INTERNAL FRICTION IN METALS AND ALLOYS

VNUTRENNEE TRENIE V METALLAKH I SPLAVAKH

ВНУТРЕННЕЕ ТРЕНИЕ В МЕТАЛЛАХ И СПЛАВАХ

Internal Friction in Metals and Alloys

Edited by
V. S. Postnikov, F. N. Tavadze, and L. K. Gordienko

Translation from Russian by
Albin Tybulewicz
Editor of the American Institute of Physics
translation of *Soviet Physics — Semiconductors*

 Springer Science+Business Media, LLC 1967

ISBN 978-1-4899-4727-7   ISBN 978-1-4899-4725-3 (eBook)
DOI 10.1007/978-1-4899-4725-3

The original Russian text was published for the A. A. Baikov Metallurgy
Institute of the Academy of Sciences of the USSR by Nauka Press, Moscow,
in 1966.

Library of Congress Catalog Card Number 67-25402

PREFACE

Real solids are never perfectly elastic, so that when they are deformed some of the mechanical energy is transformed into heat. The various mechanisms of this transformation are known, in general, as the internal friction. The internal friction is extremely sensitive to the solid's structure. Its high sensitivity to changes in the microstructure has led to the development of a new investigation method — the "internal friction" method. This method has been used to obtain extensive data which have aided in the better understanding of microprocesses taking place during the deformation of real solids, and helped to establish, in some cases, the relationship between the microstructure and the macroscopic properties of solids.

However, the complexity of the phenomenon of the internal friction and its dependence on many factors has, to a considerable degree, prevented the widespread use of this method in the investigations of various physical properties of materials and of the microprocesses taking place during deformation. Because of this, recent years have seen intensive theoretical and experimental investigations whose aim has been to establish the most general relationships for the internal friction, necessary to provide a firm basis for the method and a satisfactory theory of the phenomenon.

In this publication are collected together the results of the majority of investigations on the internal friction in metals carried out in the last 2–3 years at leading Soviet research institutes and higher educational establishments.

Many of the papers in this volume were discussed at the 1964 Tbilisi Conference on Internal Friction. The whole collection ranges over a wide variety of relevant problems.

Regretfully, we had to exclude many interesting papers because of the limitations of space. The editors apologize to teams and individuals whose papers are not included in this collection for this reason.

The Editors

PUBLISHER'S NOTE

The following Soviet journals cited in this book are available in cover-to-cover translations:

Russian title	English title	Publisher
Atomnaya énergiya	Soviet Journal of Atomic Energy	Consultants Bureau
Doklady Akademii Nauk SSSR	Soviet Physics — Doklady	American Institute of Physics
Fizika metallov i metallovedenie	Physics of Metals and Metallography	Acta Metallurgica
Fizika tverdogo tela	Soviet Physics — Solid State	American Institute of Physics
Metallovedenie i termicheskaya obrabotka metallov	Metal Science and Heat Treatment	Consultants Bureau
Plasticheskie massy	Soviet Plastics	Rubber and Technical Press, Ltd.
Uspekhi fizicheskikh nauk	Soviet Physics — Uspekhi	American Institute of Physics
Zavodskaya laboratoriya	Industrial Laboratory	Instrument Society of America
Zhurnal éksperimental'noi i teoreticheskoi fiziki	Soviet Physics — JETP	American Institute of Physics
Zhurnal neorganicheskoi khimii	Russian Journal of Inorganic Chemistry	Consultants Bureau
Zhurnal tekhnicheskoi fiziki	Soviet Physics — Technical Physics	American Institute of Physics

CONTENTS

INTRODUCTION

I. INTERNAL FRICTION IN PURE METALS
AND IN NONMAGNETIC ALLOYS

II. INTERNAL FRICTION IN STEELS

III. INTERNAL FRICTION IN CONSTRUCTIONAL MATERIALS AT HIGH VIBRATION AMPLITUDES

IV. INSTRUMENTS AND METHODS FOR MEASURING INTERNAL FRICTION

V. INTERNAL FRICTION THEORY

INTERNAL FRICTION IN METALS

V. S. Postnikov

The problem of the dissipation of energy by oscillating systems is the most important among the many various problems associated with oscillations of mechanical systems, particularly systems in which resonance conditions may arise. The oscillations of a system take place in some medium, and so the elastic energy losses of an oscillating system are divided into three types. The first type comprises the aerodynamic and hydrodynamic losses due to the resistance of the medium. The second type is known as the internal friction, because the energy is dissipated in the material from which the particular object is made. The third type of losses occurs in moving (bearings, guides, etc.) and fixed (riveted, slotted, screw-thread, etc.) joints or couplings. The losses in rigid joints are known as structural damping.

So far, most attention has been concentrated on the second type of losses — the internal friction. The great interest shown in this form of losses is not accidental, but is justified by many reasons. We shall mention only two of them. It is essential to be able to prepare alloys which have good damping properties and a relatively high strength at room and elevated (up to 1000°C) temperatures. The alloy Nivko-10 has such properties [1].

Another reason for the great interest in the internal friction is associated with the development of a new refined measurement method based on the measurement of the internal friction. Until now, this method has been deservedly recognized as one of the main highly sensitive methods for the investigation of the transport phenomena taking place in metals and alloys during their preliminary treatment and subsequent service life in alternating force fields (mechanical, thermal, electromagnetic, etc.). This method has been used successfully to study the physical properties associated with the multiplication and motion of various types of defect [2], phase transitions [3, 4], diffusion [4], cyclic and thermal fatigue of metals [5, 6], and many other problems. However, in spite of the great interest in the internal friction and the considerable success in the investigation and use of this phenomenon, many problems associated with the interpretation of the experimental data have not yet been solved.

We shall consider the internal friction as a whole and note some of the unsolved problems.

Methods of Measuring Internal Friction. Measures of Internal Friction

The following effects can be used to measure the internal friction:

1) changes in the temperature of a body;

2) changes in the angular phase shift between stress and deformation;

1

3) amplitudes of free and forced oscillations;

4) damping of elastic waves.

In the case of metals, the internal friction is relatively weak and if we represent it, say, by a quantity $\tan \varphi$, then $\tan \varphi < 0.1$. Because of this, the third and fourth group of effects are the most useful [4]. Using these effects, we can measure the internal friction in the frequency range from 10^{-3} to 10^{11} cps.

It is very difficult to carry out experiments at frequencies below 10^{-3} cps and above 10^{10} cps (hypersound). It is not possible artificially to excite hypersound at frequencies higher than 10^{11} cps. Natural thermal hypersonic waves are limited, even in crystals, to frequencies of the order of 10^{12}-10^{13} cps. Investigations of the velocity of propagation and of the absorption of natural hypersound in crystals can be carried out only indirectly, i.e., using optical [7] and x-ray diffraction methods [8-11].

The frequency range which can be used depends on the type of excited oscillations, the dimensions and shape of a sample, and the presence or absence of a special inertial system. Oscillations may be excited and recorded in various ways. Moreover, the quantities used as measures of the internal friction may be different. If we employ the natural damped oscillations of a system, then the usual measure of the internal friction is the logarithmic damping decrement θ of the quantity Q^{-1}, which is smaller by the factor π. If forced oscillations are used, then the usual measure of the internal friction is the quantity $B = \Delta \nu / \nu_0$, where $\Delta \nu$ is the half-width of the resonance peak and ν_0 is the resonance frequency.

The intensity of the absorption of elastic waves is usually represented by an absorption coefficient α, which is related to the wavelength λ and to the quantity $\delta = \Delta E / E$ (ΔE is the energy dissipated per cycle and E is the oscillation energy) by the simple relationship

$$\alpha = \frac{\delta}{2\lambda}.$$

(1)

We can show [12] that there is a simple relationship between the various measures of the internal friction, which is independent of the actual mechanism of energy loss, provided the internal friction is not very high (for example, $\tan \varphi \le 0.1$). This relationship has the following form:

$$\tan \varphi = \frac{\theta}{\pi} = \frac{\sigma}{2\pi} = \frac{B}{\sqrt{3}} = \frac{\lambda \alpha}{\pi}.$$

(2)

Using this relationship, we can compare the data on the internal friction obtained in various experiments in which different measures were used. However, we must remember that, in fact, the momentum is lost not only by absorption but also by scattering on inhomogeneities. The attenuation of stress waves is particularly strong if the average dimensions of individual inhomogeneities are comparable with the wavelength. Under these conditions, stress waves are attenuated not so much by absorption, whose mechanism is similar to the internal friction mechanism, but by the diffuse scattering of stress waves on such inhomogeneities.

We must also remember that the relationship of Eq. (2) does not apply when $\tan \varphi > 0.1$, and that it depends on the mechanism of energy dissipation [12]. When the internal friction is very high (as in the case of many plastics [13] and alloys in the phase-transition region [14]), the logarithmic decrement cannot be used as a reliable measure, and in the case of aperiodic motion the logarithmic decrement concept loses its meaning. Therefore, a different quantity, for example the absorption coefficient α [15], should be used to characterize the internal friction in the case of strong damping of oscillations.

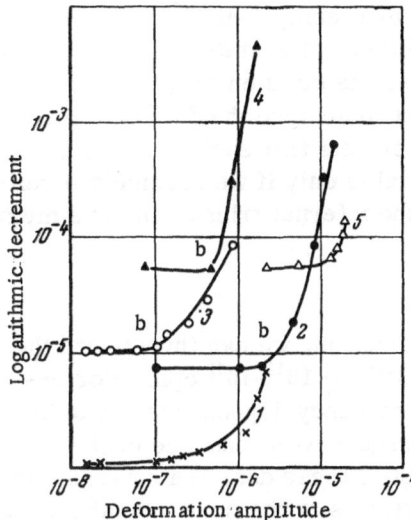

Fig. 1. Amplitude dependence of the internal friction of tin (5) and copper (1-4) crystals. 1) ν = 40 kc, t = 4.2°K; 2) 1450 cps, 27°K; 3) 40 kc, 21°K; 4) 1450 cps, 380°K; 5) 40 kc, 20°K [72].

It should be mentioned that there is no single method which could be used to measure the internal friction throughout the whole range from 10^{-3} to 10^{10} cps. It would be impossible to develop such a method and, therefore, many methods with overlapping frequency ranges are employed.

The most widely used are the low-frequency methods (10^{-3}-10^1 cps), which are closely followed by the high-frequency methods (10^4-10^8 cps). The methods employing the middle range of frequencies (10^1-10^4 cps) and hypersonic frequencies (10^9-10^{11} cps) are not yet used sufficiently widely.

Low temperatures, particularly helium temperatures, and high temperatures — above $0.8T_S$ — are not yet used sufficiently widely in internal friction measurements. No internal friction measurements have been carried out in the region of the transition from the liquid to the solid state. Metal physicists, investigating the general relationships governing the internal friction, regard this as a serious omission.

Amplitude — Frequency — Temperature Dependence of the Internal Friction

It is difficult to interpret the curve Q^{-1} (ε, ν, T). Therefore, we shall consider the three special cases of this curve, which are the ones investigated in practice.

1. Amplitude Dependence

The range of deformation amplitudes is governed by the method selected for measuring the internal friction and by the type of problem to be solved. The high-frequency methods make it possible to carry out measurements of the internal friction in the range of relative deformations beginning from 10^{-9}-10^{-8}, which cause practically no distortions in the crystal structure, right up to deformations of 10^{-3}-10^{-2}, which are limited by the strength of a crystal. The range of deformations which can be used at low frequencies is somewhat less: from 10^{-7} to 10^{-2}. For example, for a torsional pendulum, the relative deformation can in practice vary from 10^{-6} to 10^{-3}.

The nature of the $Q^{-1}(\varepsilon)$ curves of some single crystals and polycrystalline aggregates is shown in Fig. 1. The shape of the curves indicates two groups of energy dissipation mechanisms acting at relatively weak deformations (up to the point b) and at high deformations (above the point b). Therefore, the total internal friction is usually divided into two components:

$$Q^{-1} = Q_I^{-1} + Q_{II}^{-1},$$

where Q_I^{-1} is the internal friction observed at low amplitudes (up to the point b) and Q_{II}^{-1} is the amplitude-dependent contribution to the total internal friction. This division is arbitrary because, on the one hand, the position of the point b on the curve $Q^{-1}(\varepsilon)$ of the selected material depends on the frequency, temperature, and its previous history, and, on the other, Q_I^{-1} and

Q_{II}^{-1} may be interdependent. Moreover, the amplitude-dependent component Q_{II}^{-1} always depends on time, since, in the case of large deformation amplitudes, the state of the material changes during the measurement of Q^{-1} itself. Nevertheless, the division of Q^{-1} into Q_{I}^{-1} and Q_{II}^{-1} is used widely by all investigators, since it is convenient to analyze the data on the internal friction separately into these two components. It is self-evident that such analysis requires great care. The interpretation of the data can be reliable only if we accumulate reliable experimental material on the amplitude dependence of the internal friction in pure metals and in various alloys.

2. Frequency Dependence

Using the various methods for measuring the internal friction, we can investigate the frequency dependence of the friction in the range from 10^{-4}-10^{-3} to 10^{11}-10^{12} cps. For example, a torsional pendulum can be used to investigate the frequency dependence of the internal friction from 10^{-2} to 10^2 cps. Experimental data on the frequency dependence of the amplitude-independent internal friction are very scarce — and most of these data are contradictory. Hiki [16] measured the internal friction of two lead single crystals at the fundamental frequency of 64 kc and at the third harmonic (192 kc), and found that Q_{I}^{-1} was proportional to the frequency in the temperature range from 140 to 220°K. Hikata and Truell [17] reported that the internal friction in aluminum at 5 and 10 Mc was almost proportional to the oscillation frequency, while Hasiguti et al. [18] concluded that the internal friction was inversely proportional to the frequency in the same range of frequencies. They also found that the internal friction of aluminum was almost proportional to the vibration frequency between 20 and 200 Mc. Nowick [19] measured the amplitude-independent internal friction of copper single crystals at the fundamental frequency (39 kc) and at the first harmonic (78 kc). The results were fairly widely scattered, but they did not exhibit any marked frequency dependence. Alers and Thompson [20] found that, in the megacycle region (5–50 Mc) at 195°K, Q_{I}^{-1} of copper single crystals decreased with increasing frequency. Takahashi [21] conducted measurements on polycrystalline copper at room temperature between 1 and 10 kc and found that Q_{I}^{-1} was independent of the frequency. Bordoni et al. [22] observed little change in the internal friction of polycrystalline copper in the frequency range from 1.8 kc to 6.5 Mc.

We should mention also the measurements of Granato and Truell [23] on single crystals of germanium, and the measurements of Lamb et al. [24] on germanium and silicon. Granato and Truell measured the internal friction of germanium at frequencies from 5 to 300 Mc; the internal friction was approximately proportional to the frequency. Lamb et al. found, in the frequency range from $1 \cdot 10^2$ to $8 \cdot 10^2$ Mc, that germanium and silicon exhibited an almost quadratic dependence of the absorption coefficient on the frequency. However, germanium and silicon are semiconductors and the results obtained for them are probably not representative of metals.

From these unsystematic and uncorrelated measurements, we cannot draw any reliable conclusions about the nature of the frequency dependence of the internal friction Q_{I}^{-1}. Undoubtedly, there is such a dependence.

Careful experiments using overlapping methods are needed to determine the nature of the frequency dependence of Q_{I}^{-1} for various materials.

The data on the frequency dependence of the amplitude-dependent internal friction are even more limited and contradictory. Read [25] measured the amplitude-dependent internal friction of a zinc single crystal at 31 and 76 kc and found an approximately inversely proportional dependence on the frequency. Nowick [26] measured Q_{II}^{-1} of several copper single crystals at the fundamental frequency (39 kc) of the samples and at the second harmonic. The scatter of the data for different samples was large, but the results nevertheless indicated that the

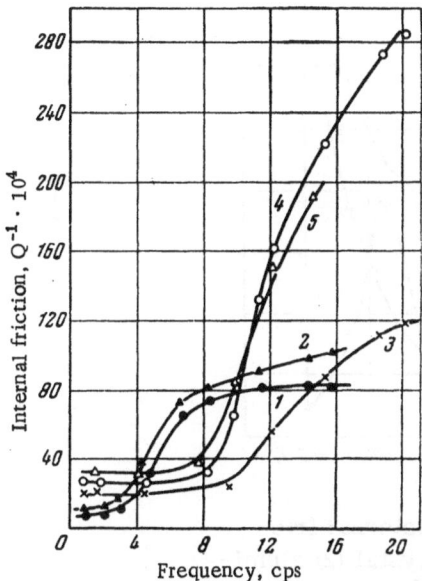

Fig. 2. Frequency dependence of the internal friction of steel of the SN-3 type (normalized from 1100°C; curves 1, 2) and of the 50 type (quenched from 740°C; curves 3-5) at 20°C (1-4) and 50°C (5). The deformation amplitude for curves 1-4 was 2.5 kg/mm² and, for curve 5, 6.6 kg/mm².

damping was independent of the oscillation frequency. Kamenetsky [27] investigated copper single crystals at various harmonics in the kilocycle region and also found some scatter, but there was a tendency for Q_{II}^{-1} to increase with the frequency. Hiki [16] measured the internal friction of annealed lead single crystals at the fundamental frequency (64 kc) and at the third harmonic (192 kc) and found no marked frequency dependence of Q_{II}^{-1}. Zhuravlev [28] investigated the internal friction of high-carbon and low-carbon steels at room temperature in the frequency range from 285 to 1180 cps. The internal friction Q_{II}^{-1} of the steels decreased slightly when the frequency increased.

Analyzing the results obtained in 1937 on the frequency dependence of Q_{II}^{-1}, Davidenkov [29] concluded in 1938 that the internal friction Q_{II}^{-1} was independent of the frequency. This conclusion was later accepted and applied by a number of investigators [30, 31].

In 1958, Davidenkov [32], in a discussion of G. S. Pisarenko's review, "Methods of Calculating Vibrations of Elastic Bodies Allowing for the Energy Dissipation in a Material," noted that, in particular, "it would be necessary to allow also for the additional dependence of the damping on the frequency, which is frequently observed in practice." This remark was perfectly correct.

A special experiment carried out by I.M. Sharshakov and V.V. Usanov indicated that Q_{II}^{-1} does depend on the frequency (Fig. 2).

It is perfectly obvious that the establishment of the nature of the frequency dependence of Q_{II}^{-1} is of basic importance. Because of this, and in view of the fact that the available data on the nature of the frequency dependence of Q_{II}^{-1} are few and contradictory, it is necessary to carry out an extensive and systematic investigation of this dependence for pure metals and alloys over a wide range of frequencies and amplitudes, as well as to study the dependence on the type of deformation (twisting, bending, elongation) used in the measurements of the internal friction.

3. Temperature Dependence

The temperature dependence of the internal friction has been investigated more fully than other dependences [27, 33-35, etc.]. Nevertheless, there are at least two serious deficiencies in the investigations of the temperature dependence of the internal friction. First, many investigations failed, until recently, to report whether the measured quantity Q^{-1} was the amplitude-dependent component. Secondly, no investigation has yet been carried out using one and the same method between helium temperatures and the melting point. Such an investigation can, in fact, be carried out.

The reported investigations [16, 17, 19, 33-39] show that Q^{-1} increases monotonically with temperature between 4°K and 0.5-0.6 T_S, and that the increase is almost linear. At

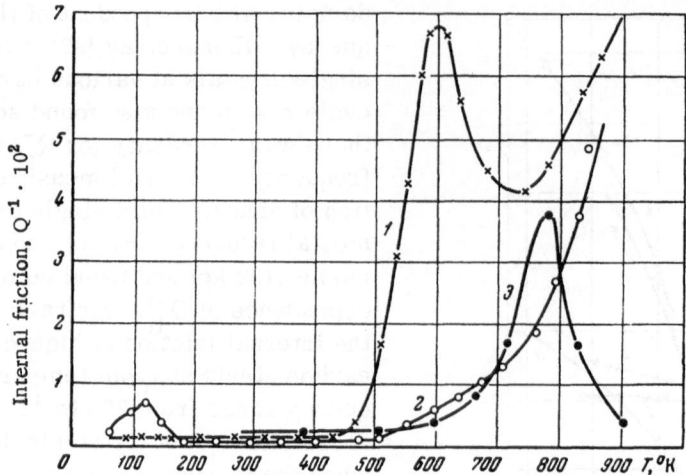

Fig. 3. Temperature dependence of the internal friction of polycrystalline (1, 3) and single-crystal (2) aluminum. 1, 2) $\varepsilon = 10^{-5}$, $\nu = 1$ cps; 3) $\nu = 10^4$ cps.

higher temperatures, Q^{-1} grows exponentially (Fig. 3). In some cases, "peaks" (whose nature
is not always clear) are superimposed on a uniformly increasing "background."

The temperature dependence of the amplitude–dependent internal friction has been investigated insufficiently fully, particularly in the kilocycle frequency range [18, 36, 38–44]. The
nature of the temperature dependence of Q_{II}^{-1} is more complex than in the case of Q_I^{-1} (see Fig. 3).

Principal Types of Internal Friction

Three types of internal friction are distinguished: relaxational, resonant, and hysteretic.
Sometimes, a fourth type is identified; it is called the "viscous" friction [45]. The introduction of this form of friction is justified only at very high deformation amplitudes when whole
regions move relative to one another in the interior of a material.

The relaxational internal friction is characterized by a dependence on the frequency and
temperature, and by an independence of the vibration amplitude. The resonance internal friction depends on the frequency and is practically independent of the deformation amplitude and
the temperature. The hysteretic internal friction is due to irreversible changes in the state
of a sample subjected to periodic deformation. The hysteretic internal friction is assumed to
be independent of frequency [40]. This is because the relationship between stress and deformation is assumed to be independent of time. Since hysteresis is a nonlinear process, energy
losses per vibration period usually increase with increase in the deformation amplitude.

The phenomenological theory of these three types of internal friction is quite well developed. As far as the microscopic theory and actual mechanisms are concerned, the situation is not so good. Until recently, the most authoritative theory had been that of Granato and
Lücke [46], which described the resonant and hysteretic internal friction within a certain approximation. However, this theory had many limitations. First, it considered only one form
of lattice imperfection, i.e., a dislocation network. Other well-known forms of imperfection,
whose irreversible motion could cause losses, were ignored by this theory. Secondly, the
theory of Granato and Lücke was based on many simplifying assumptions. Consequently, the
results of this theory frequently disagreed with experiment, particularly at low frequencies.
A modification of the dislocation theory of absorption has been put forward by Swartz and

Weertman [47]. This modification gave an amplitude dependence of Q^{-1} which was different from the Granato–Lücke dependence and in better agreement with experiment. However, even this modified theory did not solve all the difficulties of the Granato–Lücke theory at low and megacycle range of frequencies in the case of superpure metals. Moreover, the hysteretic mechanisms are exceptionally variegated and are difficult to describe mathematically.

The relaxational internal friction, caused by dislocations, was described first by Seeger [48] and Donth [49]. A slightly different theory was proposed by Lothe [50]. This theory describes the relaxation effect discovered by Bordoni [51] in various *fcc* metals. Somewhat later, Seeger [52] wrote another paper on this problem. In addition to the Bordoni effect, there is also the Snoek relaxation effect, associated with the presence of carbon atoms in α-iron; the Finkel'shtein effect, caused by the presence of carbon in γ-iron; the relaxation effects due to the presence of both dislocations and impurity atoms (a peak near 220°C in cold-worked iron, containing carbon and nitrogen), the relaxation along grain boundaries, etc. Only some of these effects have been accounted for theoretically by Zener [53].

Peaks and Background of Internal Friction

Most of the investigations have dealt with the temperature dependence of internal friction. This bias is quite understandable and does not call for comment. However, it should be mentioned that there have been very few Soviet investigations of the internal friction at low temperatures [67-71].

Investigations of $Q^{-1}(T)$ at low temperatures and low pressures are of very great theoretical and practical importance. Equally important are the investigations of $Q^{-1}(T)$ at high temperatures, above $0.7T_S$, which are also not numerous [55]. Any $Q^{-1}(T)$ curve of a metal or an alloy has a background with peaks superimposed on it. For example, the first peak, shown in Fig. 3, is the Bordoni peak [51]. The second is the "grain-boundary peak" [35, 54], and the third small peak is due to the relaxation of stresses along block boundaries [55]. The background is associated with the migration of defects in the stress field [33, 56]. The phenomenological theory [57-59] predicts the existence of the peaks and of the background in the $Q^{-1}(T)$ and $Q^{-1}(\nu)$ curves. However, there is, as yet, no microscopic description of these curves, with the exception of the first peak [53]. Moreover, the nature of the second and third peaks has not yet been investigated experimentally in sufficient detail. One of the published investigations [6] deals with the complex nature of the "grain-boundary peak."

Factors Influencing the Magnitude of Internal Friction and the Nature of Its Temperature Dependence

1. Dimensions and Shape of Sample

There have been no systematic investigations of the influence of the dimensions and sample shape on the magnitude of the internal friction. The author is aware of only a few papers (see, for example, [60]) which have demonstrated the dependence of the magnitude of the internal friction on the dimensions and shape of a sample (Fig. 4). This dependence must be allowed for in comparisons of the internal friction data obtained by different investigators. Interesting results have been obtained by Postnikov et al. [61] for copper whiskers.

It was found that the temperature dependence of a copper whisker was similar to the $Q^{-1}(T)$ curve (Fig. 5) for ordinary wire samples, although the cross sections differed by a factor of more than ~5000.

2. Type of Deformation

The energy dissipated by a vibrating sample per period should depend on the nature of the deformation of a sample during the measurement of Q^{-1}, because the nature of deformation

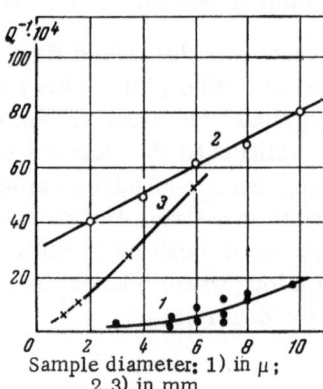

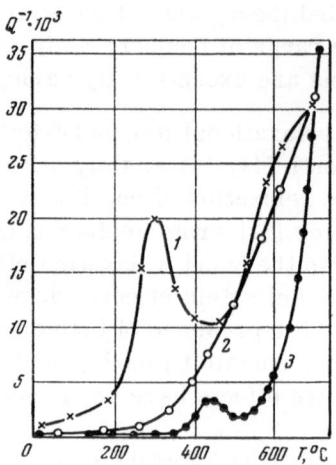

Fig. 4. Dependence of the internal friction on sample diameter. 1) Copper whisker; 2) steel 45 (diameter multiplied by 2); 3) polycrystalline aluminum. 1,3) $\nu \sim 1$ cps; 2) $\nu = 30$ cps [73].

Fig. 5. Temperature dependence of the internal friction of polycrystalline (1) and single-crystal (2) copper, and of a copper whisker (3). Frequency ~ 1 cps.

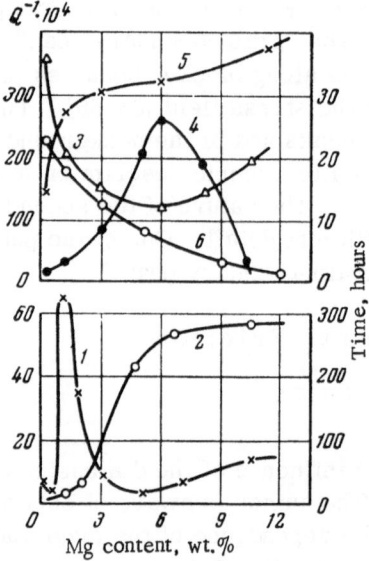

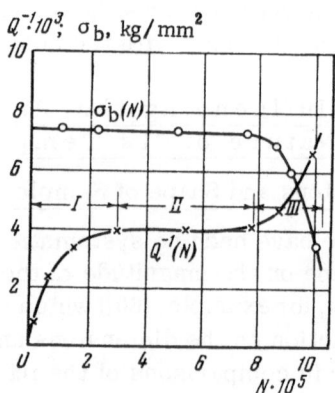

Fig. 6. Dependence on composition of the internal friction (1, 3, 5) and of the time necessary to reach a certain deflection (2, 4, 6) of Al—Mg alloys. (The value of Q^{-1} for curve 5 should be multiplied by 2.) 1, 2) 20°C; 3, 4) 300°C; 5, 6) 400°C.

Fig. 7. Dependence of the internal friction and the strength of aluminum on the number of deformation cycles. Maximum cyclic stress $\tau = 5.3$ kg per mm².

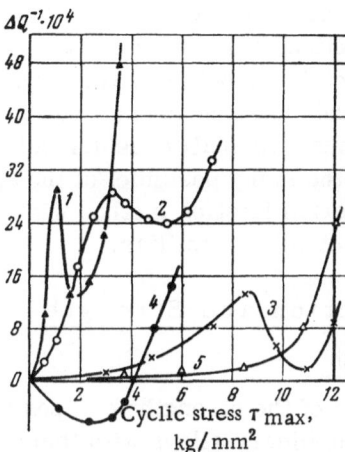

Fig. 8. Dependence of the value of ΔQ^{-1} for cadmium, aluminum, and copper in annealed (1, 2, 3) and deformed (4, 5) states on the cyclic deformation stress. $\Delta Q^{-1} = Q_C^{-1} - Q_0^{-1}$, where Q_C^{-1} is the value after a critical number of cycles, and Q_0^{-1} is the value of Q^{-1} before deformation.

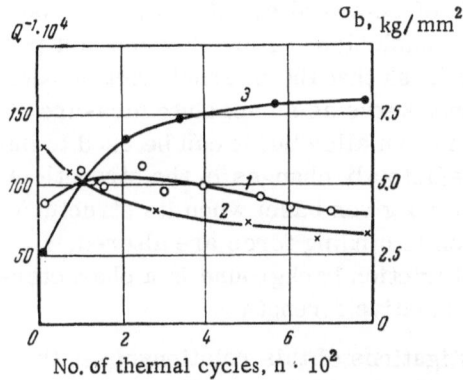

Fig. 9. Dependence of the internal friction (1, 3) and of the strength (2) of cadmium on the number of thermal cycles. Cycle parameters: $T_{max} = 250°C$, $T_{min} = 50°C$; cycle period: 2 min. 1, 2) 25°C; 3) $0.85\,T_S$.

(shear, tension, compression, bending) determines the nature of the action of the sources which generate defects and the nature of their motion in the lattice of a solid. It is difficult to observe the differences between the types of deformation because the magnitude of the internal friction depends on a large number of factors, which may be different for two different deformation methods. However, experiments on the influence of the type of deformation should be carried out because the theory predicts differences between the types of deformation [58].

3. Structure

It is well known why the internal friction is structure-sensitive. The high sensitivity of the internal friction to the structural changes which take place in a metal during its manufacture and service life is the basis of the "internal friction method," which is currently used widely by metallurgists and metal physicists. However, the method is frequently applied carelessly and this leads to errors. In the author's opinion, those investigators who obtain experimental data on the internal friction at low temperatures [2] are correct because, in this range of temperatures, the influence of the thermal motion of atoms on the structure of a metal is unimportant. Unfortunately, the available information on the internal friction in single crystals is limited to helium temperatures. There are no data on the internal friction at temperatures below 4°K. Moreover, the majority of the results on the internal friction have been obtained on imperfect single crystals containing a fairly high percentage of impurities. It is essential to continue improving the experimental technique of the preparation of single crystals to obtain the most perfect structure possible, containing only traces of impurities. Then, by disturbing the crystal structure in a controlled manner (slight cold working, irradiation, the introduction of impurities from the atmosphere, etc.) and measuring Q^{-1} simultaneously, we should obtain the fundamental information that is so essential for the theoretical justification of the internal friction method. Very valuable information can be obtained by investigating the internal friction of ionic crystals (NaCl, LiF, and others), whose internal structure is sufficiently perfect. Moreover, most of these crystals are transparent, and this makes it possible to use, in addition to other methods, optical methods of observing changes in the structure caused deliberately by the experimenter.

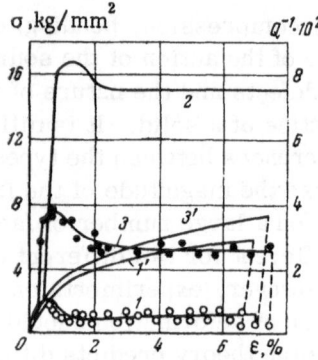

Fig. 10. Dependence of the internal friction on the magnitude of deformation (1-3) and the stress—strain diagrams (1',3') for aluminum at room temperature. 1, 1') $\dot{\varepsilon}$ = 0.00012 min^{-1}; 2) $\dot{\varepsilon}$ = 0.0018 min^{-1}, Ke's curve [65]; 3, 3') $\dot{\varepsilon}$ = 0.002 min^{-1}.

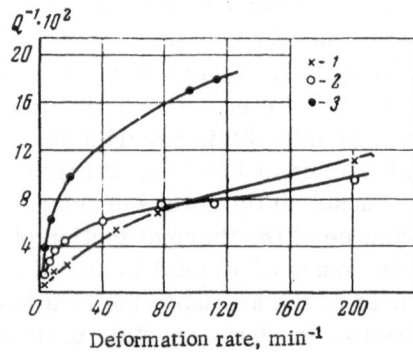

Fig. 11. Dependence of the internal friction on the deformation rate, determined for ε = 3%. 1) Al; 2) Cu; 3) Zn.

Without such fundamental investigations, it is difficult to provide sufficiently reliable experimental and theoretical bases to the internal friction method, which is undoubtedly very promising, and which is already used successfully. Because of space limitations, we shall consider only two examples (out of the many possible) of the applications of the internal friction method with which the reader may not yet be familiar.

Internal Friction and Strength

1. Static Strength

Comparison of the strength—composition isotherms of some binary alloys with their internal friction—composition isotherms has led to the conclusion that there is a one-to-one correspondence between the strength of an alloy and its high-temperature internal friction background, i.e., the greater the strength, the weaker is the internal friction background. In the temperature range from room temperature to $0.6\,T_S$, no reliable correlation between the magnitude of the internal friction and the strength has yet been established (Fig. 6). Investigations carried out on a series of refractory alloys [62, 63] have confirmed this conclusion. They have made it possible to establish that the internal friction background cannot serve as an absolute measure of the strength of an alloy but it can be used to describe satisfactorily changes in the strength of an alloy (with a given base) when its structure and interatomic binding force are altered, i.e., the internal friction background is a characteristic of the relative strength.

Unfortunately, there have been no systematic investigations of this relationship in the temperature range $(0.6-0.95)\,T_S$ using modern refractory alloys. No investigations have been carried out either in the $(0-0.2)\,T_S$ range, i.e., at low temperatures. The investigations carried out in the $(0.2-0.6)\,T_S$ range show that the correlation between Q^{-1} and the strength, referred to earlier, is not observed because, in this range of temperatures, there are many phase transitions, relaxation effects, and other processes which alter greatly the magnitude of the internal friction but have little effect on the strength.

2. Cyclic Strength

Investigations have shown [5] that the cyclic (fatigue) strength and the phenomenon of fatigue fracture can be investigated very successfully by the internal friction method. Figure 7 illustrates changes in the strength and internal friction as the number of deformation cycles is increased. The internal friction curve $Q^{-1}(N)$ has three definite regions corresponding to three stages of fatigue fracture [5]. The first, most interesting, range of the dependence of Q^{-1} on N, corresponding to the first fatigue stage, is almost absent from the $\sigma_B(N)$ curve. Consequently, investigations of the fatigue by the internal friction method give much more useful information than direct investigations of the strength during cycling.

An equally interesting result is obtained in the investigation of changes in the ΔQ^{-1} of a metal due to cyclic deformation at various torsional amplitudes. It follows from Fig. 8 that the $\Delta Q^{-1}(\tau)$ curves of the investigated metals have a maximum which represents the fatigue limit of these metals [5].

Consequently, using the internal friction method, we can easily establish the fatigue limit of a metal which is notoriously difficult to find by direct experiments.

Using the internal friction method, we can also follow (Fig. 9) the development of the stages of the thermal fatigue of a metal [6].

Internal Friction and Plasticity

According to current ideas, when a metal is deformed, various defects (vacancies, dislocations, etc.) are generated and set in motion. Since the internal friction is governed mainly by these defects, it is interesting to follow changes in the internal friction during the process of deformation of a metal. Such an experiment was first carried out by Maringer [64] on molybdenum and by Ke [65] on aluminum, copper, and low-carbon steels. Both these workers used a torsional pendulum to measure the internal friction. El'kin has continued these investigations. Figures 10 and 11 show the results of investigations of the dependences of Q^{-1} on ε and on $v = \dot{\varepsilon}/\dot{\varepsilon}_0$, where $\dot{\varepsilon} = 0.00008$ min^{-1} is the lowest rate of tensile deformation [66].

Figure 10 shows that, in the second half of the so-called elastic region, structural changes have already taken place in a metal and these changes have led to a sharp increase in the magnitude of the internal friction. Near the elastic limit, the internal friction reaches its maximum value, then it decreases somewhat and remains constant during further deformation. When the deformation is stopped ($\dot{\varepsilon} = 0$, $\dot{\varepsilon} = $ const), the internal friction drops rapidly. After the complete removal of the load, the magnitude of the internal friction is somewhat higher than at the beginning of the experiment and represents the internal friction of a sample deformed by an amount corresponding to ε.

Figure 11 shows that the internal friction of aluminum depends almost parabolically ($Q^{-1} \propto v^{1/2}$) on the rate of deformation, but for copper and zinc this dependence is somewhat more complex.

There have been no theoretical investigations of these phenomena. The first steps in this direction have been made in [55].

Conclusions

This review confirms the high sensitivity, universality, and importance of the internal friction method for metallurgists, but it also shows the need for a careful justification of this method. It is necessary to carry out, very soon, a number of fundamental experimental investigations referred to earlier and use them to develop the theoretical basis of the method.

We are speaking here not of the phenomenological theory, which is already developed to a considerable extent, but of the atomic-kinetic basic theory. The atomic-kinetic (microscopic) theory of the internal friction is still in its infancy.

The future of the internal friction method depends wholly on the successful development of the microscopic theory and on our efforts in attracting theoretical physicists to this problem.

Literature Cited

1. A.W. Cochardt, Collection: Magnetic Properties of Metals and Alloys [Russian translation], IL, Moscow (1961), p.328.
2. Collection: Internal Friction and Defects in Metals [Russian translation from English and German], V.S. Postnikov, ed., Izd. "Metallurgiya" (1965).
3. G.K. Mal'tseva, Relationships Governing Internal Friction in Metals and Alloys During Solid-State Phase Transitions, Dissertation for Candidate's Degree, Voronezh (1963).
4. M.A. Krishtal, Yu.V. Piguzov, and S.A. Golovin, Internal Friction in Metals and Alloys, Metallurgizdat (1964).
5. G.A. Gorshkov; Investigation of the Fatigue of Metals by the Internal Friction Method, Dissertation for Candidate's Degree, Voronezh (1964).
6. I.V. Zolotukhin, Some Problems of the Thermal Fatigue of Metals and Alloys, Dissertation for Candidate's Degree, Voronezh (1964); Dokl. Akad. Nauk SSSR, 158(3):590 (1964).
7. M.V. Vol'kenshtein, Molecular Optics, GITL (1951).
8. G.N. Ramachandran and W.A. Wooster, Acta Cryst., 4:335,451 (1951).
9. E. Prince and W.A. Wooster, Acta Cryst., 6:450 (1953).
10. W.P. Binnie and A.M. Liebschutz, Proc. Conf. Nucl. Eng., E-26 (1953).
11. S.C. Prasad and A.W. Wooster, Acta Cryst., 8:361,507,614,682 (1955).
12. S.I. Meshkov, V.S. Postnikov, and T.D. Shermergor, Izv. Akad. Nauk SSSR, No.3:104 (1964).
13. V.S. Postnikov, Plasticheskie Massy, No.11:60 (1960).
14. G.K. Mal'tseva, I.V. Zolotukhin, and V.S. Postnikov, Fiz. Metal. i Metalloved., 16:754 (1963).
15. S.I. Meshkov, T.D. Shermergor, and V.S. Postnikov, Collection: Energy Dissipation in Vibrations of Elastic Systems, Izd. AN UkrSSR (1963), Vol.4, p. 357.
16. Y. Hiki, J. Phys. Soc. Japan, 13:1138 (1958).
17. A. Hikata and R. Truell, J. Appl. Phys., 28:522 (1957).
18. R.R. Hasiguti, N. Igata, and K. Tanaka, J. Phys. Soc. Japan, 18:102 (1963).
19. A.S. Nowick, Phys. Rev., 80:249 (1950).
20. G.A. Alers and D.O. Thompson, J. Appl. Phys., 32:283 (1961).
21. S. Takahashi, J. Phys. Soc. Japan, 11:1253 (1956).
22. P.G. Bordoni, M. Nuovo, and L. Verdini, Nuovo Cimento, 14:273 (1959).
23. A. Granato and R. Truell, J. Appl. Phys., 27:1219 (1956).
24. J. Lamb, M. Redwood, and Z. Steinshliefer, Phys. Rev. Letters, 3:28 (1959).
25. T.A. Read, Phys. Rev., 58:371 (1940).
26. A.S. Nowick, Carnegie Institute of Technology Symposium on the Plastic Deformation of Crystalline Solids (1950).
27. L.A. Kamenetsky, Cornell University, AF-OSR-TN-425 (1956).
28. V.A. Zhuravlev, Zavodsk. lab., No. 5 (1948).
29. N.N. Davidenkov, Zh. Tekhn. Fiz., 8:483 (1938).
30. G.S. Pisarenko, Energy Dissipation in Mechanical Vibrations, Izd. AN UkrSSR (1962).
31. Ya.G. Panovko, Internal Friction in Vibrations of Elastic Systems, Fizmatgiz (1960).
32. N.N. Davidenkov, Zh. Mekhan., No. 7:93 (abstract) (1958).
33. V.S. Postnikov, Zh. Tekhn. Fiz., 94:1599 (1954).
34. V.S. Postnikov, Usp. Fiz. Nauk, 66:43 (1958).
35. V.S. Postnikov, I.M. Sharshakov, and É.M. Maslennikov, Collection: Relaxation Phenomena in Metals and Alloys, Metallurgizdat (1963), p. 165.
36. H.L. Caswell, J. Appl. Phys., 29:1210 (1958); Thesis, Cornell University, AF-OSR-TR-57-69 (1957).
37. D.H. Niblett and J. Wilks, Proc. Phys. Soc. (London), 73:95 (1959).

38. D.O. Thompson and D.K. Holmes, J.Appl.Phys., 30:525 (1959).

39. C.W. Wert, J.Appl.Phys., 20:29 (1949).

40. R.H. Chambers, Carnegie Inst.Tech.Rept., AT(30-1)-1193 (1957).

41. R.H. Chambers and R. Smoluchowski, Phys.Rev., 117:725 (1960).

42. J. Weertman and E.J. Salcovitz, Acta Met., 3:1 (1955); J.Appl.Phys., 27:1251 (1956).

43. D.H. Niblett and J.Wilks, Proc.Intern.Congr.Refrig., Copenhagen (1959).

44. R.S. Barnes, N.H. Hancock, and E.C.H. Silk, Phil.Mag., 3:519 (1958).

45. H.G. van Bueren, Imperfections in Crystals [Russian translation], IL (1962), Chap.17. [English edition: 2nd ed. (Interscience) Wiley, New York.]

46. A. Granato and K. Lücke, J.Appl.Phys., 27:583 (1956).

47. J.C. Swartz and S. Weertman, J.Appl.Phys., 32:1860 (1961).

48. A. Seeger, H. Donth, and F. Pfaff, Discussions Faraday Soc., 23:19 (1957).

49. H.Donth, Z. Physik, 149:111 (1957).

50. J. Lothe, Phys.Rev., 117:704 (1957).

51. P.G. Bordoni, Ric. Sci., 19:851 (1949); J. Acoust. Soc.Am., 26:495 (1954).

52. A. Seeger and P. Schiller, Acta Met., 10:348 (1962).

53. C.M. Zener, Elasticity and Anelasticity of Metals, University of Chicago Press, Chicago,Illinois (1948).

54. T.S. Ke, Phys.Rev., 71:533 (1947).

55. V.S. Postnikov, Internal Friction in Pure Metals and Alloys at High Temperatures, Doctoral Dissertation, Tomsk (1959).

56. V.S. Postnikov, Fiz.Metal.i Metalloved.,7:777 (1959; Collection: Investigations of Refractory Alloys, Izd.AN SSSR (1959), Vol.4, p.181; Collection: Relaxation Phenomena in Metals and Alloys, Metallurgizdat (1960), p.264 [English translation: B.M. Finkel'shtein, ed., Consultants Bureau, New York (1963),p.199.]

57. V.S. Postnikov, Fiz.Metal.i Metalloved., 6:522 (1958).

58. S.I. Meshkov, Some Problems of the Phenomenological Theory of Relaxation Internal Friction in Solids, Dissertation for Candidate's Degree, Voronezh (1964).

59. S.I. Meshkov and T.D. Shermergor, Zh.Prikl. Mekhan.i Tekhn.Fiz., 6(3) (1962).

60. V.S. Postnikov and A.M. Belyaev, Collection: Relaxation Phenomena in Metals, Metallurgizdat (1963), p.159.

61. V.S. Postnikov, S.A. Ammer, and A.M. Belyaev, Fiz.Metal.i Metalloved.,19(2) (1965).

62. V.S. Postnikov, Collection: Fatigue Strength of Metals, Izd.AN SSSR (1962),p.207.

63. I.M. Sharshakov, Internal Friction and Mechanical Problems of Austenitic and Martensitic Steels, Dissertation for Candidate's Degree, Voronezh (1965).

64. R.E. Maringer, J.Appl.Phys., 24:1525 (1953).

65. T.S. Ke,P.T.Yung,and C.C. Chang, Sci. Record, 1:231 (1957).

66. V.S. Postnikov and Yu.M. El'kin, Collection: Investigation of Steels and Alloys, Izd. "Nauka" (1964),p.376.

67. V.A. Pavlov, N.F. Kryuchkov, and I.D. Fedotov, Fiz.Metal.i Metalloved., 5:371 (1957).

68. V.A. Pavlov, Fiz.Metal. i Metalloved., 6:122 (1958).

69. Yu.Kh. Vekilov and Yu.V. Piguzov, Fiz. Tverd. Tela, 4:1099 (1962).

70. P.L. Gruzin and A.N. Semenikhin, Fiz.Metal.i Metalloved., 15:791 (1963).

71. Collection: Relaxation Phenomena in Metals and Alloys, Metallurgizdat (1963).

72. E.G. Shvidkovskii and A.A. Durgaryan, Nauchn.Dokl. Vysshei Shkoly,Fiz.-Mat.Nauki, No.5:217 (1958).

73. N.M. Mukhin and G.S. Pisarenko, Collection: Energy Dissipation in Vibrations of Elastic Systems, Izd. AN UkrSSR (1963), p. 214.

INVESTIGATION OF THE BEHAVIOR OF INTERSTITIAL ATOMS IN MOLYBDENUM BY THE INTERNAL FRICTION METHOD

Yu. V. Piguzov, V. D. Verner, V. I. Shulepov, and I. Ya. Rzhevskaya

Interstitial impurities dissolved in Mo govern, in many respects, its technological properties. As is known, the internal friction can be used successfully to analyze the behavior of interstitial impurity atoms in metals and alloys with the bcc lattice. This application of the internal friction method is based on the Snoek peak [1], which can be used to determine the concentration of interstitial atoms in a solid solution. However, in the case of such metals as W, Cr, and Mo, the Snoek peak has not yet been reported and, therefore, our first problem was to attempt to detect this peak. We investigated fired and deformed molybdenum containing Fe, P, C, O_2, and N_2 impurities. Samples 1 mm wide were cut from a strip 0.35 mm thick; they were heated to various temperatures in a directly heated electric furnace and were quenched in an argon blast.

The temperature dependence of the internal friction was determined using plate-shaped samples in a vacuum relaxator (torsional pendulum) of the RKF-MIS type. The frequency of vibrations of the samples was varied within the limits 0.5-2.1 cps.

For quenched samples we found a broad internal friction of complex form, located in the 60-400°C range (Fig. 1). The maximum amplitude of the peak increased with increase in the quenching temperature.

Figure 2 compares the quenching-temperature dependences at the peak amplitude and of the concentration in interstitial solutions of C, N_2, and O_2. The correlation between the internal friction and the concentration of the interstitial impurity can be seen most clearly in Fig. 3, where the abscissa represents the relative concentration C/C_{max} or the relative height of the peak Q^{-1}/Q^{-1}_{max}. The internal friction curve lies between the solubility curves of C or N_2 and O_2; it is almost parallel to the solubility line of oxygen [2]. Thus, the detected peak was undoubtedly associated with the presence of interstitial atoms in the solid solution and is probably directly connected with the presence of oxygen. However, the peak was complex; in many cases, it had a central component and two satellites. Even in the apparent absence of one of these satellites, the half-width of the peak was about three times as large as that expected for a single relaxation time. Figure 4 demonstrates the internal friction peak as a function of $1/T$ and shows, by graphical analysis, that the peak can be represented as the superposition of at least three components: I, II, III. Using the Marx—Wert method, we can ascribe the activation energies of 26, 32, and 39 kcal/mole, respectively, to these three peaks. After

15

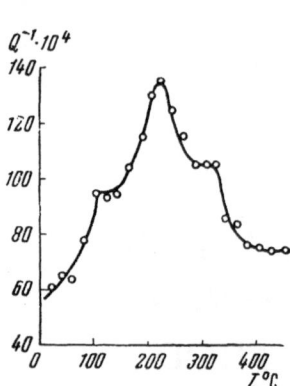

Fig. 1. Temperature dependence of the internal friction of molybdenum (sample 1) after quenching from 2000°C.

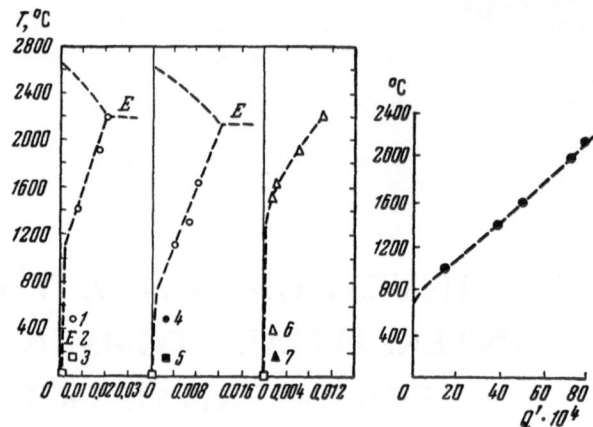

Fig. 2. a: Quenching-temperature dependence of the solubility (in %) of C, O_2, and N_2 in a molybdenum solid solution [2]; 1-7) data of different workers. b: quenching-temperature dependence of the amplitude of an internal friction peak of molybdenum.

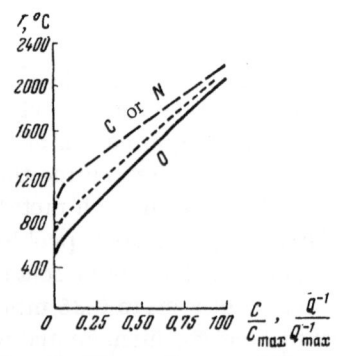

Fig. 3. Dependence of the relative concentration and of the relative internal friction on the quenching temperature.

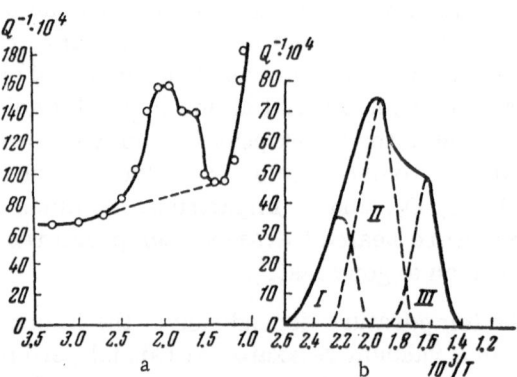

Fig. 4. a: Temperature dependence of the internal friction of molybdenum (sample 2) quenched from 2200°C. b: The same dependence after subtracting the internal friction background.

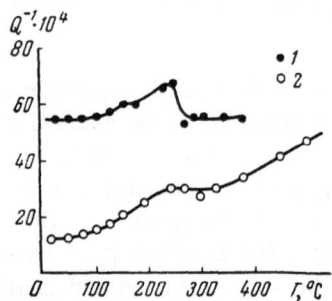

Fig. 5. Temperature dependence of internal friction of molybdenum. 1) After quenching from 1000°C; 2) after quenching from 2200°C and tempering at 600°C for 30 min.

quenching from 1000°C, the component III disappears (Fig. 5). Bearing in mind that 1000°C is below the temperature (1200°C) at which the concentration of carbon in the solid solution becomes noticeable, we may conclude that the component III is associated with carbon atoms. This is also supported by the value of the activation energy of the component III which, within the limits of the experimental error, agrees with the value of the activation energy of the diffusion of carbon (44.5 ± 8 kcal/mole), determined by Hartley and Wilson [3] from the measurements of the Young's modulus during aging.

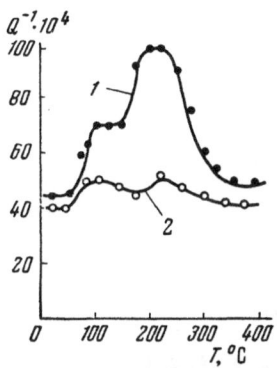

Fig. 6. Temperature dependence of the internal friction of molybdenum. 1) After quenching from 1600°C; 2) after annealing in hydrogen and quenching from 1600°C.

Moreover, Schnitzel [4] has recently reported an internal friction peak for carbon-bearing molybdenum in the region of 300°C. The amplitude of this peak was much less than that of the component III ($2 \cdot 10^{-4}$ and $50 \cdot 10^{-4}$, respectively), but their positions coincided quite well. The differences in the amplitudes of the peaks were probably due to the fact that Schnitzel [4] investigated a single crystal with the (100) main axis. In the ideal case, there should be no Snoek peak when such a crystal is subjected to torsion.

The central component II may be associated with oxygen, if we bear in mind that the variation in the amplitude of this peak correlates best of all with the variation in the solubility of oxygen (see Fig. 3). The component I is probably associated with nitrogen. This is supported by, in particular, the activation energy of this component, which is identical with the activation energy of the diffusion of nitrogen, found by Hartley and Wilson [3] to be 25.1 ± 2.7 kcal per mole.

Tempering at 600°C for 30 min reduced sharply the peak amplitude and the low-temperature background; moreover, it shifted the high-temperature component to the left, which could be associated with the process of aging of the solid solution supersaturated by quenching (see Fig. 5). A similar change in the peak amplitude took place when the solution was annealed in hydrogen (Fig. 6).

All these reported observations indicate that the observed peak was the Snoek peak of molybdenum. Deformation of annealed samples by rolling caused a shift of the high-temperature component to the left without any marked regular change in the background at 60-350°C. It has been reported [5] that a broad internal friction peak appeared on deformation of molybdenum. The shape of this peak resembled the peak which appeared on quenching, but the maximum of the peak was shifted toward high temperatures and the peak itself was broader. There are two possible explanations of this peak: it may be associated with the breaking away of dislocations from Cottrell atmospheres and with transitions of impurity atoms into solid solutions, or it may be due to differences between the conditions of deformation and measurement of the internal friction in our case and in [5].

Literature Cited

1. J. Snoek, Physica, 9:862 (1942).
2. Molybdenum (collection of papers, A.K. Natanson, ed.) [Russian translation], IL (1962).
3. C.S. Hartley and R.J. Wilson, Acta Met., 11:835 (1963).
4. R. Schnitzel, J.Appl.Phys., 30:2011 (1959).
5. S.Z. Bokshtein, M.B. Bronfin, and V.A. Marichev, Collection: Relaxation Phenomena in Metals and Alloys, Moscow (1963), Vol. 3, p. 123.

MAGNETIC DAMPING OF
ULTRASOUND IN NICKEL AND COBALT

V. F. Taborov and V. F. Tarasov

Magnetic damping of elastic vibrations in ferromagnetic materials has been investigated in a wide range of frequencies. It has been shown [1] that the maximum value of the logarithmic decrement, associated with the motion of domain boundaries, is observed at frequencies of the order of 0.1 Mc. Further increase in the frequency increases the damping much less than would be expected from the motion of domain boundaries only. Therefore, it is assumed that the rotational processes make a considerable contribution at higher frequencies. However, West [2], who investigated the temperature dependence of the damping of ultrasound in nickel single crystals, concluded that the modern theories of the damping of elastic vibrations in ferromagnets could not explain satisfactorily the dependence of the damping of ultrasound on the value of the magnetic anisotropy. Moreover, no explanation has as yet been offered for the monotonic reduction or the maximum of the elastic vibration damping observed when the magnetizing field is increased. This has been observed by us in the measurements of the damping of ultrasound in the megacycle frequency range.

The present paper reports a study of the damping of ultrasound in nickel and cobalt in a magnetic field, in the range of temperatures in which the magnetic anisotropy constants changed considerably. The measurements were carried out by a pulse method. A two-coordinate automatic recorder was used to determine the change in the damping due to the magnetization of a sample. The magnetic field intensity was measured with a Hall probe. The damping in nickel was measured using single-crystal samples at 30 Mc in the temperature range 77–375°K. Single crystals were grown by the Czochralski method. The samples were in the form of cylinders. The orientation and preparation of the samples were carried out using standard methods. A longitudinal wave was propagated along the <110> direction, coinciding with the cylinder axis. The magnetic field direction coincided with the base of the cylinder. By rotating a sample about its axis, we could magnetize it along various crystallographic directions. The damping in cobalt was measured using polycrystalline samples, cut from a cobalt ingot produced in a high-frequency furnace. The grain dimensions were 1–3 mm.

The following results were obtained in the measurements of the damping of ultrasound in nickel single crystals. When a sample was magnetized along the <111> direction, lying in the plane of the cylinder base, the damping decreased monotonically with increase in the magnetizing field. The dependence of the damping on the magnetic field intensity for magnetization along the <001> direction is shown in Fig. 1. At a certain value of the field intensity, the damping passed through a maximum. The maximum shifted toward stronger fields when tem-

18

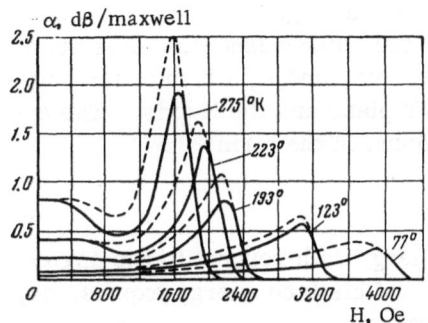

Fig. 1. Dependence of the damping
of ultrasound in a nickel single crystal on the magnetic field intensity.

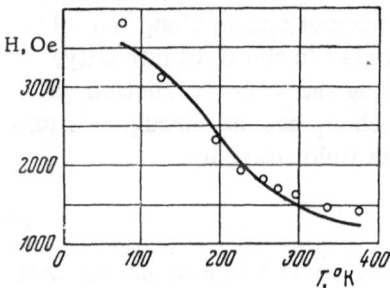

Fig. 2. Temperature dependence of
the field H corresponding to the damping maximum.

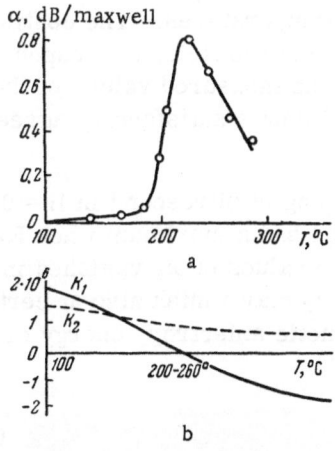

Fig. 3. Temperature dependence of
the damping of ultrasound in cobalt (a)
and the temperature dependences of
K_1 and K_2 (b).

perature was lowered. The continuous curves in
Fig. 1 represent the damping when the field was increased, while the dashed curves represent damping
when the field was reduced. The damping maximum
amplitude decreased on cooling. The damping at
saturation was assumed to be zero.

The results obtained on the damping of ultrasound in nickel single crystals can be explained satisfactorily using some of the theoretical investigations
of the interaction of spin and elastic waves and the
calculations of the ferromagnetic resonance frequency, which allow for the magnetic anisotropy energy.
Kittel [3] has considered the interaction of elastic and
spin waves and obtained an expression for the wave
number of transverse waves. His formula can be
used in the qualitative sense for longitudinal waves
as well. The damping per wavelength can be written
in the form

$$\frac{k_2}{k_1} = \frac{A\omega_i}{(\omega_r - \omega)^2 + \omega_i^2},$$ (1)

where k_1 and k_2 are coefficients of the real and imaginary components of the wave number; $A = \gamma b_2^2/2\alpha M_S$;
ω_r is the real component of the ferromagnetic resonance frequency, and ω_i is the imaginary component
of this frequency; ω is the frequency at which measurements are carried out; γ is the magnetomechanical ratio for spin; b_2 is the magnetoelastic coupling
constant [4]; M_S is the saturation magnetization; α is
an elastic constant.

Bearing in mind that γ = 2.8 Mc/Oe, we find
that the field corresponding to 30 Mc is about 11 Oe,
and, since the observed width of the maximum is
much greater than this value, we may assume that
$\omega = 0$. Then, Eq. (1) shows that the damping will increase with decrease in ω_r and with decrease in ω_i
(provided $\omega_r = 0$). Thus, in order to obtain the dependence of the damping on the magnetizing field intensity, it is necessary to know the dependences of
ω_r and ω_i on the field intensity. According to Kittel's
estimates [3], the damping per wavelength should be
of the order of unity, if ω_r and ω_i are of the order of
10^8 sec^{-1}. At a fixed temperature and frequency ω,
the value of ω_i does not vary greatly with the magnetic field intensity. As far as ω_r is concerned, there
are published data on the calculation of the ferromagnetic resonance frequency, in the presence of magnetic anisotropy, as a function of the magnitude and
direction of the magnetizing field. Artman [5] calcu-

lated the dependence of the ferromagnetic resonance frequency (for some special cases of crystals with cubic and hexagonal symmetries, allowing for the first magnetic anisotropy constant and for the domain structure) on the magnitude and direction of the magnetizing field. The calculation is usually carried out using coordinates such that the independent variables are θ, which is the angle between the Z axis and the magnetization vector, and φ, which is the angle between the projection of the magnetization vector onto the XY plane and the X axis. The expression for the ferromagnetic resonance frequency is then found in the form

$$\omega_r = \gamma \, (F_{\theta\theta} F_{\varphi\varphi} - F_{\theta\varphi})^{1/2} M_s^{-1} \mathrm{cosec}\ \theta, \tag{2}$$

where $F_{\theta\theta}$ and $F_{\varphi\varphi}$ are the second derivatives of the free energy with respect to the angles θ and φ; $F_{\theta\varphi}$ is a mixed second derivative. In the derivation of his free energy expression, Artman took into account the appropriate demagnetization factors.

For nickel, whose first magnetic anisotropy constant is negative, it is found that, when a sample is magnetized along the <111> direction, ω_r increases monotonically with increase in the field intensity. In the case of magnetization along the <100> direction, ω_r vanishes at some value of the magnetizing field. This variation of ω_r may explain the observed variation of the damping of ultrasound in nickel single crystals subjected to magnetization. If we substitute ω_r obtained by Artman into Eq.(1), then, in the case of magnetization along the <111> direction, the damping should decrease monotonically with increase in the field intensity, which was indeed observed, and in the case of magnetization along the <100> direction, a maximum of the damping should be observed at the value at which ω_r passes through a minimum. From Artman's calculations, it follows that $\omega_r = 0$ at the field intensity

$$H = 2K_1 M_s^{-1} + 4\pi h M_s, \tag{3}$$

where K_1 is the first magnetic anisotropy constant (absolute value), and h is a demagnetization factor.

For the sample on which our measurements were carried out, h = 0.2. The above formula could be used to determine the field intensity at which the damping maximum should be observed, because all the quantities in this expression were known. We knew also the dependences of K_1 and M_s on temperature [6, 7]. Therefore, we measured the dependence of the damping of ultrasound on the magnetic field intensity at various temperatures. The continuous curve in Fig. 2 shows the temperature dependence of the value of the field, corresponding to $\omega_r = 0$, calculated using Eq.(3). The circles in Fig. 2 denote the measured values of the field intensity corresponding to the damping maximum. Figure 2 shows satisfactory agreement between theory and experiment.

We can also relate the temperature dependence of the damping of ultrasound in H = 0 to the variation of ω_r. West [2] noted that the damping of ultrasound had a maximum when K_1 had a minimum. This was expected because, in H = 0, one of the two values of ω_r vanished in the absence of magnetic anisotropy [5]. In the demagnetized state, ω_r may vanish also at certain values or ratios of the magnetic anisotropy constants. The magnetic anisotropy energy of cobalt may be represented satisfactorily by the expression

$$F = K_1 \sin^2 \alpha + K_2 \sin^4 \alpha, \tag{3a}$$

where K_1 and K_2 are, respectively, the first and second magnetic anisotropy constants; α is the angle between the hexagonal axis and the magnetization vector. If $K_1 \geq 0$ and $K_2 > 0$, the magnetization vector is directed along the hexagonal axis, i.e., $\alpha = 0$. In this case,

$$\omega_r = \gamma F_{\alpha\alpha} M_s^{-1} = \gamma \, (2K_1 + 12K_2\alpha^2) M_s^{-1}, \tag{4}$$

where $F_{\alpha\alpha}$ is the second derivative of F with respect to α. It follows that if K_1 is equal to zero, then, irrespective of the value of K_2, ω_r vanishes.

The measurements of the damping of ultrasound in cobalt were carried out in the temperature range in which K_1 passed through zero. Figure 3b shows the temperature dependences of K_1 and K_2 [4, 8]. According to different authors, K_1 vanishes at temperatures of 200-260°C. A maximum of the damping of the ultrasound should occur in the same region. Figure 3a shows the results of the measurements of the magnetic damping of ultrasound in cobalt as a function of temperature in the range 100-290°C. The measurements were carried out at 5 Mc. A damping maximum was observed near 230°C. Thus, the experimental data were in agreement with the theoretical predictions.

If $K_1 < 0$, then ω_r depends also on the anisotropy in the basal plane. However, we shall not consider this point here. We can mention that the damping maximum can also be observed when $-K_1 = 2K_2$.

We note also that a maximum of the damping of ultrasound should correspond also to a magnetic susceptibility maximum. The terms which follow γ on the right-hand side of Eq. (2) represent an effective field H_{eff}, which is "sensitive" to the magnetization vector. The equation for forced harmonic vibrations of small amplitude, for example, vibrations of the component M_x, has the form

$$\gamma^2 H_{eff}^2 M_x - \omega^2 M_x = \gamma^2 H_{eff} M_s h_x - i\omega M_s h_y,$$

(5)

where h_x and h_y are the corresponding components of the real alternating field; $h_x, h_y \ll H_{eff}$; H_{eff} is directed along the z axis. If we neglect the second terms on the left-hand and right-hand sides of Eq. (5) [which is equivalent to $\omega = 0$ in Eq. (1)], then

$$M_x = h_x \chi,$$

(6)

where $\chi = M_s / H_{eff}$ is the magnetic susceptibility which will have a maximum when H_{eff} has a minimum.

Our discussion has so far ignored eddy currents. However, these currents can alter considerably the conditions under which the ferromagnetic resonance frequency can fall to values necessary to explain the observed magnetic damping of ultrasound. If the conditions for zero frequency of the ferromagnetic resonance, associated with an external field and anisotropy are satisfied, the expression for the ferromagnetic resonance frequency associated with the eddy currents has the form [3]

$$\omega_e = 4\pi\gamma M_s \left(1 - i\frac{c^2 k^2}{4\pi\sigma\omega}\right)^{-1},$$

(7)

where c is the velocity of light; σ is the conductivity; ω is the elastic wave frequency. For $k = \omega/v$, where v is the elastic wave velocity, the separation of ω_e into the real and imaginary components gives

$$\operatorname{Re}\omega_e = 4\pi\gamma M_s (1 + B^2\omega^2)^{-1},$$
$$\operatorname{Im}\omega_e = 4\pi\gamma M_s B\omega (1 + B^2\omega^2)^{-1},$$
$$B = c^2/4\pi\sigma v^2.$$

(8)

For nickel at room temperature at 30 Mc, $(\omega_e)_r$ and $(\omega_e)_i$ are of the order of 10^{11} sec^{-1}. Substituting these values of $(\omega_e)_r$ and $(\omega_e)_i$ into Eq. (1), we find that the magnetic damping, even in the most favorable case (a transverse wave along the field), is much less than the observed value. The reason for this discrepancy may be the fact that k in Eq. (7) is governed not by the acoustic wavelength but by inhomogeneities in the sample. Consequently, we can use Döring's model [9], which he has proposed for the explanation of the microeddy currents. In this model it is assumed that a ferromagnet has an inhomogeneity giving rise to such a stress field that the application of an alternating stress of frequency ω may rotate the magnetization vector in opposite senses at points separated by a distance λ. Such a motion is equivalent to a standing wave of frequency ω and wavelength λ, i.e., we should substitute $k = 2\pi/(2 - \lambda)$ and not $k = \omega/v$ into Eq. (8); here $2\pi/(2 - \lambda) \gg \omega/v$. To account for the observed values of the magnetic damping, we should have $\lambda \simeq 10^{-4}$ cm or less. It is natural to assume that these inhomogeneities represent the domain structure itself. This may be of some importance in the region where hysteresis is observed (see Fig. 1). However, the damping is also strong in fields in which hysteresis is not observed and in which the domain structure vanishes. This means that there are other stable inhomogeneities. Some information on the dimensions and nature of these inhomogeneities can be obtained by investigating the frequency dependence of the amplitude of the magnetic damping maximum. Investigation of this dependence will help also in the determination of the cause of the fall of the damping maximum amplitude when temperature is reduced.

Literature Cited

1. W.P. Mason, Physical Acoustics and the Properties of Solids, D. Van Nostrand Co., Inc., Princeton, New Jersey (1958), pp. 217-220.
2. F.G. West, J. Appl. Phys., 29(3):480-482 (1958).
3. C. Kittel, Phys. Rev., 110(4):836-841 (1958).
4. C. Kittel, Rev. Mod. Phys., 21(4):553-556 (1949).
5. J.O. Artman, Phys. Rev., 105(1):74-84 (1957).
6. W.J. Carr, J. Appl. Phys., 29(3):436-437 (1958).
7. R.M. Bozorth, Ferromagnetism [Russian translation], IL (1956), p. 343. [English edition: D. Van Nostrand Co., Inc., Princeton, New Jersey.]
8. G. Asch, Compt. Rend., 248:784 (1958).
9. W. Döring, Z. Physik, 114:597 (1939).

INFLUENCE OF ANNEALING TEMPERATURE
AND OF PURITY ON HIGH-TEMPERATURE INTERNAL
FRICTION OF NICKEL

O. A. Belous, V. N. Gridnev, A. I. Efimov,
and N. P. Kushnareva

The high-temperature internal friction of nickel has been investigated on several occasions. A grain-boundary peak at a vibration frequency of 1 cps has been observed in the region 420-450°C [1, 2, 3]. Moreover, the heating of samples (annealed first at 700-900°C) shows [4] that, in addition to its grain-boundary peak, nickel has one more internal friction peak in the range 630-720°C, associated with the relaxation of stresses along block boundaries. However, the available data on the high-temperature internal friction of nickel are not systematic and the measurements have been carried out on samples of various degrees of purity, which makes it difficult to compare them.

For these reasons, we carried out an investigation of the internal friction of nickel by heating in the temperature range 200-900°C deformed and annealed samples of various purities.

Materials and Investigation Method

We investigated electrolytic nickel of 99.9 and 99.99% purities. Nickel, initially in the form of slabs, was remelted in vacuum. Ingots were forged at 800°C into rods from which a wire of 0.8-mm diameter was prepared by cold drawing. We also investigated nickel of 99.99% purity after melting with an electron beam, which, according to the chemical analysis data, reduced considerably the content of Mn, Si, P, and other impurities. We could assume that the material subjected to the electron-beam melting was purer than 99.99%.*

A wire was produced from this extra-pure nickel by cold drawing. In all cases, the degree of deformation was 93-95%. The samples were annealed at various temperatures in a vacuum furnace for 1 h.

The measurements were carried out using a torsional pendulum [5] at frequencies of 1.7-2 cps. The rate of heating was 4 deg/min. We used samples 150 mm long. The maximum deformation of the sample surface was $3.5 \cdot 10^{-5}$. The measurements were made oscillographically every 10°C; in the region of a peak they were made in steps of 5°.

---

*Nickel subjected to electron-beam remelting was supplied by A.L. Tikhonovskii of the Electric Welding Institute, Academy of Sciences of the USSR.

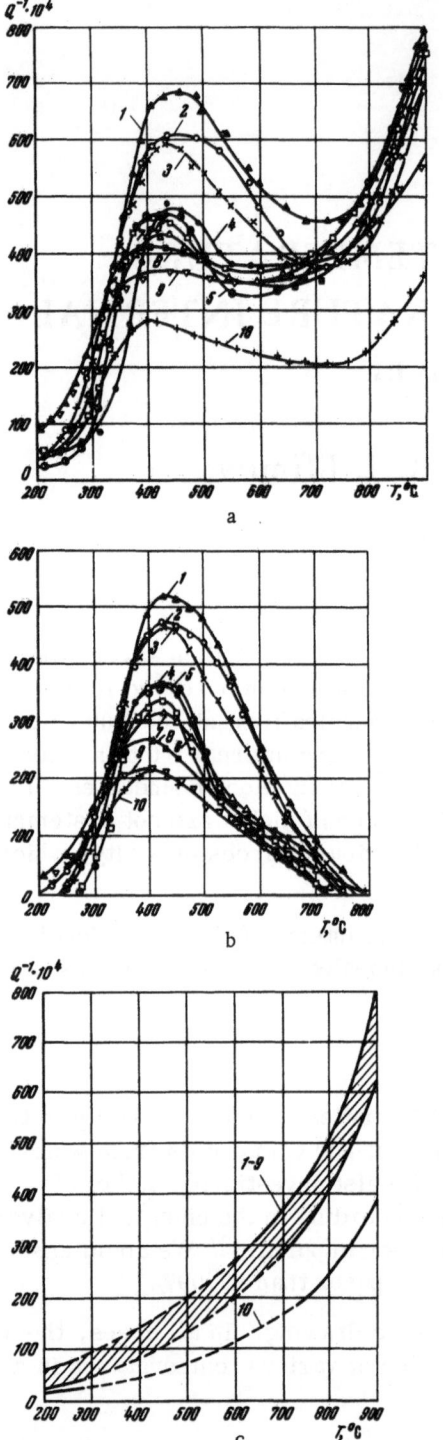

Fig. 1. Internal friction in 99.9% nickel.
a) General nature of variation in internal
friction of samples subjected to preliminary annealing at temperatures (°C): 1)
300; 2) 400; 3) 500; 4) 600; 5) 700; 6) 800;
7) 900; 8) 1000; 9) 1100; 10) 1200. b) Grain-
boundary relaxation. c) Internal friction
background.

To eliminate losses associated with the
magnetoelastic damping in the ferromagnetic
state of nickel, we recorded the necessary
curves in a saturating alternating field, produced by passing a current through the furnace
winding. We established that the alternating magnetic field had no influence on the high-temperature branch of the internal friction curve (above
350°C).

Results and Discussion

The internal friction curves of 99.9% pure
nickel, annealed in the temperature range 300–
1200°C, are shown in Fig. 1a. It is evident that
the internal friction background at 200°C was
strongest for samples annealed at low temperatures and that it decreased with increase in the
annealing temperature. At present, it is com-
monly assumed [6] that the higher values of the
internal friction of deformed samples, compared
with annealed samples, are associated with crystal lattice imperfections, especially dislocations.
Increase in the annealing temperature reduces
the number of lattice defects and decreases the
internal friction background. The beginning of
the internal friction rise shifts from 200 to 250°C.
It is necessary to mention that the slight
plastic deformation which could arise in an annealed sample affects the temperature at which
the internal friction begins to rise, by displacing
it toward lower temperatures.

The internal friction has a peak at 410–
430°C. As the preliminary annealing temperature is increased (i.e., the grain dimensions
are increased), the magnitude of the internal
friction in the region of the peak decreases. The
peak may be associated with the relaxation of
stresses along grain boundaries. Beginning
from 700°C, the internal friction background increases again. The magnitude of this background
decreases with increase in the preliminary annealing temperature. Figure 1c shows the internal friction background of 99.9% nickel extrapolated to low temperatures. The differences
between the values of the high-temperature internal friction of 99.9% nickel samples annealed
at various temperatures in the 300–1100°C range
are slight (Fig. 1c — shaded region). A considerable reduction in the background is observed
after a preliminary heating to 1200°C (curve 10).

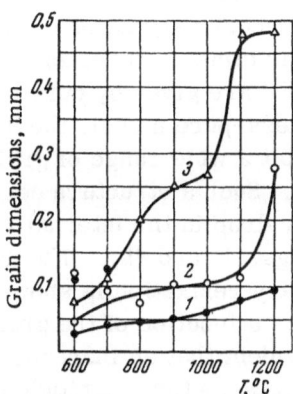

Fig. 2. Influence of the annealing temperature on the average grain diameter in nickel of various purity grades. 1) 99.9%; 2) 99.99%; 3) electron-beam treated material.

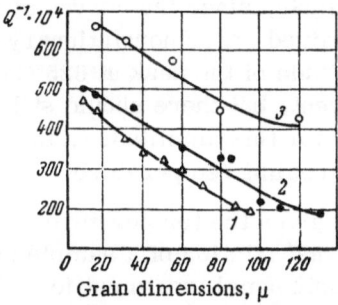

Fig. 3. Dependence of the amplitude of the grain-boundary peak on the grain dimensions in nickel of various purity grades. 1) 99.9%; 2) 99.99%; 3) electron-beam treated material.

The grain-boundary relaxation curves, obtained from Fig. 1a by subtracting the background, are given in Fig. 1b. We can see that the peak amplitude decreases with increase in the grain dimensions, but the temperature of the peak remains approximately the same. The profile of the peak is asymmetric. We may assume that the slow decrease in the 550-700°C region, observed for these samples, is associated with additional relaxation processes.

Microstructure investigations show that the recrystallization temperature of nickel of this purity is 350°C. The change in the grain dimensions on heating in the temperature range 600-1200°C is shown in Fig. 2 (curve 1). At 600-700°C there is a considerable range of grain dimensions. The dimensions of large grains are shown in Fig. 2. We note the relatively slight change in the grain dimensions due to heating up to 1200°C, which is evidently due to the retarding influence of impurities on the collective recrystallization processes.

The degree of the grain-boundary relaxation is shown in Fig. 3 as a function of the grain dimensions (curve 1).

The temperature dependence of the internal friction of 99.99% pure nickel, deformed plastically and annealed at 300-1200°C, is shown in Fig. 4a. An increase in the annealing temperature in the case of nickel of this purity leads to a reduction in the internal friction at 200°C, which is stronger than in the less-pure nickel. Moreover, the beginning of the rise of the internal friction curves shifts to 250-280°C.

In contrast to 99.9% nickel, the curves representing 99.99% nickel samples, deformed and annealed up to 700°C, exhibit two peaks at 400-440°C and 620-630°C. Their amplitude decreases with increase in the annealing temperature. Heating above 730-740°C leads to a further enhancement of the internal friction background.

The temperature dependence of the background, extrapolated to low temperatures, is shown in Fig. 4c. It is evident that a sharp reduction in the high-temperature internal friction of 99.99% nickel occurs at lower annealing temperatures (1100°C). The value of the background above 700°C is less than for 99.9% nickel (Fig. 1c).

The curves in Fig. 4b were obtained by subtracting the background from the curves shown in Fig. 4a. The peak at 400-440°C (see Fig. 4b) may be associated with the relaxation of the stresses along the grain boundaries. The temperature at which this peak occurs does not remain constant when the annealing temperature is increased. However, it has not been possible to establish a definite dependence on the grain dimensions. A considerable shift, to 500°C, is observed only for a sample annealed at 1200°C. The peak amplitude depends on the

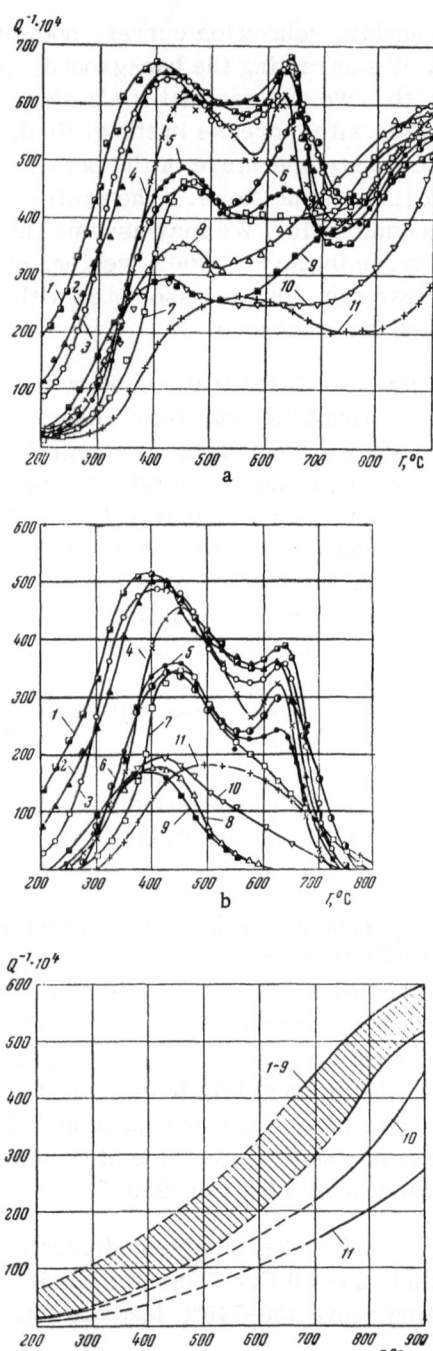

Fig. 4. Internal friction in 99.99% pure nickel. a) General nature of variation in the internal friction of samples deformed first and then annealed at temperatures (°C): 1) deformed; 2) 300; 3) 400; 4) 500; 5) 600; 6) 700; 7) 800; 8) 900; 9) 1000; 10) 1100; 11) 1200. b) Grain-boundary relaxation; c) internal friction background.

grain dimensions. Microstructure investigations show that the recrystallization temperature is 300°C. The grain dimensions of samples annealed at 600-1200°C are given by curve 2 in Fig. 2. As in the less-pure nickel, the 99.99% pure samples exhibit a wide range of grain dimensions at 600°C. Such a structure corresponds to a greater drop in the internal friction (cf. curves 5 in Figs. 4a and 4b). The degree of the grain-boundary relaxation is shown by curve 2 in Fig. 3 as a function of the grain dimensions. It is evident that an increase in the purity of nickel increases the magnitude of the internal friction for the same size of grain. The peak at 620-630°C is unstable and can be destroyed by heating to 800°C or above (Fig. 5a). The amplitude of this peak depends on the rate of heating used to record the curve. When the rate of heating is increased, the peak at 625°C increases markedly in amplitude (Fig. 5b). Because of this, the rate of heating of the samples was kept the same. Nevertheless, since the curves of Fig. 4a and 4b were obtained under nonstationary conditions, the amplitude of the peak at 625°C could have been different, but there should still have been a tendency for this amplitude to fall with increase in the annealing temperature.

Figure 5c gives the temperature dependence of the internal friction of a sample annealed at 600°C. An isothermal treatment for 1 h at the temperature of the peak reduces this amplitude monotonically. It is interesting to note that the rate of change of the internal friction decreases considerably after 40-min treatment. Further increase in temperature gives rise to a new, larger fall of the value of the damping. At 700°C an increase in the damping is observed, associated with the high-temperature background.

The behavior of the 625°C internal friction peak of 99.99% pure nickel leads to the conclusion that this peak may be associated with secondary recrystallization processes taking place in samples during heating in the measurements of the internal friction. Since the recrystallization peak is observed at temperatures higher than the grain-boundary peak, the activation energy of the recrystallization process is higher than the activation energy of the boundary relaxation [7]. The absence of such a peak in the less-pure

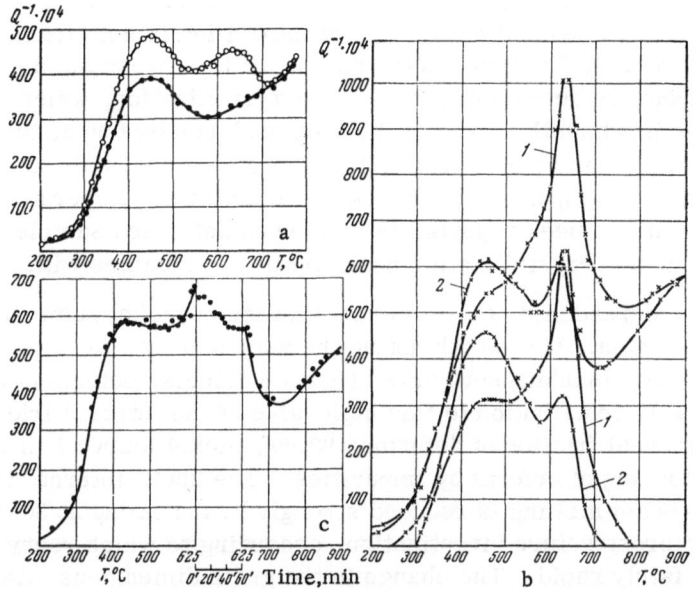

Fig. 5. Behavior of the recrystallization peak of the internal friction as a function of: a) heating a sample to 800°C; b) rate of heating: 1) 2.5-3°C per min; 2) 5-6°C per min; c) isothermal treatment at 625°C.

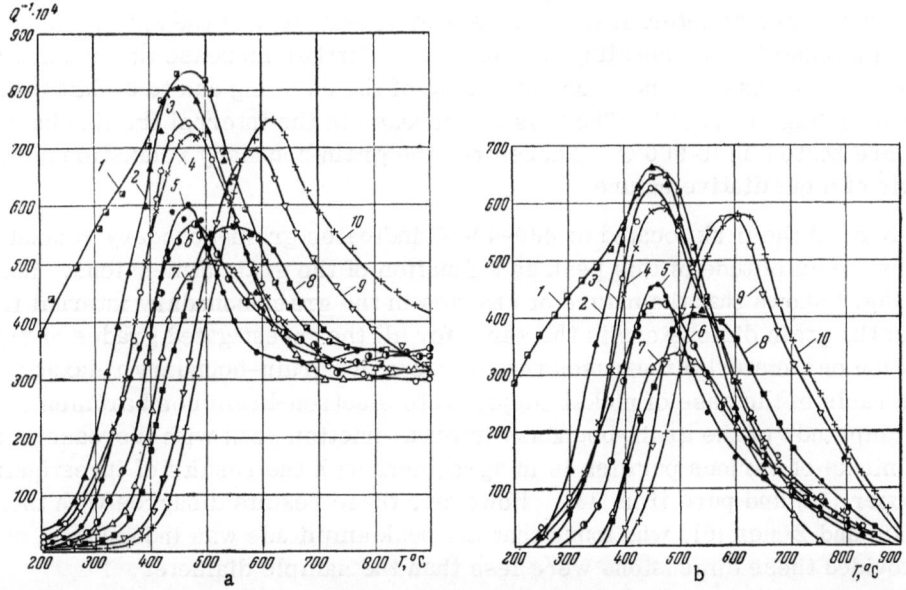

Fig. 6. Internal friction of nickel subjected to electron-beam remelting. a) General nature of variation in the degree of relaxation in samples deformed first and then annealed at temperatures (°C): 1) deformed; 2) 300; 3) 400; 4) 500; 5) 600; 6) 700; 7) 900; 8) 1000; 9) 1100; 10) 1200. b) Grain-boundary relaxation.

nickel and, as we shall show later, in nickel subjected to electron-beam remelting, indicates that the role of impurities in the appearance of the recrystallization peak may be important. Since there is, as yet, no information on the mechanism of the appearance of the recrystallization-type internal friction peak, the influence of impurities is not clear. It should be mentioned that secondary recrystallization takes place in all the investigated grades of nickel. However, we may expect the activation energy of recrystallization, which governs the possible range of temperatures in which the recrystallization peak can be found, to depend on the purity of the material. Evidently, in 99.99% nickel, the collective recrystallization and the peak are found in the same range of temperatures, which is the optimum temperature range for the appearance of the peak. It has been reported [8] that when deformed samples of tungsten of the VA-3 grade are heated, a recrystallization peak appears in samples with a certain structure.

The temperature dependence of the internal friction of nickel subjected to electron-beam remelting is given in Fig. 6a. It is worth noting the very high values of the damping for a deformed sample (curve 1). In addition to crystal lattice defects, defects of the slip-band type may make a considerable contribution to this high value of the internal friction [6, 9]. The lower values of the internal friction of deformed 99.99% nickel (curve 1 in Fig. 4a) may be due to the processes of blocking of defects by impurities. The 200°C internal friction of nickel subjected to electron-beam melting is reduced strongly by annealing at 300°C. This is in agreement with the microstructure investigation, according to which recrystallization takes place at 200°C and is fairly rapid. The change in the grain dimensions after annealing in the 600-1200°C range is represented by curve 3 in Fig. 3.

Comparison of the 200°C values of the internal friction of the investigated nickel samples, subjected to annealing at the same temperatures, indicates that the degree of relaxation decreases with increase in the purity of the material. This may be associated with the different degree of perfection of the crystal lattice of grains reached at a given annealing temperature.

Nickel subjected to electron-beam bombardment has two internal friction peaks. Samples annealed up to 900°C have an internal friction peak at 460-490°C and the amplitude of this peak decreases with increase in the annealing temperature. Further increase in the annealing temperature leads to an increase in the maximum value of the damping and to a shift in the peak to 625°C after annealing at 1200°C. There is no increase in the internal friction background when samples are heated up to 900°C. Therefore, the grain-boundary relaxation curves of Fig. 6b are basically of a qualitative nature.

The behavior of the peak located at 460-490°C indicates grain-boundary relaxation. Curve 3 in Fig. 3 gives the amplitude of this peak as a function of the grain dimensions. Inspection of the curves in Fig. 3 shows that the nature of changes in the grain-boundary internal friction with changes in the grain dimensions is the same for all the investigated grades of nickel. However, purification of a material increases the degree of its grain-boundary relaxation, which is very considerable in the case of nickel subjected to electron-beam bombardment. The reduction in the amplitude of the grain-boundary internal friction peak with increase in the grain dimensions of nickel of various purities is in agreement with the results of investigations carried out on copper [10] and pure iron [11]. However, these results disagree with those obtained by Ke [12] and Zener [6], who found that the peak amplitude was independent of the grain dimensions provided these dimensions were less than the sample diameter.

The internal friction peak found in samples annealed at or above 1000°C cannot be associated with the grain-boundary relaxation. Its behavior is anomalous. Moreover, it appears in samples with very large grains, which have a characteristic "bamboo" structure in which the overall area of the grain boundaries is relatively small. It has been suggested [13] that this peak may be due to the relaxation of stresses along block boundaries. Further investigations have shown that it may be associated with structural changes taking place under the action of axial loads (25 g/mm²) applied at elevated temperatures during the recording of the curve. Numerous investigations have shown that, under certain conditions, a substructure is formed in nickel during creep tests [14, 15].   The absence of enhancement of the high-temperature internal friction of nickel subjected to electron bombardment supports this conclusion.

Literature Cited

1. T. Ikhiyama, J. Japan Inst. Metals, 24(3) (1960);   Ref. Zh. Met., Vol. 2 (1961).
2. V. S. Postnikov, I. M. Sharshakov, and É. M. Maslennikov, Collection: Relaxation Phenomena in Metals and Alloys, Metallurgizdat (1963).
3. I. B. Kekalo and B. G. Livshits, Fiz. Metal. i Metalloved., 13(1) (1962).
4. O. I. Datsko and V. A. Pavlov, Collection: Relaxation Phenomena in Metals and Alloys, Metallurgizdat (1960), p. 234. [English translation: B. N. Finkel'shtein, ed., Consultants Bureau, New York (1963), p. 174.]
5. A. I. Efimov, this volume, p. 217.
6. C. M. Zener, Elasticity and Anelasticity of Metals [Russian translation], IL (1954). [English edition: University of Chicago Press, Chicago, Illinois (1948).]
7. C. Wert and I. Marx, Acta Met., 1(2) (1953).
8. S. V. Elyutin, A. K. Natanson, E. I. Mozzhukhin, and O. A. Vasil'ev, Collection: Relaxation Phenomena in Metals and Alloys, Metallurgizdat (1963).
9. W. Koster, Z. Metallk., B53: 1 (1962).
10. W. Koster, L. Bangett, and W. Lang, Z. Metallk., B46(2) (1955).
11. G. W. Miles and G. M. Leak, Proc. Phys. Soc. (London), 78: 6 (1961).
12. Ke Ting-sui, Phys. Rev., 72: 41 (1947).
13. O. A. Belous, A. I. Efimov, and N. P. Kushnareva, Collection: Problems in the Physics of Metals and Metallurgy, No. 20, Izd. 'Naukova dumka" (1964).
14. G. Ya. Kozyrskii, Investigations of Refractory Alloys, Izd. AN UkrSSR (1963).
15. G. Ya. Kozyrskii and P. N. Okrainets, Collection: Problems in the Physics of Metals and Metallurgy, No. 16, Izd. AN UkrSSR (1962).

INVESTIGATION OF THE INTERNAL
FRICTION OF TECHNICAL-GRADE NICKEL

I. A. Azizov and K. V. Popov

The internal friction of nickel has been much investigated [1-10]. However, most investigations have been concerned with the temperature dependence of the internal friction of very pure nickel and have been carried out at above room temperatures.

The internal friction is sensitive to structure and, therefore, the results of individual investigations, concerned with different materials and different experimental conditions, are frequently very different. However, the following characteristics of the temperature dependence of the internal friction of nickel can be regarded as established.

1. Annealed pure nickel exhibits, in the temperature range 70-150°C, a broad internal friction peak associated with the ferromagnetism of nickel and magnetoelastic losses.

2. If nickel is cold-worked, the internal friction peak (or, more exactly, the internal friction background) increases strongly and the magnetoelastic peak disappears.

3. Variation in temperature and duration of annealing, as well as in the degree and method of preliminary deformation, affects the amplitude and position of this peak.

4. The application of an external constant or alternating magnetic field also affects the amplitude of the peak: in weak fields it increases, while in saturation fields it disappears.

5. In the temperature range 400-500°C pure polycrystalline nickel has a second internal friction peak, which is due to the relaxation of stresses along the grain boundaries when the grains slip in a viscous manner.

6. Nickel containing an excess of carbon in a solid solution has an internal friction peak of the relaxation type in the region of 250°C [11]. This peak is due to the diffusion of carbon into a nickel lattice having a high vacancy concentration. This peak is destroyed by tempering because of the precipitation of the supersaturated solid solution and dispersal of vacancies.

The purpose of the investigation described here was to determine the features introduced into the internal friction of technical-grade nickel by the presence of a considerable number of impurities, particularly those capable of giving rise to interstitial solid solutions.

Materials and Investigation Method

We investigated technical-grade nickel of the NP-3 type, containing 0.04% C and 0.0065% N, in the form of cold-drawn rods of 8-mm diameter.

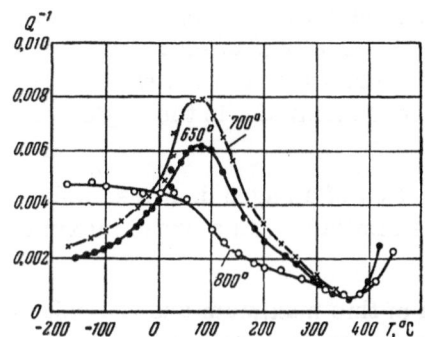

Fig. 1. Influence of the annealing temperature on the internal friction of the NP-3 nickel.

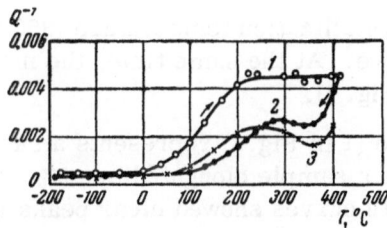

Fig. 2. Internal friction of cold-worked nickel. 1) Deformation (60%) by twisting during heating; 2) the same sample after 1 h at 420°C, curve recorded during cooling; 3) deformation by drawing, followed by aging for 1 year.

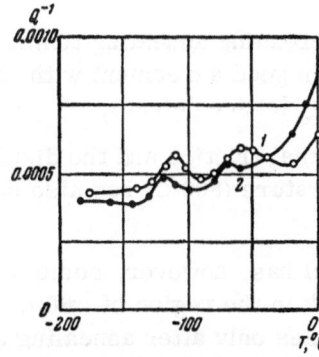

Fig. 3. Internal friction of cold-worked nickel below 0°C. 1) 10% deformation by elongation; 2) 40% deformation by twisting.

The internal friction was measured by the low-frequency method using apparatus described in [12] and cylindrical samples of 5 × 75 mm dimensions. Such samples were sufficiently rigid and they could be used repeatedly to carry out heat treatment, preliminary deformation, etc.

The measurements were carried out in the temperature range from −196 to +500°C. Liquid nitrogen was used for cooling below room temperature. Subsequent heating was carried out at a rate of 2-3°C/min. The measurements above 0°C were carried out during continuous heating or cooling, as well as at fixed temperatures.

Some of the measurements were carried out in a magnetic field of up to 250 Oe intensity. The vibration frequency was 1-3 cps. The maximum relative shear deformation did not exceed $2 \cdot 10^{-5}$.

Results of the Investigation

Internal Friction of Annealed Nickel. To investigate the influence of the annealing temperature on the internal friction, we annealed samples for 3 h a little above the recrystallization temperature − at 650°C and then at 700 and 800°C.

The results of the measurements are given in Fig. 1. It is evident from the curves in Fig. 1 that an increase in the annealing temperature led first to an increase in the 80°C magnetic peak and then this peak degenerated into an inflection on the internal friction curve.

Internal Friction of Cold-Worked Nickel. The temperature dependence of the internal friction of cold-worked nickel was investigated using samples subjected to a preliminary annealing at 800°C, followed by torsion or extension, as well as samples prepared from a cold-drawn rod in the as-supplied state. The latter were aged at room temperature for one year. We shall call them the drawn samples.

The measurements at room temperature began one hour after the deformation of a sample. The internal friction was about twice as high as in the case of the aged samples. The internal friction decreased considerably during measurements, falling in one hour by a factor of 1.5 (with the background subtracted) and the fall was nearly exponential.

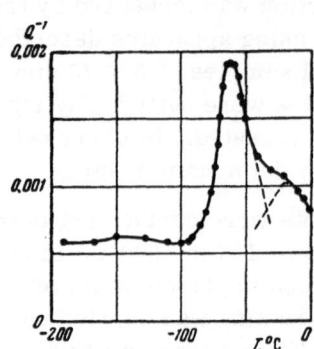

Fig. 4. Internal friction of drawn nickel after annealing for 24 h at 450°C.

The results of the measurements during continuous heating of a sample deformed by 60% twisting are given in Fig. 2 (curve 1). Cooling, after 1 h at maximum temperature, gave curve 2, which did not have a magnetic peak but had a clear maximum in the region of 280°C.

The internal friction of an aged drawn sample (curve 3 in Fig. 2) was similar, with a broad maximum near 250°C.

The large width of this maximum and the relatively narrow range of frequency variation in the measurements gave only a rough estimate of the activation energy of the process responsible for the increase in the internal friction peak (H ≈ 16 kcal/mole).

Subsequent 3-h annealing at temperatures of 400, 500, and 600°C caused a gradual decrease in the amplitude of this peak.

An increase in the annealing temperature, to the recrystallization temperature (650°C), destroyed the peak and altered it into an inflection in the curve. At the same time, the magnetic peak appeared with its typical amplitude-dependence (Fig. 1).

The low-temperature part of the internal friction curve (1 in Fig. 3) represents a sample deformed by 40% twisting, while curve 2 in Fig. 3 represents a sample elongated by 10%. The measurements were carried out 24 h after deformation. Both curves showed clear peaks in the region of ⁻(100-120) and ⁻(50-70)°C.

The cold-drawn aged samples did not exhibit such peaks. However, after 24-h annealing at 400-450°C, a fairly strong asymmetric "two-humped" peak appeared with a well-defined maximum at ⁻60°C and a broad maximum at ⁻20°C (Fig. 4).

Discussion of Results

The internal friction of annealed technical-grade nickel has no features to distinguish it from high-purity nickel.

The appearance of a magnetic peak and its growth with increasing annealing temperature to 700°C, as well as its disappearance after cold-working, are in good agreement with [2, 3], and are due to the blocking of the motion of domain boundaries by lattice defects.

The reduction of the magnetoelastic component of the internal friction and the displacement of the peak to the left after annealing at a very high temperature (800°C) has also been reported in [3].

The internal friction of cold-worked technical-grade nickel has, however, some features which distinguish it from pure nickel. Thus, the relaxation peak in the region of ⁻60°C has not been reported before, and we have found it for aged drawn samples only after annealing at 450°C. We may assume that this peak is associated with the formation of clusters at dislocations.

The broad maximum in the region of 250-300°C, which is observed both for recently cold-worked as well as for aged samples, and which disappears on annealing, must be considered in more detail.

As mentioned earlier, a peak in this region has been reported by Ke for carbon-bearing nickel [11], and it is associated with the microdiffusion of carbon in the host lattice under the action of stresses. The necessary asymmetric lattice distortions are provided by excess vacancies generated by quenching. The peak is destroyed completely by annealing or reduction of the carbon content (to 0.04%) [11].

In the nickel investigated by us, the concentration of carbon did not exceed 0.04%, but nitrogen was present in our samples.

A similar maximum has been found also for slightly purer but also cold-worked technical-grade nickel of the N-1 type [13]. This peak was reduced by low-temperature annealing until it disappeared at the recrystallization temperature.

We can therefore conclude that the 250-300°C peak is of the deformation type, associated with the interaction between dislocations and the interstitial atoms of carbon and nitrogen.

Literature Cited

1. W. Köster, Z. Metallk., 35 : 246 (1943).
2. I. B. Kekalo and B. G. Livshits, Fiz. Metal. i Metalloved., 13(1) : 54 (1962); Fiz. Metal. i Metalloved., 13(4) : 599 (1962).
3. O. I. Datsko, Fiz. Metal. i Metalloved., 16(3) : 416 (1963).
4. V. S. Postnikov, G. K. Mal'tseva, and V. I. Razumov, Izv. Vysshikh. Uchebn. Zavedenii, Chernaya, Metallurgiya, No. 7 : 148 (1963).
5. O. I. Datsko and V. A. Pavlov, Collection: Relaxation Phenomena in Metals and Alloys, Metallurgizdat (1960), p. 236. [English translation: B. N. Finkel'shtein, ed., Consultants Bureau, New York (1963), p. 174.]
6. B. Ya. Pines and Teng Ko-seng, Collection: Relaxation Phenomena in Metals and Alloys, Metallurgizdat (1960), p. 295. [English translation: B. N. Finkel'shtein, ed., Consultants Bureau, New York (1963), p. 222.] Fiz. Metal i Metalloved., 8(4) : 599 (1959).
7. V. S. Postnikov, Fiz. Metal. i Metalloved., 4(2) : 344 (1957).
8. K. Mishek, Fiz. Metal. i Metalloved., 18(3) : 373.
9. A. Cisman, B. Rothenstein, and I. Hrianca, Stud. Cercetari Stiint. Tehn., 5(1-2) : 47 (1958).
10. I. L. Mirkin, V. Z. Tseitlin, and G. G. Morozova, Collection: Investigation of New Refractory Materials for Power-Engineering Applications, Mashgiz, Moscow (1961), p. 32.
11. Ting-sui Ke, Sci. Sinica, 4(4) : 519 (1955).
12. I. A. Azizov, K. V. Popov, and V. F. Vinogradov, Collection: Relaxation Phenomena in Metals and Alloys, Metallurgizdat (1963), p. 72.
13. V. F. Sukhovarov, N. A. Aleksandrov, and L. A. Kudryavtseva, Fiz. Metal. i Metalloved., 6 : 895 (1962).

INTERNAL FRICTION OF BORON

F. N. Tavadze, I. A. Bairamashvili,
V. Sh. Metreveli, and G. V. Tsagareishvili

At the Georgian Metallurgy Institute, experiments have been carried out on the preparation of pure boron single crystals by the method of zone-refining and the properties of such single crystals have been investigated. Boron has been known as an element for a long time, but its properties have not often been investigated. Data on the physicochemical properties of boron have been accumulated over the last 150 years, but the earlier data are not sufficiently accurate because pure crystalline boron has become available only in the last 3–5 years. For this reason, many physicochemical properties of pure boron have not yet been investigated. One such property is the internal friction of boron. Investigation of the internal friction will undoubtedly give additional information on relaxation processes taking place in crystals having the covalent type of binding. We investigated the temperature dependence of the internal friction of boron by a low-frequency method.

To measure the internal friction we prepared single-crystal boron filaments ~0.7 mm in diameter and up to 110 mm in length by the floating-zone method, in which an electron beam was used, followed by chemical thinning.

The internal friction of crystalline boron, containing 0.002% metallic impurities, 0.0001% hydrogen, and 0.01% oxygen, was investigated by the method of damping of torsional vibrations in the temperature range from room temperature to 850°C at a frequency of about 1 cps using apparatus of the RKF-MIS type. The internal friction was measured in 15° steps with 5 min at a given temperature. The continuous curve in the figure gives the temperature dependence of the internal friction of boron. We can see that at about 260°C there is a well-defined internal friction peak with an unusually large amplitude of about $1400 \cdot 10^{-4}$ units. The background amounts to $\sim 8 \cdot 10^{-4}$ (the internal friction of boron was measured at deformation amplitudes of the order of 10^{-5}).

The temperature dependence of the square of the natural frequency (proportional to the shear modulus) is shown dashed. At first, the square of the frequency decreases with increase in temperature. Beginning from 160°C, this decrease becomes rapid, corresponding to the beginning of the

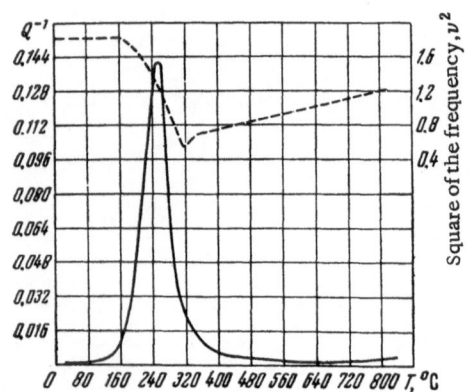

Temperature dependence of the internal friction and of the modulus of rigidity of boron.

34

rapid rise in the internal friction. At 320°C the curve representing the frequency squared has a minimum and above this temperature it increases. In the present stage of our investigation it is difficult to interpret the nature of the observed peak. All we can say is that a polymorphic transition is not involved in the described relaxation effect, because dilatometric and thermal analyses of microsamples showed no such transition in remelted crystalline boron.

The peak is not due to grain-boundary relaxation processes either, because the experiments were carried out on boron single crystals. It follows that the nature of the internal friction peak on boron requires further investigation.

INFLUENCE OF BORON ON THE
INTERNAL FRICTION OF PURE IRON

F. N. Tavadze, I. A. Bairamashvili,
and V. Sh. Metreveli

It is known that when boron is dissolved in α-iron its anelastic properties are affected.

Hasiguti and Kamoshita [1] investigated the internal friction of boron steel and discovered, at 20 kc, three peaks located at temperatures of 130, 160, and 180°C; they ascribed these peaks to the effects of interstitial atoms of boron, nitrogen, and carbon, respectively.

Pridantsev, Meshcherinova, and Piguzov [2] investigated the influence of boron on low-carbon iron and found a peak at 30°C and 1 cps, which increased in amplitude with increase of the boron content in the sample.

Thomas and Leak [3] found, for α-iron at 1 cps, a broad peak near 40°C, which they ascribed to the simultaneous effect of carbon and boron. The amplitude of this peak increased with increase in the boron content in the iron. They were unable to prepare a carbon-free boron-doped sample, since the iron became saturated with carbon during the introduction of boron. They ascribed the peaks observed at 15.5 cps to boron and carbon: the 65°C peak they ascribed to carbon and the 80°C peak to boron. They found the activation energy of the diffusion of boron in iron to be 15 kcal/mole.

All the investigators of the anelastic effects in iron, caused by the presence of boron atoms, concluded that boron dissolved in iron to form, like carbon and nitrogen, a solid solution of the interstitial type. However, the nature of the solid solution of boron in iron is still in doubt.

Wever and Muller [4] concluded, from x-ray structure data, that boron formed a substitutional solid solution with α-iron since they observed a reduction in the lattice constant of α-iron. Investigating Armco iron containing various amounts of boron, Shvelev [5] observed a reduction in the crystal lattice period, alloy density, and interference lines in the x-ray diffraction patterns and he came to the same conclusion. Busby, Warga, and Wells [6] investigated the diffusion of boron in iron and obtained the activation energy values of 62 and 21 kcal per mole for α- and γ-iron, respectively, and, thence, they concluded that boron formed a substitutional solid solution in α-iron, but an interstitial solid solution in γ-iron.

Thus, the problem of the nature of the solid solution of boron in boron steel remains unsolved. We must mention that each of the cited investigations has been carried out using insufficiently pure materials.

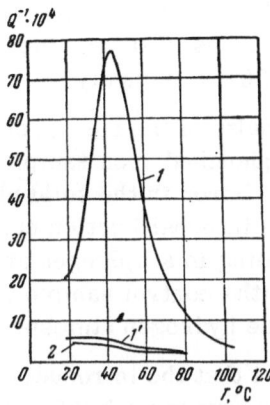

Fig. 1. Temperature dependence of the internal friction of iron before and after introduction of boron; the samples quenched from 720°C after being kept at this temperature for 10 min. 1) Sample containing boron; 1') sample before introduction of boron; 2) control sample.

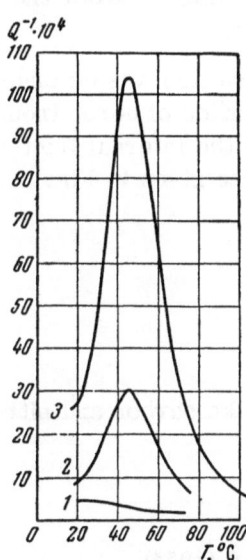

Fig. 2. Temperature dependence of the internal friction of iron samples containing different amounts of boron and quenched from 720°C. 1) Free of boron; 2) 0.004% B; 3) 0.011% B.

We attempted to shed some light on this problem.

As the initial material we used carbonyl iron, whose chemical composition (in wt.%) is given below:

C	N	S	P	Mn	Si	Cr	Ni
0.001—0.002	0.005	0.002	0.001	traces	traces	traces	traces

The iron was purified by zone-melting in an atmosphere of moist hydrogen and this was followed by annealing in a stream of moist hydrogen at 750–800°C for 50, 100, and 200 h. Iron samples purified in this way showed no internal friction peak due to carbon.

Iron prepared in this way was then doped with high-purity boron, prepared by pyrolysis of diborane, whose chemical composition (in wt.%) is given below:

O$_2$	H$_2$	Si	Fe	Al	Ca	Mg	Pb	C
0.1	0.18	$1.5 \cdot 10^{-4}$	$2.5 \cdot 10^{-4}$	$1 \cdot 10^{-4}$	$2 \cdot 10^{-4}$	$5 \cdot 10^{-5}$	$5 \cdot 10^{-5}$	$1 \cdot 10^{-3}$

Doping with boron was carried out in two ways: in the solid state and by melting.

The introduction of boron in the solid state was carried out using iron wires (0.75 mm in diameter and 110 mm long), whose internal friction was recorded first. The samples were placed in a powder of pure boron. The process was carried out in an atmosphere of hydrogen applying 900°C for 2.5 h. Then, the samples were taken out of the boron powder and made homogeneous by heating for 6 h at 900°C in a hydrogen atmosphere.

In the second method, we introduced boron by melting in an atmosphere of pure helium. We melted in the same furnace a control sample made of pure iron in order to check the effect of the atmosphere and of the substrate (Al$_2$O$_3$). After melting, the samples were kept at 1100°C for 4 h in a helium atmosphere. The resultant ingots were in the form of spherical pellets, 10 g in weight. The material was melted in the form of a pellet in order to reduce the area of contact with the substrate.

The ingots were subjected to cold rolling and drawing to reduce the diameter to 0.75 mm. The cold working of an ingot, right down to producing a wire of the final dimensions, was carried out without intermediate annealing. The samples finally obtained were annealed at 860°C for 1 h. The final heat treatment of all the samples consisted of quenching in oil from a temperature of 720°C after keeping them at this temperature for 10 min. This heat treatment was carried out in a vacuum furnace at a residual pressure of 10^{-3} mm Hg.

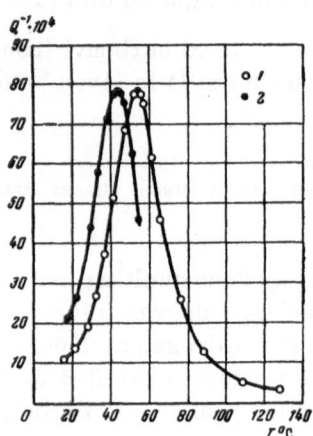

Fig. 3. Internal friction peak of boron–doped iron after quenching. 1) Vibration frequency 0.62 cps; 2) 1.57 cps.

The internal friction was measured using a vacuum relaxator (torsional pendulum) of the RKF–MIS type, vibrating at a frequency of ~ 1 cps, and applying an alternating magnetic field of 120-Oe intensity.

Figure 1 shows the results of the measurements of the internal friction of a quenched iron sample before and after introduction of boron in the solid phase. The same figure includes the internal friction curves of a control sample after annealing in a hydrogen atmosphere for 8 h and quenching (the control sample was used to check the purity of the hydrogen atmosphere).

It is evident from Fig. 1 that the introduction of boron into iron gave rise to a strong peak (75 · 10^{-4} units) at 44°C.

Figure 2 shows the temperature dependence of the internal friction of two samples doped with boron by melting in a helium atmosphere, and of a sample of pure iron melted in the same atmosphere; all these samples were quenched. According to chemical analysis data, the samples contained 0.004 and 0.011 wt.% boron.

It is evident from Fig. 2 that, as the content of boron in the iron was increased, the internal friction peak increased in amplitude. A sample of pure iron, melted under the same conditions and tested after the same heat treatment, showed no peak.

To determine the activation energy of the peak, the internal friction was determined at two different frequencies (0.62 and 1.57 cps). To avoid the precipitation of boron from the solid solution in a quenched sample, the temperature dependence of the internal friction was investigated at relatively low temperatures. The curves obtained are given in Fig. 3.

The activation energy, calculated from the formula

$$H = \frac{R \cdot T_1 \cdot T_2}{T_2 - T_1} \ln \left(\frac{\omega_2}{\omega_1} \right),$$

was 18.8 kcal/mole.

Thus, the data indicated that boron dissolved in iron forms, like carbon and nitrogen, a substitutional solid solution.

Literature Cited

1.   R.R. Hasiguti and O. Kamoshita, J.Phys.Soc.Japan, 9(4):646 (1954).
2.   M.V. Pridantsev, O.N. Meshcherinova, and Yu.V. Piguzov, Dokl.Akad.Nauk SSSR, 111(1) (1956).
3.   W.R. Thomas and G.M. Leak, Nature, 29:176 (1955).
4.   F. Wever and A. Muller, Mitt. Kaiser-Wilhelm-Inst. Eisenforsch., Düsseldorf, 2:152 (1930).
5.   A.K. Shvelev, Dokl.Akad.Nauk SSSR, 123(3) (1958).
6.   P.E. Busby, M.E. Warga, and C. Wells, Trans. AIME, 197:1463 (1953).

INVESTIGATION OF THE INTERNAL FRICTION
AND YOUNG'S MODULUS OF
ANTIMONY AND BISMUTH

A. Z. Zhmudskii, P. A. Maksimyuk,
and V. D. Mikhalko

It has been suggested [1–3] that an increase in the conduction electron density reduces Young's modulus. This suggestion is based on the observation that an increase in the conduction electron density leads to an increase in the repulsion forces between atoms in a crystal lattice [4, 5] and thus reduces Young's modulus.

It is known that antimony and bismuth have metallic and semiconducting properties [5, 6]. The metallic properties are exhibited at elevated temperatures; at low temperatures they are much weaker. This characteristic behavior of antimony and bismuth is due to the complex interatomic binding: the metallic and covalent bonds coexist.

Since the role of the metallic bonding of bismuth and antimony increases with increase of temperature, it was of interest to investigate the temperature dependence of Young's modulus and of the internal friction of these "semimetals."

Experimental Data and Discussion

The internal friction and Young's modulus were measured using apparatus of the UIMD-2 type. The dynamic investigations were carried out using small deformations, which made it possible to measure Young's modulus at elevated temperatures. The internal friction was determined from the formula

$$Q^{-1} = \frac{\ln 2}{n\pi},$$

where n is the number of vibrations necessary for the amplitude of the vibrations to fall to half its original value.

Young's modulus was determined from the formula

$$E = 1.6388 \cdot 10^{-8} \frac{P}{l} \left(\frac{l}{d}\right)^4 f^2 \ \text{kg/mm}^2,$$

where P is the weight of the sample (g), l is the length (cm), d is the diameter (cm), and f is the resonance frequency (cps).

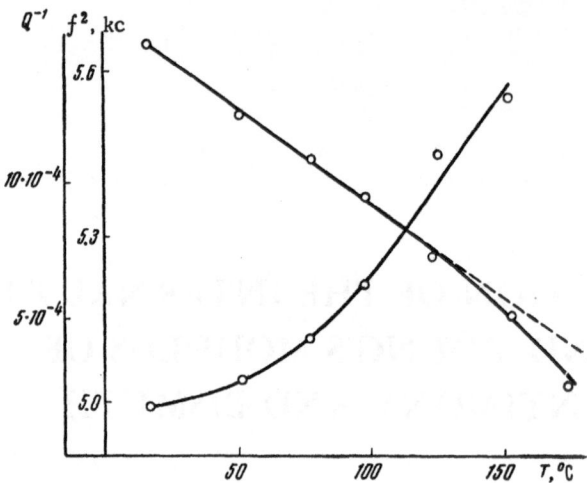

Fig. 1. Temperature dependence of the internal friction Q^{-1} and of the square of the frequency f^2 for antimony.

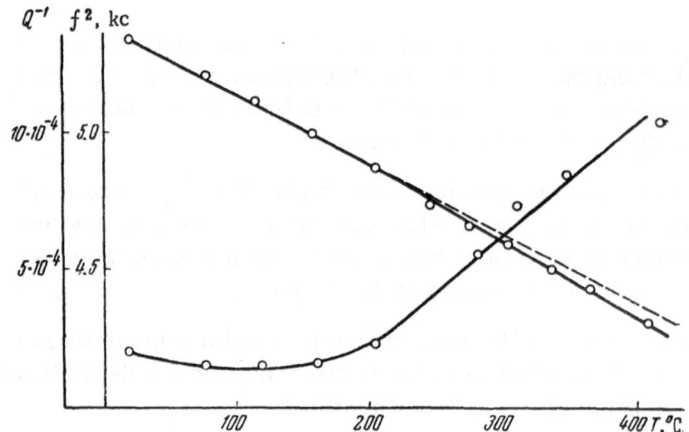

Fig. 2. Temperature dependence of the internal friction Q^{-1} and of Young's modulus (square of the frequency f^2) for a bismuth single crystal.

In the investigation of the relative change in Young's modulus with temperature, we can plot the dependence of the square of the frequency on temperature. In plotting the temperature dependence of Young's modulus itself, a correction has to be made for the change in the linear dimensions of the sample, bearing in mind a possible temperature dependence of the linear thermal expansion coefficient of the investigated material.

We investigated cylindrical samples of antimony and bismuth of the Analar grade. Since antimony is easily oxidized, the melting was carried out in vacuum. Variation in the rate of cooling of molten antimony made it possible to obtain samples having grains of various sizes: ranging from single crystals to fine-grained polycrystalline samples.

Bismuth was melted in glass tubes, as well as in a graphite demountable mold, and single-crystal samples were obtained. To determine the grain structure, we used an etchant of the following composition: 3 parts HF, 5 parts HNO_3, 3 parts CH_3COOH, and 1 part Br_2.

Figure 1 shows the temperature dependence of the internal friction and of the square of the frequency for antimony (the sample was virtually a single crystal; there were several grains along its length). It is evident from Fig. 1 that the internal friction increased slightly with temperature, without any singularities. As the grain size increased, the internal friction increased with temperature.

The temperature dependence of the square of the frequency had the following features. Up to 200°C, Young's modulus decreased linearly with increase in temperature. On further increase in temperature there was a deviation from the linear dependence. For the curve shown in Fig. 1, this deviation appeared as an additional decrease in Young's modulus. For polycrystalline samples we found a more complex dependence.

Figure 2 shows the temperature dependence of the internal friction and of Young's modulus of a single-crystal sample of bismuth. The axis of this cylindrical sample lay in the (111) plane. The internal friction of the bismuth single crystal increased slightly with temperature. The square of the frequency of free vibrations of the sample decreased linearly with increase of temperature. When the temperature was increased further, there was a deviation from the linear dependence.

At sufficiently high vibration frequencies (of the order of kilocycles), the internal friction of antimony and bismuth was obviously due to resonance phenomena. The additional decrease in Young's modulus in the investigated range of temperatures was much less than the possible considerable fall in the modulus due to the appearance of defects in the crystal lattice. The observed additional reduction in Young's modulus was probably caused by a weakening of the interatomic interaction, due to an increase in the number of conduction electrons, leading to an increase in the repulsion forces between atoms in the crystal lattice.

This assumption was in agreement with the data obtained in an investigation of the temperature dependence of the electrical resistivity and of the relative elongation of antimony [6]. Thus, the electrical resistivity of antimony decreased, instead of the usual increase in the electrical resistivity observed for metals on departure from the linear law. The linear expansion coefficient, which, in turn, depended on the interatomic interaction, increased with increase in temperature. This preliminary investigation of the internal friction and of Young's modulus of antimony and bismuth led to the following conclusions.

1. The internal friction in these two materials was due to resonance phenomena.

2. The additional decrease of Young's modulus was caused by the weakening of the interatomic interaction due to an increase in the conduction electron density.

Additional investigations are needed to refine these conclusions.

Literature Cited

1.   B. N. Finkel'shtein and A. I. Yamshchikova, Dokl. Akad. Nauk SSSR, 98(5): 781 (1954).
2.   L. J. Brunor and R. W. Reyes, Phys. Rev. Letters, 7(2): 15-16 (1961).
3.   P. A. Maksimyuk, Relaxation Phenomena in Metals and Alloys, Metallurgizdat (1963), p. 233.
4.   Ya. I. Frenkel', Introduction to the Theory of Metals, Metallurgizdat (1950).
5.   W. Delinger, Theoretical Metallurgy [Russian translation], Metallurgizdat (1960).
6.   Gmelins Handbuch der anorganische Chemie, Antimonium, B1, 18: 120-124 (1943).

INFLUENCE OF CRYSTAL STRUCTURE IMPERFECTIONS ON THE ELASTIC AND ANELASTIC PROPERTIES OF ALUMINUM

T. Ya. Benieva, L. N. Larikov, and I. G. Polotskii

Anelastic effects, associated with imperfections of the crystal structure of metals, have recently attracted more interest [1-5].

We investigated the temperature dependence of Young's modulus (E) and of the logarithmic decrement (δ) of single-crystal and polycrystalline aluminum of various degrees of perfection. Cylindrical single crystals of aluminum (100 mm long and 5 mm in diameter) were grown by the Bridgman method. The initial material used to prepare single crystals was 99.99% pure aluminum. The axes of the investigated single crystals were along the [110] direction. The lateral surfaces of the samples were ground and the deformed layers were removed by etching. Before measurements, single crystals of aluminum were annealed in vacuum for 1 h at 600°C.

The investigation of the temperature dependence of E and δ of aluminum in various structural states was carried out on the same sample. A polygonized structure was obtained by compression (by 1%) along the single-crystal sample axis in a special steel yoke, which ensured that the required degree of deformation was achieved and which prevented the sample from bending. Then, the aluminum sample was annealed in vacuum at 600°C for 1 h. The degree of polygonization was estimated by the x-ray diffraction method. A polycrystalline coarse-grained structure was obtained by annealing polygonized samples, deformed additionally by bending, and then straightened again.

To investigate E and δ, we used the resonance method [6]. To compare the data obtained on the temperature dependence of E and δ for various structural states, and to eliminate the influence of the deformation amplitude, we studied the amplitude dependence of E and δ over the whole investigated temperature range. From the measurements of the dependence of the vibration amplitude on the exciting stress, we determined the maximum relative deformation of an aluminum sample and found that its value was within the limits $1.5 \cdot 10^{-6}$-$5.8 \cdot 10^{-5}$. The value of E for zero vibration amplitude was determined by graphical extrapolation of E from a given deformation amplitude to the zero amplitude. From the measured values of the frequency we determined the relative change in the modulus, which was assumed to be equal to the ratio of the squares of the corresponding frequencies. In the investigation of the temperature dependence of E and δ, the sample was kept at a given temperature for 1.5-2 h before measurements.

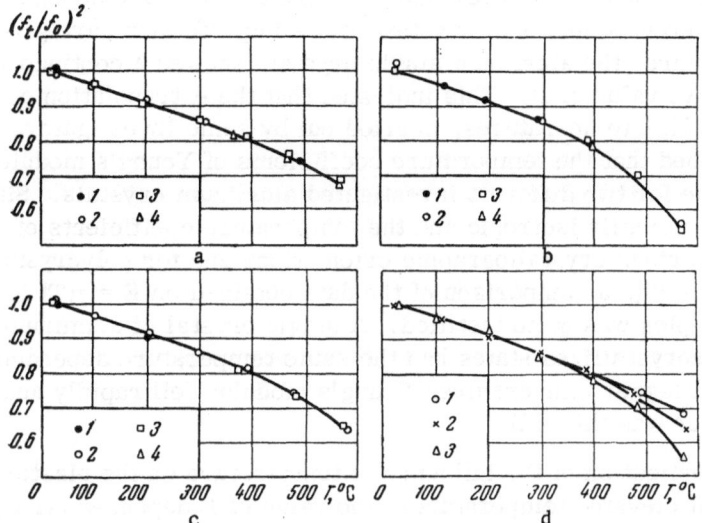

Fig. 1. Temperature dependence of Young's modulus of an aluminum single crystal: a) initial state; b) polygonized; c) recrystallized. 1) First measurement; 2) second; 3) third; 4) fourth. d) Average values for a single crystal: 1) initial state; 2) polygonized; 3) recrystallized.

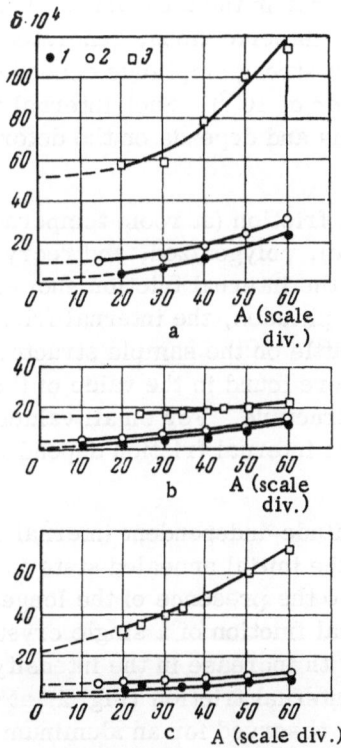

Fig. 2. Amplitude dependence of logarithmic decrement of vibrations in aluminum single crystal: a) initial state; b) polygonized; c) recrystallized. 1) 20°C; 2) 200°C; 3) 300°C.

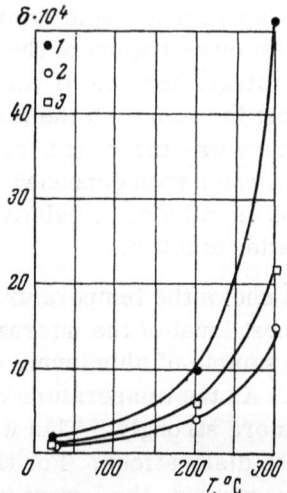

Fig. 3. Temperature dependence of amplitude-independent logarithmic decrement of vibrations in aluminum single crystal: 1) initial state; 2) polygonized; 3) recrystallized.

The results obtained in the investigation of the temperature dependence of E of aluminum samples having different amounts of crystal-lattice imperfections are given in Fig. 1. With increasing temperature, the aluminum single crystals showed a continuous but nonlinear decrease in the relative value of E. This indicated that the extrapolation of the low-temperature dependence E(T) to high temperatures, carried out by some investigators [7], was not justified. It should be mentioned that the temperature coefficients of Young's modulus ($\Delta E/\Delta T \cdot E_0$) were found to be the same for two different investigated aluminum crystals. Since aluminum crystals were almost elastically isotropic and the temperature coefficients of Young's modulus for single crystals of various crystallographic orientations and for polycrystalline samples did not differ too greatly [8], a comparison of the data obtained on $E = f(T)$ for single crystals and polycrystalline samples was quite justified. A single crystal of aluminum in the annealed, polygonized, and recrystallized states had the same temperature dependence of Young's modulus up to 300°C. At higher temperatures Young's modulus fell rapidly and the less perfect the crystal, the stronger was this fall.

Thus, an investigation of the influence of temperature on the elastic properties of aluminum showed that, at elevated temperatures, the value of E depended strongly on the structure of the sample. At temperatures close to the melting point, the largest value of the modulus defect was found for polycrystalline samples, and this effect increased in magnitude as the grain dimensions got smaller, i.e., as their boundaries increased. Consequently, Young's modulus could not be regarded as a definite measure of the interatomic binding forces at temperatures exceeding $0.5\,T_{melt}$ of a metal.

The internal friction of single crystals was very sensitive to the structure and was affected even by slight deformation. The reduction of the internal friction of aluminum single crystals, observed after some time from the moment of placing it in the measuring chamber, was due to a recovery process, associated with some unstable imperfections, generated during the mounting of the crystal. It is known that metals have a strong amplitude dependence of the internal friction even at very low deformations (of the order of 10^{-7}). Such internal friction is due to the unpinning of dislocations from impurity atoms and depends on the deformation amplitude [9].

Figure 2 shows the amplitude dependence of the internal friction (at room temperature, at 200 and at 300°C) for a single crystal in the initial (annealed), polygonized, and recrystallized states. An investigation of the amplitude dependence of the internal friction showed that, at room temperature, for initial values of the deformation amplitudes, the internal friction level was almost the same for the three states and depended little on the sample structure. When temperature was increased, considerable differences were found in the value of the internal friction, which then depended strongly on the sample structure. For small values of the deformation amplitudes at relatively low temperatures, the internal friction depended little on the deformation amplitude.

Figure 3 shows the temperature dependence of the amplitude-independent internal friction. The highest level of the internal friction was found for the initial annealed state in a single-crystal sample of aluminum, which was evidently due to the presence of the longest dislocation loops. As the temperature was increased, the internal friction of a single crystal increased even more strongly. This was probably associated with increase in the intensity of vibrations of the dislocations. For the amplitude-dependent internal friction (Fig. 2) at moderate temperatures, the lowest internal friction level was observed for an aluminum sample with the polygonized structure, i.e., with the structure characterized by the shortest dislocations.

Changes in the internal friction at high temperatures could be deduced from the change in Young's modulus of the samples caused by a change in the crystal structure. The smallest change in Young's modulus was found for a single crystal, and the largest for a recrystallized sample. This was because, at temperatures close to the melting point of the metal, the role of the anelastic phenomena along grain boundaries increased.

Literature Cited

1. A.S. Nowick, Progr.Metal Phys., 4:1 (1953).
2. K. Lücke and A. Granato, in: Dislocations and Mechanical Properties of Crystals, J.R.C. Fisher (ed.), Wiley, New York (1957).
3. W.P. Mason, Physical Acoustics and the Properties of Solids, D.Van Nostrand Co., Inc., Princeton, New Jersey (1958).
4. E.G. Shvidkovskii, M.P. Shaskol'skaya, N.A. Tyapunova, A.A. Predvoditelev, and A.A. Durgaryan, Paper presented at an International Regional Congress on Single Crystals, Turnov (1961).
5. D.H. Niblett and J. Wilks, Advan. Phys., 9(33):1 (1960).
6. I.G. Polotskii and V.F. Taborov, Zavodsk.Lab., No.8:986 (1957).
7. T.S. Ke, Metals Technol., 15:4 (1948).
8. P.M. Sutton, Phys.Rev., 91:4 (1953).
9. A. Granato and K. Lücke, J.Appl.Phys., 27:583 (1956).

INFLUENCE OF DIMENSIONS, IMPURITIES, AND DEFORMATION ON THE INTERNAL FRICTION AND STRENGTH OF WHISKERS

S. A. Ammer, A. T. Kosilov, and V. S. Postnikov

Whiskers have been intensively investigated since 1952 [1], after discovery of their unusual mechanical properties. In the last two years, several papers have been published on the relaxation properties of whiskers.

The change in the elastic properties and the relaxation of stresses in iron whiskers 30-70 μ in diameter have been investigated at low temperatures (down to 20°K) [2]. Torsional oscillations, manually applied, have been used to measure the internal friction at 20°C in 10^{-6} mm Hg vacuum.

The logarithmic decrement of vibrations has been found to be of the order of 10^{-2}. A more accurate measurement of the internal friction has been carried out using a torsional pendulum in 10^{-6} mm Hg vacuum [3]. The data obtained at room temperature are given in the table below.

The considerable weight of the torsional system, amounting to 80 mg, and the resultant strong axial stress in the sample ($\sim$225 g/mm^2 in a whisker of 10-μ diameter) resulted in an overestimate of the value of Q^{-1} for whiskers 6 μ in diameter and made it impossible to investigate thinner whiskers.

We proposed a low-frequency torsional micropendulum [4], by means of which the temperature dependence of the internal friction and of the shear modulus of whiskers of 3- to 10-μ diameter, or thicker whiskers, could be investigated in the temperature range from 20 to 800-900°C in 10^{-5} mm Hg vacuum. The axial load on a sample did not exceed 120 g/mm^2 and the tangential stress was 10 g/mm^2.

The present investigation is a continuation of earlier studies [4, 5] intended to determine the physical nature of the unusual properties of whiskers. It deals with the internal friction and strength of pure copper whiskers and whiskers made of copper with iron as an impurity.

Internal Friction in Whiskers

Material	Diameter, μ	Internal friction, $Q^{-1} = \theta/\pi$
Iron whisker	6	$2.1 \cdot 10^{-3}$
» »	8	$6.7 \cdot 10^{-4}$
» »	16	$7.6 \cdot 10^{-4}$
» »	20	$9.4 \cdot 10^{-4}$
» »	40	$4.0 \cdot 10^{-3}$
Copper whisker	30	$1.5 \cdot 10^{-3}$
Tungsten whisker	20	$1.4 \cdot 10^{-3}$
Quartz whisker	10	$1.5 \cdot 10^{-3}$

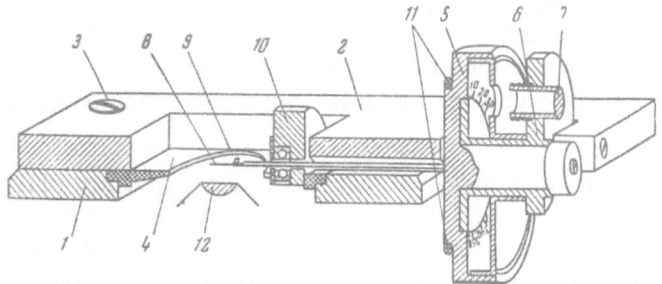

Fig. 1. Schematic diagram of the attachment to an MIM-7
microscope, used to study the surface of a whisker by rotating it. 1) Microscope stage; 2) pressure plate; 3) screw
for fixing and removing the attachment from the stage; 4)
diaphragm; 5) rotating graduated drum; 6) fixed disk; 7)
ocular; 8) whisker; 9) magnetic filings; 10) fixed disk with
a bearing; 11) drive from a motor; 12) microscope objective.

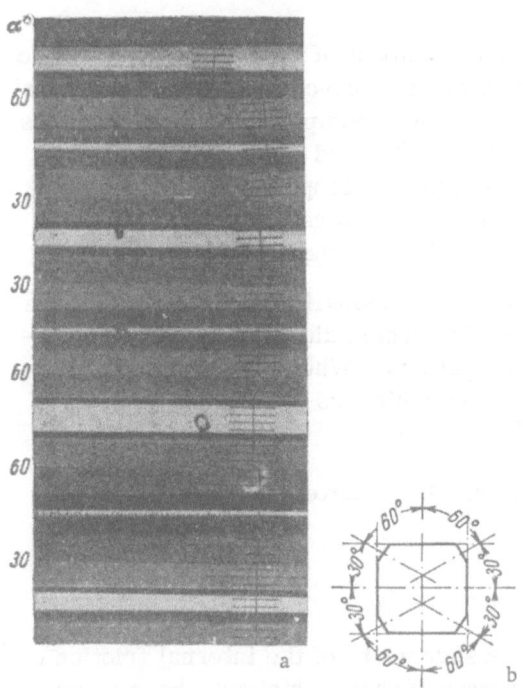

Fig. 2. Surface of an irregular whisker
with random defects: α is the angle of rotation of the whisker under a microscope
(a). The cross section of this whisker, deduced from the measured faces and angles,
is shown in (b).

Experimental Method

Pure copper whiskers were prepared at
560-570°C by the hydrogen reduction of copper
chloride anhydrate of Analar grade [6].

Copper−iron whiskers were grown at 800-
850°C in quartz boats by the simultaneous reduction of copper and iron halides in proportions of
10:1, 5:1, and 1:1 in a stream of hydrogen
mixed with argon [7]. The whiskers obtained
were of various lengths, diameters, and cross
sections. To determine more exactly the shape
of the cross section, and to find defects on the
whisker surface, we constructed a special
rotating device which was used as an attachment
to an MIM-7 microscope. This attachment made
it possible to investigate a whisker by rotating it
in front of an objective. The attachment was
simple, compared with others [8], and reliable.
It is shown schematically in Fig. 1.

The attachment was fixed directly to the
microscope stage 1, and moved together with the
stage. A diaphragm 4 inserted in the stage
and fixed rigidly to a plate 2, made it possible to
locate an object exactly with respect to an objective 12. A whisker 8 was placed on a special
recess in a steel rod and held to the rod by mag-
netic filings 9 (iron, nickel, and other strongly magnetic whiskers were self-adhering). The
rod, attached to a rotating drum 5, was rotated by a reversible motor DSDR-2, through a
transmission system 11. The drum was graduated in degrees in order to determine the angle
between the faces of a whisker by looking through an eyepiece 7, which could be moved along
its length in a fixed disk 6. The whisker grain dimensions were determined using the eyepiece

scale. A photographic camera ("Zorkii-S") was mounted on the microscope tube of the microscope and the investigated object was photographed. The attachment made it possible to avoid gross errors in the determination of the cross-sectional area, particularly when it had irregular geometrical shape and when the defects were distributed nonuniformly on its surface (Fig. 2).

This treatment did not damage the whisker, which distinguished our method from the labor-consuming preparation of microsections [9]. The measurements of the internal friction and of the strength of whiskers were carried out using a method described in [4]. The quantity $Q^{-1} = \theta/\pi$ was used as the measure of the internal friction. We investigated samples with ideally smooth and with defective surfaces. The measurements were carried out at room temperature in $2 \cdot 10^{-5}$ mm Hg vacuum.

To measure the internal friction of whiskers of a diameter less than 5μ, we first used torsional systems of 0.9-1.5 mg, parts of which were made of thick copper whiskers. The frequency of oscillations of the micropendulum was then of the order of 1 cps and the axial load on the sample did not exceed 100 g/mm^2.

In determining the strength of the whiskers, the length of the working part ranged from 1 to 5 mm and the diameter did not exceed 50μ for pure copper whiskers, and 100μ for copper—iron whiskers.

The structure of the Cu—Fe whiskers depended on the amount of the salt $FeCl_2$ added to CuCl. For the ratio of the salts 1:1 in the boats we obtained copper-colored whiskers as well as iron-colored whiskers. Metallographic investigations showed complex laminar structures similar to that described in [10, 11]. Interpretation of the rocking and Laue diffraction patterns showed that the central core of such whiskers consisted of a copper single crystal and the outer shell consisted of polycrystalline iron. For thicker whiskers we found traces of polycrystalline copper, which was evidently produced in the latter stages of growth.

Copper-colored whiskers, without an external shell, had a small amount of iron (of the order of the equilibrium amount at growth temperature [12]), which did not appear as a separate phase in metallographic and x-ray diffraction investigations. When the ratios of the copper and iron salts were 10:1 and 5:1, only copper-colored whiskers containing iron as an impurity were obtained.

We investigated whiskers with and without an iron shell (the latter representing a pure copper whisker).

Experimental Results and Discussion

It is evident from Fig. 3 that the results of the measurements of the internal friction of pure copper whiskers, as well as the results on the strength of these whiskers, have a considerable scatter. Moreover, the value of the internal friction depends strongly on the whisker diameter. The internal friction of the whiskers of more than $20-\mu$ diameter was of the same order as that of normal (large) single crystals. The minimum value of the internal friction was found for whiskers of diameter less than 10μ, i.e., those whiskers which had high strength and exhibited a yield point.

Assuming that the internal friction at room temperature is purely of a dislocation nature [13], the scatter of the experimental data on the internal friction and on the strength of whiskers, tested under the same conditions, can only be due to the different degrees of perfection of the whiskers. The least imperfect whisker should have the lowest value of internal friction and the highest strength. However, some whiskers were an exception to this rule. Among them were those of 3- to $12-\mu$ diameter (denoted by squares in Fig. 3), which, in spite of their

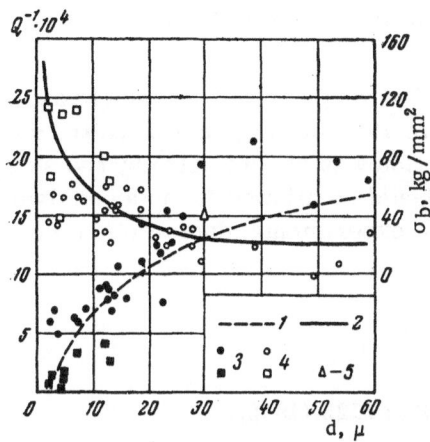

Fig. 3. Internal friction and tensile strength of pure copper whiskers of various diameters: 1) internal friction; 2) strength; 3) minimum value of the internal friction of whiskers and the corresponding strength (4); 5) results from [3].

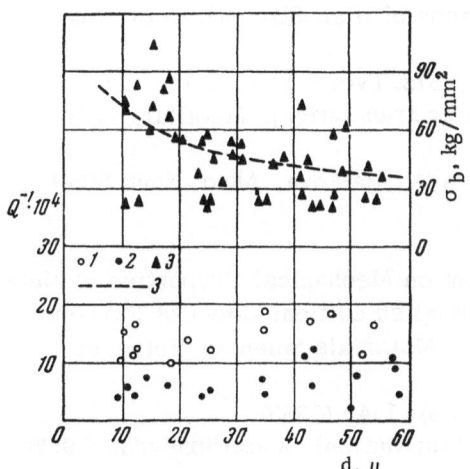

Fig. 4. Internal friction and tensile strength of copper whiskers with iron impurity: 1) internal friction of whiskers prepared from CuCl and $FeCl_2$ salts taken in the ratio 10 : 1; 2) the same, for the ratio 1 : 1; 3) dependence of the strength of impure whiskers on their diameter.

small internal friction, had a low tensile strength. Such a disagreement was obviously due to the statistical nature of the value of the strength of whiskers [14], because of the presence of random surface or volume defects in whiskers with a perfect lattice. The contribution of these defects to the internal friction was small. However, during deformation they may have acted as dislocation sources or helped to nucleate dislocations, and this reduced the strength. The surface defects are particularly effective in this way, since they can generate dislocations under much lower stresses than the defects in the interior [15, 16].

The addition of an iron impurity to copper whiskers (Fig. 4) (CuCl and $FeCl_2$ salts in the ratio 10 : 1) increased the internal friction of whiskers of less than 20-μ diameter to the value for the thicker whiskers, which indicated a considerable departure from the perfection of the crystal lattice and easier nucleation of dislocations at impurities. The strength of thin whiskers was strongly reduced by the iron impurity, while the strength of whiskers more than 10 μ in diameter increased, which was in good agreement with the well-known data [17].

Further increase in the impurity content (CuCl/$FeCl_2$ ratio 1 : 1) decreased the logarithmic decrement of whiskers by a factor greater than 2. Such behavior of the internal friction was probably associated with further increase in the density of structure defects and with the locking of dislocations.

The whiskers consisting of a mixture of metals with an iron shell exhibited the same damping of vibrations as the copper whiskers without such a shell, which were obtained using the 1 : 1 ratio of salt concentrations.

The internal friction of the pure copper whiskers was sensitive to cold working. After deformation of a whisker of 4.8-μ diameter to the stage of easy slip, its internal friction increased by a factor of almost 20. Subsequent annealing at 700°C for 1 h re-established fully the initial value of Q^{-1}. The effect of annealing may be explained by the emergence of dislocations on the surface of a crystal under the action of internal force fields.

It is known that the stresses generated by dislocations in a crystal lattice are very high: of the order of 10^7 dyn/cm^2 at a distance of 1 μ from a dislocation axis [18]. In very thin whiskers they may emerge on the surface without difficulty. This is confirmed also by the re-establishment of the initial shape and mechanical properties of deformed whiskers by annealing [19].

The copper—iron whiskers were less sensitive to cold working. Deformation of such whiskers made up to the point of appearance of slip bands increased the internal friction by a factor of only 2.

The value of the logarithmic decrement also depended strongly on the degree of vacuum. A change in vacuum conditions, for example from $2 \cdot 10^{-5}$ to 10^{-3} mm Hg, increased the damping by a factor of almost 10. The value of the internal friction was affected also by the strength of adhesion of a whisker to the micropendulum. All this should be allowed for in an analysis of the experimental data, particularly in comparisons with the experimental data of other investigators.

Literature Cited

1. C. Herring and J.K. Galt, Phys. Rev., 85:1060 (1952).
2. R. Conte, B. Dreyfus, and L. Weil, Acta Met., 10(12):1125 (1962).
3. E. Schurerovó, Czech. J. Phys., A14(2):151 (1964).
4. V.S. Postnikov, S.A. Ammer, and A.M. Belyaev, Fiz. Metal. i Metalloved., 19(2):268 (1965).
5. V.S. Postnikov and S.A. Ammer, Izv. Vysshikh Uchebn. Zavedenii, Chernaya Met. (in press).
6. S.S. Brenner, Acta Met., 4(1):62 (1956).
7. T.S. Ke and Y.K. Wan, Sci. Sinica, 10(3):301 (1961).
8. M. Schtreckenbach, Exptl. Tech. Physik, 11(1):67 (1963).
9. S.Z. Bokshtein and I.L. Svetlov, Zavodsk. Lab., 28(5):595 (1962).
10. T.S. Ke, Y.H. Chuang, and Y.K. Wan, Sci. Sinica, 10(3):318 (1961).
11. I.A. Oding and I.M. Kop'ev, Collection: Investigations of High-Strength Alloys and Whiskers, Izd. Akad. Nauk SSSR (1963), p. 3.
12. V.G. Kostyuk, K.K. Ziling, and A.V. Serebryakov, Fiz. Tverd. Tela, 5(11):3060 (1963).
13. D. McLean, Mechanical Properties of Metals [Russian translation], Metallurgizdat (1965). [English edition: Wiley, New York.]
14. V.N. Geminov and I.M. Kop'ev, Tr. Inst. Met. im A.A. Baikova, Akad. Nauk SSSR, No. 10:202 (1962).
15. M.A. Adams, Acta Met., 6(5):327 (1958).
16. I.R. Kramer and L.J. Demer, Effect of Environment on Mechanical Properties of Metal [Russian translation], Izd. "Metallurgiya" (1964). [English edition: Effect of Environment on Mechanical Properties of Metal (Progress in Materials Science, Vol. 9, Pt. 3), Pergamon Press, Inc., New York.]
17. S.Z. Bokshtein and I.L. Svetlov, Fiz. Tverd. Tela, 5(6):1749 (1963).
18. W.T. Read, Jr., Dislocations in Crystals [Russian translation], Metallurgizdat (1957). [English edition: McGraw-Hill, New York.]
19. F.N.R. Nabarro, Phil. Mag., 6(7) (1961).

INTERNAL FRICTION AND SHEAR MODULUS OF HEAT-RESISTANT NICHROME-BASE ALLOYS

V. N. Gridnev and A. I. Efimov

Heat-resistant materials used in modern gas turbines must satisfy a number of require-ments, among which the damping capacity of the alloy is important. It has been reported [1] that a cobalt-base alloy having a high value of the logarithmic decrement has found wide appli-cation in the manufacture of the fixed blades in steam turbines. Therefore, an investigation of the internal friction of heat-resistant alloys is definitely of interest. The determination of the internal friction at high temperature is also important in the investigation of viscous prop-erties of grain boundaries, and for the establishment of a possible relationship between the strength, heat-resistance, and the internal friction [2,3].

We investigated, over a wide range of temperatures (20-1000°C), the internal friction (the background and the grain-boundary relaxation component), as well as the shear modulus of a number of industrial heat-resistant Nichrome-base alloys. The results of the internal friction measurements were compared with the microstructure of the alloys.

Materials and Investigation Method

The chemical compositions of the investigated alloys are given in Table 1. The alloys Kh22N78 and VZh98 were heat-resistant. The strength of these alloys was increased by alloy-ing nickel with chromium (Kh22N78), as well as with chromium, tungsten, and carbon (VZh98). The strength of the alloys ÉI437B, ÉI607A, ÉI617 was increased by increasing the binding forces in the solid solution lattice using alloying elements (Cr, Nb, Fe, Mo, W, V), as well as by the precipitation of particles of the intermetallic γ'-phase on introduction of Ti and Al. The grain boundaries were further strengthened by boron.

To obtain the optimum heat resistance, the alloys were subjected to typical heat treat-ments listed in Table 2. The damping curves and the shear modulus (square of the frequency) were recorded in the temperature range 20-1000°C using a torsional pendulum [4]. In this in-vestigation the samples were 150 mm long and 0.8 mm in diameter. With the exception of the alloy ZhS6-K, wire-shaped samples were obtained by cold drawing. The degree of deforma-tion amounted to about 93%. Samples of the cast alloy ZhS6-K were prepared by centerless grinding. To reduce the error in the measurements at high temperatures, in the range of fre-quencies used (~2.5 cps), the damping curves and time marks were recorded on an oscillogram. The logarithmic decrement was determined at intervals of 10°C.

Table 1

Alloy designations	Alloy composition, wt.%							
	C	Si	Mn	P	S	Cr	Ni	Ti
Kh22N78	—	—	—	—	—	22	Base	—
VZh98	⩽0.1	⩽0.8	⩽0.5	—	—	23.5— 26.5	»	0.3—0.7
ÉI437B	⩽0.08	⩽1.0	⩽0.5	⩽0.02	⩽0.015	19—23	»	2.0—2.9
ÉI607A	⩽0.08	⩽0.8	⩽1.0	⩽0.02	⩽0.02	15—17	»	1.8—2.3
ÉI617	0.08	⩽0.5	⩽0.6	⩽0.02	⩽0.02	15	»	2
ZhS6-K	0.14	0.1	0.1	0.009	0.007	10.8	»	2.8

Alloy designations	Alloy composition, wt.%							
	Al	Nb	Fe	Mo	W	V	Co	B
Kh22N78	—	—	⩽4.0	—	—	—	—	—
VZh98	0.5	—	⩽4.0	—	13.0—16.0	—	—	—
ÉI437B	0.4—1.1	—	—	—	—	—	—	0.008
ÉI607A	0.5—1.0	1.0—1.5	⩽3.0	—	—	—	—	—
ÉI617	2	—	⩽5.0	3	7	0.3	—	0.008
ZhS6-K	5.3	—	0.2	3.8	4.9	—	4.5	0.008

Results of Investigation and Discussion

Figure 1 shows the internal friction curves of the investigated alloys. With the exception of the alloy VZh98, no anomalies were found in the temperature range 20-500°C. The alloy VZh98 had an internal friction peak in the temperature range 160-170°C. The amplitude of this peak depended on the tempering temperature of the VZh98 alloy. We could assume that this peak was associated with anelasticity, due to carbon atoms in the fcc lattice of the solid solution [5]. The value of the logarithmic decrement at room temperature is given in Table 3 for the various alloys. The same table includes the values of the tensile strength and of the yield point, in accordance with [6, 7]. We were unable to establish a definite relationship between the tensile strength and the internal friction background, and this was in agreement with the conclusions reported in [3].

It follows from Fig. 1a that, beginning from 500°C, the internal friction increased, exhibiting a peak or plateau in the region of 650-750°C. We may assume that this peak was due to the viscous properties of the grain boundaries. Increase in temperature to above 800°C increased the internal friction again. Figures 1b and 1c show the grain-boundary peak and the background of the internal friction, deduced from Fig. 1a. It is evident that the temperature of the peak of the Kh22N78 alloy was 640-650°C. It has been reported [8-11] that the grain-boundary peak for pure nickel was at 410-450°C. The shift of the peak of the Nichrome alloys Kh22N78 could be associated with the strengthening of grain boundaries by chromium atoms and by increase in the activation energy [12]. If, during the process of heat treatment, the grain boundaries were enriched with chromium atoms, this could also have a definite effect on an increase of temperature of the peak. The amplitude of the grain-boundary peak for the alloy Kh22N78 did not exceed $35 \cdot 10^{-4}$, although the grain dimensions were sufficiently small (0.05-0.06 mm) (Fig. 2a).

It is known that the amplitude of the grain-boundary peak of pure metals depends on the grain dimensions [10, 11, 13, 14] and, in the case of a fine-grain sample, it may be very high

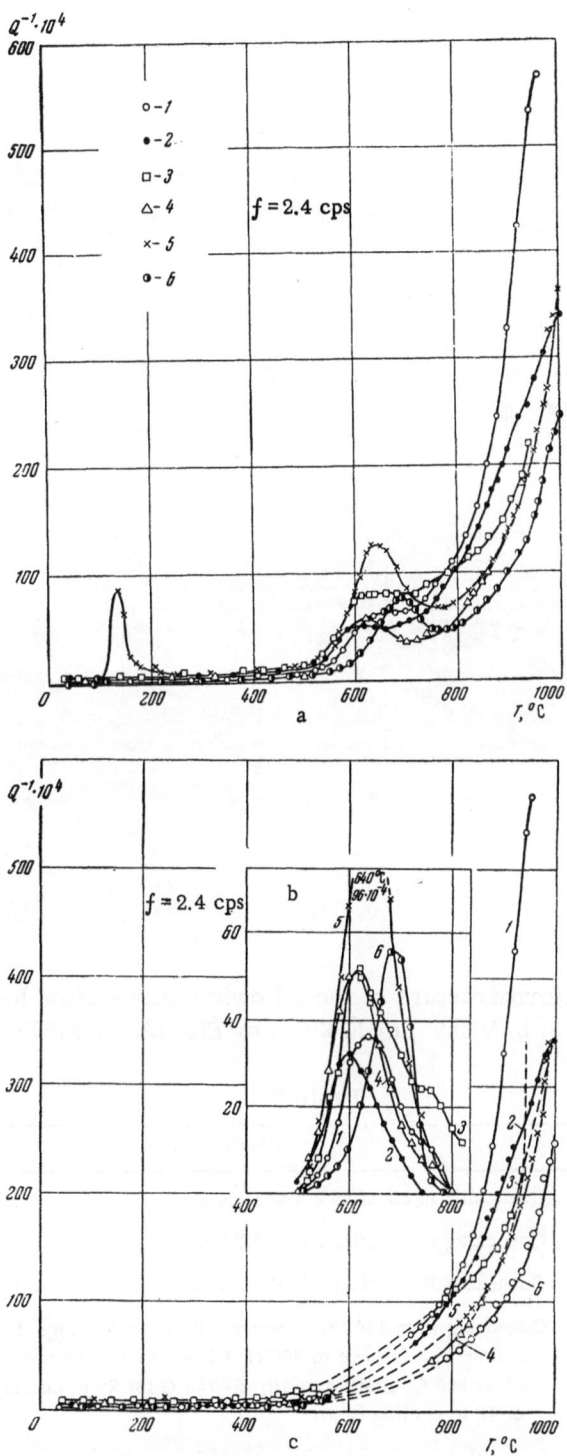

Fig. 1. Temperature dependences of the internal friction of heat-resistant alloys (a), of the grain-boundary relaxation (b), and of the internal friction background (c). 1) Kh22N78; 2) ÉI607A; 3) ÉI617; 4) ÉI437B; 5) VZh98; 6) ZhS6-K.

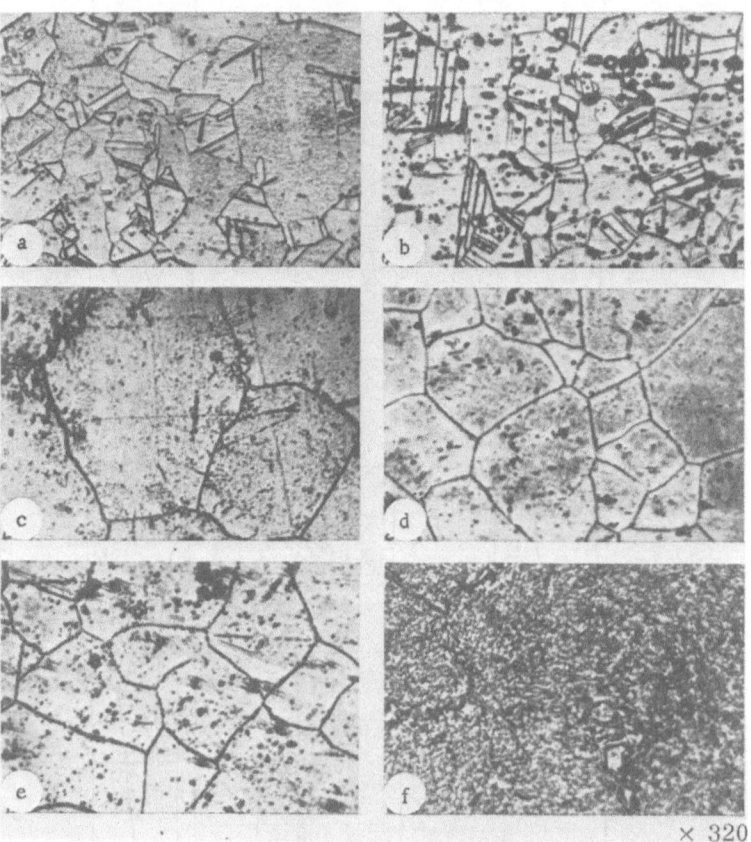

× 320

Fig. 2. Microstructure of the following alloys after heat treatment:
a) Kh22N78; b) VZh98; c) ÉI437B; d) ÉI607A; e) ÉI617; f) ZhS6-K.

Table 2

Alloy	Heat treatment
Kh22N78	Quenching in air after 2 h at 1080°C.
VZh98	Quenching in air after 2 h at 1200°C.
ÉI437B	Quenching in air after 8 h at 1080°C, aging at 700°C for 16 h, cooling in air.
ÉI607A	Quenching from 1100°C in water, three-stage aging: 1) 1000°C for 2 h, cooling in a furnace to 900°C, 1 h at 900°C, cooling in a furnace to 800°C, 2 h at 800°C, cooling in air; 2) 750°C for 20 h, cooling in air; 3) 700°C for 48 h, cooling in air.
ÉI617	Quenching in air after 2 h at 1200°C; heating for 4 h at 1050°C, quenching by cooling in air; aging at 800°C for 16 h, cooling in air.
ZhS6-K	Quenching in air after 4 h at 1200°C.

Table 3

Alloy	Temperature, °C								
	20			600			700		
	$\sigma_{s\,0.2}$, kg/mm^2	σ_b, kg/mm^2	$Q^{-1}\cdot 10^4$	$\sigma_{s\,0.2}$, kg/mm^2	σ_b, kg/mm^2	$Q^{-1}\cdot 10^4$	$\sigma_{s\,0.2}$, kg/mm^2	σ_b, kg/mm^2	$Q^{-1}\cdot 10^4$
Kh22N78	—	—	2.24	—	—	28	—	—	52
VZh98	—	80	20	—	61	20	—	54	36
ÉI437B	66	100	0.8	55	88	12	53	84	28
ÉI617	49	94	1.0	45	79	22	40	55	50
ZhS6-K	175	114	2.0	70	100	50	65	90	62
	—	—	2.26	—	—	10	—	—	24

Alloy	Temperature, °C								
	800			900			1000		
	$\sigma_{s\,0.2}$, kg/mm^2	σ_b, kg/mm^2	$Q^{-1}\cdot 10^4$	$\sigma_{s\,0.2}$, kg/mm^2	σ_b, kg/mm^2	$Q^{-1}\cdot 10^4$	$\sigma_{s\,0.2}$, kg/mm^2	σ_b, kg/mm^2	$Q^{-1}\cdot 10^4$
Kh22N78	—	—	112	—	—	320	—	—	—
VZh98	—	40	75	—	23	140	—	14	366
ÉI437B	47	53	50	—	40	140	—	—	—
ÉI607A	—	—	110	—	—	220	—	—	340
ÉI617	50	70	110	40	50	160	—	—	—
ZhS6-K	—	90	50	—	75	100	—	65 (950° C)	246

$(5 \cdot 10^{-2}\text{–}1 \cdot 10^{-1})$. Evidently, the small amplitude of the grain-boundary peak of the Kh22N78 alloy could be due to the alloying and enrichment of the grain boundaries with chromium atoms. Additional alloying of Nichrome with tungsten and carbon (the VZh98 alloy) did not shift greatly the position of the peak compared with the Kh22N78 alloy. The amplitude of the peak amounted to $96 \cdot 10^{-4}$ and was the highest of all the investigated alloys. Microstructure investigations showed (Fig. 2b) that the average grain dimensions did not exceed 0.04-0.05 mm. As in the case of Nichrome, the low values of the grain-boundary peak amplitude of the VZh98 alloy could be associated with the processes of alloying and enrichment of grain boundaries with chromium, tungsten, and other elements. The decrease in the peak amplitude could be due to carbides, precipitated at grain boundaries. The microstructure of the alloys Kh22N78 and VZh98 was characterized by a large number of annealing twins.

The increase in strength of the alloys due to the intermetallic γ'-phase (ÉI437B, ÉI607A, ÉI617) did not have much influence on the temperature of the internal friction peak. We observed only some shift of the peak into the 600-620°C region (Fig. 1b). Hence, it followed that the additional strengthening of the boundaries by the precipitation of the γ'-phase and of borides did not increase the activation energy of the grain-boundary relaxation.

Investigation of the microstructure (Figs. 2c, 2d, 2e) showed that the grain dimensions of the investigated alloys were always much less than the sample diameter. The small value of the grain-boundary relaxation $(30-50) \cdot 10^{-4}$ was evidently due to the processes of alloying and grain-boundary blocking by precipitates, consisting of the γ'-phase, carbides, and borides.

In contrast to the alloys considered so far, the highly heat-resistant alloy ZhS6-K had a maximum shifted toward the region of 700°C. This could be associated with further strengthening of the grain boundaries by elements forming a solid solution (Cr, Mo, Co, W, and others).

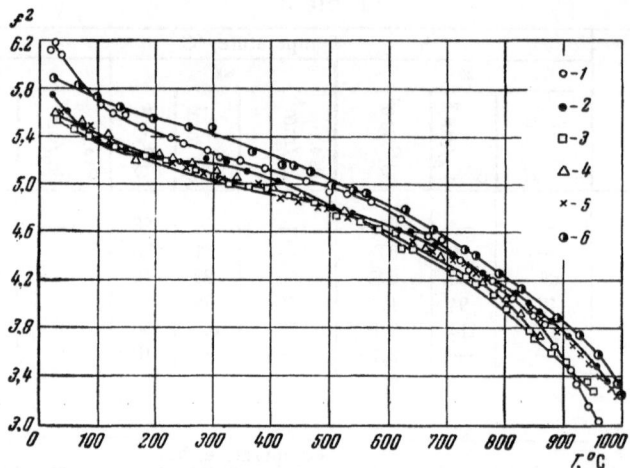

Fig. 3. Temperature dependence of the shear modulus (square of the frequency) of heat-resistant alloys. The notation is the same as in Fig. 1.

The small amplitude of the peak ($55 \cdot 10^{-4}$) of the cast alloy ZhS6-K was due to the small area of the grain boundaries and the blocking of these boundaries by impurities (Fig. 2f).

Comparison of the internal friction background in the temperature range 700-1000°C of the alloys containing the hardened solid solution (Kh22N78 and VZh98) showed (Fig. 1c) that the highest value of the background was observed for nickel alloyed with chromium. Additional alloying with tungsten and carbon reduced the internal friction background considerably. The value of the background in the temperature range 600-1000°C is given in Table 3.

We were unable to establish a clear relationship between the value of the background, the tensile strength, and the yield point in short-duration tests at high temperatures for the alloys strengthened by the intermetallic γ'-phase (Figs. 1c and Table 3).

Only the most strongly heat-resistant alloy ZhS6-K had a background which was lower, over the whole investigated range of temperatures, than the background of other heat-resistant alloys. The considerable reduction of the background in the case of the ZhS6-K alloy was probably due to the fairly strong alloying of the solid solution with chromium, molybdenum, tungsten, cobalt, and other elements.

There are, at the moment, many points of view regarding the nature of the high-temperature internal friction background. Ke [15] is of the opinion that the background is associated with the presence of dislocations in a grain, but he does not suggest a mechanism for the increase in the internal friction. Others [16] attempt to relate the background to grain-boundary creep. Still others [2, 17] explain the exponential rise of the dissipated vibration energy with increase in temperature by the participation of thermally activated vacancies in relaxation processes. The small number of experimental investigations prevents us from establishing which of the theories is to be preferred. We can suggest that, under real conditions, the internal friction background depends on many factors.

The resistance to heat is related both to the degree of strengthening of the solid solution and to the structure state of the alloy. Obviously, the mechanism of plastic deformation at high temperatures depends, to a certain degree, on the rate of deformation. In view of this, it is difficult to compare the internal friction background with the heat resistance, because it is not known under what test conditions (time-to-rupture) the background and high-temperature deformation mechanism are the same. It is possible that such a comparison is best made for strengthened solid solutions.

The temperature dependences of the shear modulus (square of the frequency) of the investigated alloys are given in Fig. 3. The shear modulus was a less-sensitive characteristic than the internal friction. The effects associated with the grain-boundary relaxation did not manifest themselves in the shear-modulus curves. Obviously, this was due to the smallness of the contribution of the grain boundaries to viscous slip processes.

The shear modulus decreased linearly when temperature was increased to 600°C. Above 600°C there was a deviation from this linear dependence. This could be associated with the influence of crystal-lattice imperfections on the shear modulus [18]. The highest degree of relaxation of the modulus in the region of 900-950°C was exhibited by the alloy Kh22N78 and the lowest by ZhS6-K.

Literature Cited

1. A.W. Cochardt, Magnetic Properties of Metals and Alloys [Russian translation], IL (1961).
2. V.S. Postnikov, Investigations of Refractory Alloys, Izd. Akad. Nauk SSSR (1959), Vol. 4, p. 181.
3. V.S. Postnikov, Fatigue Strength of Metals, Izd. Akad. Nauk SSSR (1962).
4. A.I. Efimov, this volume, p. 217.
5. K.M. Rozin and B.N. Fil'kenshtein, Dokl. Akad. Nauk SSSR, 91(4) (1953).
6. P.B. Mikhailov-Mikheev, Handbook of Metallic Materials for Turbine and Engine Construction, Mashgiz (1961).
7. F.F. Khimushin, Refractory Steels and Alloys, Izd. "Metallurgiya" (1964).
8. O.I. Datsko and V.A. Pavlov, Relaxation Phenomena in Metals and Alloys, Metallurgizdat (1960). [English translation: B.N. Finkel'shtein, ed., Consultants Bureau, New York (1963), p. 174.]
9. V.S. Postnikov, Relaxation Phenomena in Metals and Alloys, Metallurgizdat (1963).
10. O.A. Belous, A.I. Efimov, and N.P. Kushnareva, Problems of Metal Physics and Metallography, Izd. "Naukova dumka" (1965), No. 20.
11. O.A. Belous, V.N. Gridnev, A.I. Efimov, and N.P. Kushnareva, this volume, p. 23.
12. C. Wert and J. Marx, Acta Met., 1(3) (1953).
13. W. Köster, L. Bangett, and W. Lang, Z. Metallk., 46(2):84 (1955).
14. G.W. Miles and G.M. Leak, Proc. Phys. Soc. (London), 78:6 (1961).
15. T.S. Ke, J. Appl. Phys., 21:414-419 (1950).
16. L. Rotherham and S. Pearson, J. Metals, 8:881 (1956).
17. V.T. Shmatov and A.V. Grin', Fiz. Metal. i Metalloved., 12(4):600-606 (1961).
18. T.Ya. Benieva, L.N. Larikov, and I.G. Polotskii, this volume, p. 42.

INFLUENCE OF THE ANNEALING TEMPERATURE ON Q^{-1} AND G OF PURE CHROMIUM AND OF ALLOYS OF CHROMIUM WITH YTTRIUM AND GADOLINIUM

O. A. Belous, V. N. Gridnev, A. I. Efimov, Yu. V. Mil'man, and V. I. Trefilov

Chromium has a number of properties which are of practical importance: high melting point ($\sim 1930°C$), good heat resistance, and resistance to aggressive media, etc. However, the wide use of chromium is limited because of its brittleness at low temperatures. It has been shown [1-6] that the introduction of small admixtures of yttrium and of rare-earth elements (cerium, gadolinium, etc.) increases the plasticity and heat resistance of chromium. Chromium-base alloys find wide application in the manufacture of fixed and moving turbine blades, as well as in hot ducts of gas turbines. It is known that one of the important characteristics of a material suitable for fixed turbine blades is the damping capacity. However, the internal friction of plastically deformed chromium at high temperatures (up to 800°C) has been investigated only once [7], which is probably due to the difficulties met with in the preparation of wire-shaped samples, used in the torsional pendulum method. We have reported [2] the preparation of wire-shaped samples of a chromium—yttrium alloy and in further stages of this investigation we also prepared wire samples of chromium with an admixture of gadolinium. In view of this, it has become possible to use the torsional vibration method to investigate the internal friction and the shear modulus over a wide range of temperatures.

Materials and Investigation Method

We investigated zone-refined chromium as well as Cr−1% Yt and Cr−1% Gd alloys. The melting and treatment methods, as well as the purity of the materials used have been reported earlier [2]. Wires made of the zone-purified chromium were prepared by drawing pieces which were cut from an ingot. The drawing was carried out at a temperature of $\sim 300°C$. To investigate the internal friction we used samples of 0.8 mm diameter and 150 mm length. The degree of deformation during drawing was 94-96%. The damping curves were recorded using the torsional pendulum method [8]. The vibrations were recorded oscillographically every 10°C. The frequency of vibrations was 2.1 and 2.8 cps. The maximum deformation of the sample surface was $3.5 \cdot 10^{-5}$.

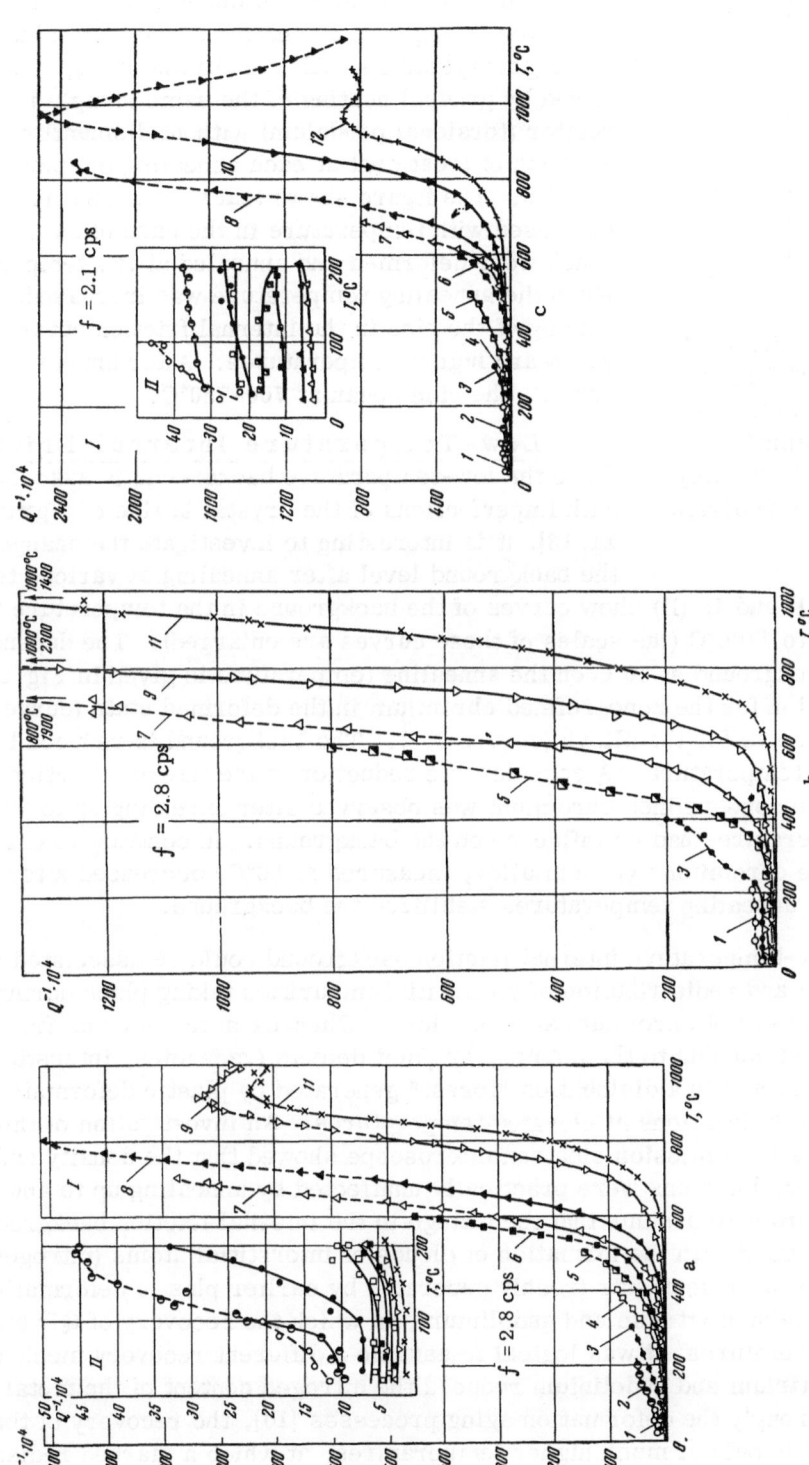

Fig. 1. Influence of annealing on the temperature dependence of the internal friction of zone-purified chromium (a), of the Cr−1% Yt alloy (b), and of the Cr−1% Gd alloy (c). 1) Deformed sample; 2) after annealing at 100°C; 3) 200°C; 4) 300°C; 5) 400°C; 6) 500°C; 7) 600°C; 8) 700°C; 9) 800°C; 10) 900°C; 11) 1000°C; 12) 1100°C.

Fig. 2. Low-temperature internal friction background (at 50°C) after annealing at various temperatures. 1) Zone-purified chromium; 2) Cr−1% Yt alloy; 3) Cr−1% Gd alloy; 4) pure chromium, results reported in [7].

The damping curves of chromium and its alloys are given in Fig. 1 as a function of the preliminary annealing temperature. All the curves in Fig. 1 were obtained by gradual heating of the same sample in a relaxator (torsional pendulum) without dismantling. The duration of treatment at each annealing temperature was 1 h. The figure shows that the internal friction increased with temperature in the case of samples which were deformed and annealed at low temperatures. When the annealing temperature was increased, the beginning of the rise in the internal friction curves shifted toward higher temperatures. After annealing at 1000°C, the rise began at 700-750°C.

Low-Temperature Internal Friction. Since the low-temperature background is associated with imperfections of the crystal lattice of a grain [9, 11, 13], it is interesting to investigate the change in the background level after annealing at various temperatures. Figures 1a (II) and 1c (II) show curves of the background in the temperature range 20-200°C after annealing to 1100°C (the scales of these curves are enlarged). The dependence of the internal friction background at 50°C on the annealing temperature is given in Fig. 2. The value of the internal friction for the zone-refined chromium in the deformed state (curve 1) was approximately half as large as for the alloys (curves 2, 3). The background is reduced by an increase in the annealing temperature. A considerable reduction in the internal friction (from $15 \cdot 10^{-4}$ to $5 \cdot 10^{-4}$) of the zone-refined chromium was observed after annealing up to 300°C. Annealing at higher temperatures had no influence on the background. In contrast to chromium, the internal friction of the chromium−yttrium alloy, measured at 50°C, decreased after annealing to 600°C. Higher annealing temperatures stabilized the background.

Reduction in the low-temperature internal friction background could be associated with the processes of recovery and redistribution of interstitial impurities taking place during the annealing of deformed samples of chromium and its alloys. Changes in the internal friction during recovery [9] were either due to the pinning, by point defects (vacancies, interstitial atoms), of dislocation segments in a dislocation "forest" generated by plastic deformation, or due to a rearrangement of dislocations at elevated temperatures. An investigation of thin foils of the deformed alloys in a transmission electron microscope showed that the density and nature of the distribution of dislocations were practically unaffected by annealing up to 400°C. Therefore, it seemed in order to assume that the change in the internal friction background of pure chromium was associated with the formation of clouds of interstitial atoms (nitrogen and, probably, carbon) at dislocation segments freshly generated by earlier plastic deformation. In the case of chromium containing yttrium and gadolinium, in which the recovery of Q^{-1} took place at much higher temperatures, it was logical to assume a different recovery mechanism. Since the admixtures of yttrium and gadolinium reduced the nitrogen content of the metal matrix and suppressed strongly the deformation aging processes [10], the recovery of the Q^{-1} background became possible only at much higher temperatures, at which a marked redistribution of dislocations took place in polygonization. The recovery temperature of the internal friction background shifted toward much higher temperatures (compared with the zone-refined chromium) in agreement with the conclusions reported in [10] on the refining effect of yttrium and rare-earth admixtures.

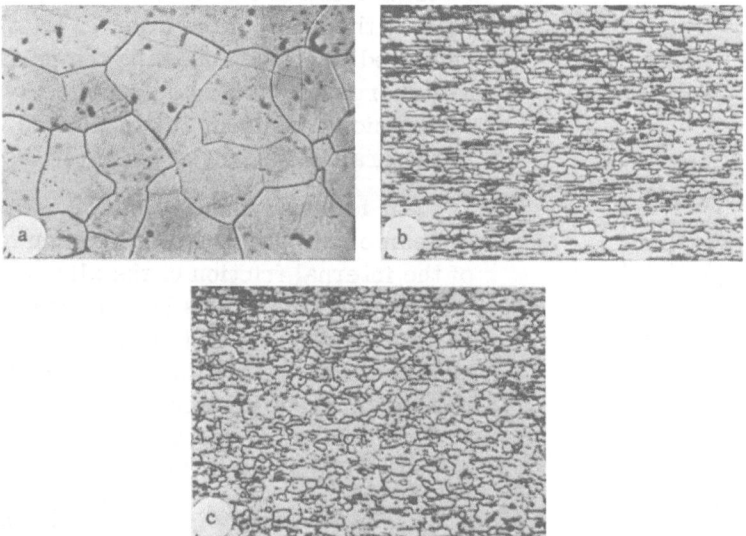

Fig. 3. Microstructure of chromium and its alloys after annealing at 1000°C for 1 h. a) Zone-purified chromium; b) Cr−1% Yt alloy; c) Cr−1% Gd alloy.

Curve 4 in Fig. 2 shows the change in the low-temperature background of the internal friction of pure chromium as a function of the annealing temperature, taken from [7]. The reported background was 2-3 times greater than the values obtained in the present investigation. However, the temperature range within which the internal friction recovery took place was the same in both cases. It was reported in [7] that the annealing of chromium above 400°C caused a new rise of the background with a maximum at 550°C. In our investigation, carried out over a wide range of annealing temperatures, a background maximum was not observed for the zone-purified chromium and for the alloys with yttrium and gadolinium.

Microstructure investigations showed that the recovery of the internal friction was not accompanied by a marked change in the microstructure of pure chromium and its alloys.

It follows from the curves in Fig. 2 that the recrystallization taking place at higher temperatures did not exert any marked influence on the magnitude of the low-temperature internal friction background.

High-Temperature Internal Friction. A strong rise in the internal friction at high temperatures was associated with the high-temperature background and with the participation of grain boundaries in relaxation processes [11, 12].

The nature of the high-temperature background is not yet sufficiently clear. It was suggested in [11, 13] that the exponential rise in the internal friction at high temperatures was associated with the migration of defects in the grain lattice. The contribution of the grain boundaries to the high-temperature relaxation depended on the grain dimensions. It had been shown [14, 15] that when the grain dimensions increased (for example, because of increase in the preliminary annealing temperature), the magnitude of the grain-boundary internal friction decreased. During heating of plastically deformed or annealed (below the recrystallization temperature) metals, slip bands [16] made a definite contribution to the value of the high-temperature internal friction, and these bands could provide an additional outlet for energy dissipation.

Comparison of the curves given in Fig. 1 shows that the alloying of chromium with small amounts of yttrium and gadolinium did not have much influence on the temperature at which the

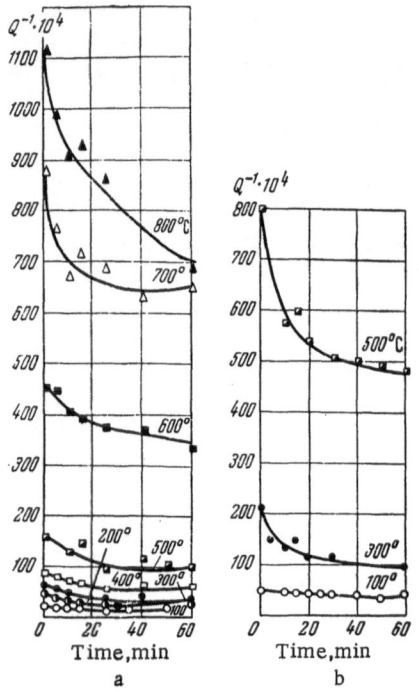

Fig. 4. Influence of isothermal heat treatment on the internal friction at the annealing temperature for zone-purified chromium (a) and for the Cr—1% Yt (b).

internal friction curves rose sharply. The observed shift of the onset of the internal friction rise, which occurred when the annealing temperature was gradually increased, could be associated with a fall in the dislocation density in slip bands and with a reduction in the grain-boundary area.

It follows from Fig. 1 that after annealing at the same temperature, the maximum measured value of the internal friction of the alloys containing yttrium and gadolinium was larger than that of pure chromium. This was in good agreement with the retarding influence of alloying on the processes of collective recrystallization. Figure 3 shows the microstructure of the zone-purified chromium and of the alloys annealed at 1000°C for 1 h.

The heat treatment of the chromium—yttrium alloy at 1000°C for 1 h caused a sharp drop in the internal friction from $2.3 \cdot 10^{-1}$ to $1.49 \cdot 10^{-1}$ (Fig. 1b). Figure 4 gives the curves representing the change in the internal friction of pure chromium and of the chromium—yttrium alloy during isothermal heat treatment at various temperatures.

At annealing temperatures higher than 300°C, an exponential fall of Q^{-1} with time was observed. However, the data obtained could not be used to specify in detail the mechanism of the processes responsible for the fall in Q^{-1}. We could assume that for the chromium—yttrium alloy the change in Q^{-1} in the 300-600°C range was associated with polygonization (the recrystallization temperature of the chromium—yttrium alloy was above 800°C) and, in the case of the zone-purified chromium above 600°C, it was associated with the recrystallization.

The heating of pure chromium to 1000°C led to the appearance of a peak in the internal friction curves (Fig. 1a). The temperature of the peak was 920°C at a vibration frequency of 2.8 cps. After annealing for 1 h, the amplitude of the peak decreased but its temperature remained practically unaffected. A similar relationship was reported in [15] for the grain-boundary internal friction in nickel. The high temperature of the peak indicated that the activation energy of the grain-boundary relaxation in pure chromium was considerably greater than the activation energies for metals of lower melting points [17].

When the chromium—yttrium alloy was heated to 1000°C, the grain-boundary internal friction peak was not observed. Evidently, the alloying with yttrium shifted the peak toward higher temperatures. Investigation of the chromium—gadolinium alloy showed that the internal friction peak was at 960-970°C at a vibration frequency of 2.1 cps (Fig. 1c). As in the case of the zone-purified chromium, the treatment at 1100°C for 1 h reduced strongly the amplitude of the grain-boundary peak (from 0.24 to 0.09) of the chromium—gadolinium alloy. The temperature of the peak was not greatly affected by this heat treatment.

Shear Modulus

An investigation of the temperature dependence of the square of the frequency (this quantity was proportional to the shear modulus) of the chromium—yttrium alloy indicated the following relationships. Deviation from the linear dependence of the modulus occurred in the same range of temperatures in which the internal friction rose sharply. The change in the shear modulus was associated with the relaxation along grain boundaries and with the influence of crystal lattice imperfections. Thus, for example, the higher absolute value of the shear modulus of an annealed metal, compared with the value of the modulus in the deformed state, was in agreement with Mott's ideas [18] on the relationship between the measured value of the modulus and the length of a dislocation segment capable of vibration (an unpinned segment). Unfortunately, the accuracy of the measurements of the square of the frequency by the method of torsional vibrations was insufficient for quantitative calculations.

In conclusion, we express our gratitude to V.G. Epifanov, member of the staff of the Institute of Metal Physics, Academy of Sciences of the UkrSSR, who supplied the zone-purified chromium, prepared by the method described in [19], involving three passes of the zone. The chromium contained ≤0.001% nitrogen.

Literature Cited

1. F.J. Collins, P.V. Calkins, and A.J. McGurty, Problems of Modern Metallurgy [Russian translation] (1960), Vol.2, p.53.
2. V.N. Gridnev, R.K. Ivashchenko, Yu.V. Mil'man, and V.I. Trefilov, Problems of Metal Physics and Metallurgy, Izd. "Naukova dumka" (1964), No. 20.
3. F.F. Khimushin, Refractory Steels and Alloys, Izd."Metallurgiya" (1964).
4. C.T. Sims, J. Metals, 15(2):127 (1953).
5. A.J. McGurty, F.J. Collins, and P.V. Calkins, USA Pat. No. 2955937, dated Nov.10, 1960.
6. A.J. McGurty and F.J. Collins, USA Pat. No. 3015559, dated Feb.1, 1962.
7. M.E. de Morton, Trans. Met. Soc. AIME, 218(2):294-299 (1960).
8. A.I. Efimov, this volume, p. 217.
9. A.S. Nowick, in: Creep and Recovery [Russian translation], Metallurgizdat (1961), p.166. [English edition: R. Maddin, ed., American Society for Metals, Novelty, Ohio.]
10. Yu.V. Mil'man, A.P. Rachek, V.I. Trefilov, et al., Metal physics, Collection: Mechanism of Plastic Deformation of Metals, Izd. "Naukova dumka" (1965).
11. V.S. Postnikov, Investigations of Refractory Alloys, Izd. AN SSSR (1959), Vol.4, p.181.
12. D.T. Peters, I.C. Bisseliches, and J.W. Spretnak, Trans. Met. Soc. AIME, 230(4): 530 (1964).
13. V.T. Shmatov and A.V. Grin', Fiz. Metal.i Metalloved., 12(4):600-606 (1961).
14. W. Köster, L. Bangett, and W. Lang, Z. Metallk., 46(2):84 (1955).
15. O.A. Belous, A.I. Efimov, and N.P. Kushnareva, Problems of Metal Physics and Metallography, Izd. "Naukova dumka" (1964), No.20.
16. C.M. Zener, Elasticity and Anelasticity of Metals [Russian translation], IL (1954). [English edition: University of Chicago Press, Chicago, Illinois, (1948).]
17. C. Wert and J. Marx, Acta Met., 1(3) (1953).
18. N.F. Mott, Phil. Mag., 43:1151 (1952).
19. V.G. Epifanov and A.G. Lesnik, Problems of Metal Physics and Metallography, Izd. "Naukova dumka" (1964), No.20.

MECHANISM OF THE DAMPING OF ELASTIC VIBRATIONS IN THE CASE OF PHASE TRANSITIONS IN COPPER AND COBALT BASE ALLOYS

I. G. Polotskii and N. S. Mordyuk

To check the conclusions of the theory of the damping of elastic vibrations in the case of phase transitions [1, 2], it is essential to obtain new experimental data.

We investigated the mechanism of damping of elastic vibrations in alloys based on copper undergoing phase transitions. We studied several copper-base alloys (Cu—Be, Cu—In, Cu—Al), whose phase transitions have been investigated quite thoroughly [3-8]. We investigated also the damping, elastic properties, and structural changes taking place during the precipitation of a supersaturated Co—W solid solution in the initial aging stages and during prolonged isothermal treatment.

The investigation was carried out using resonance apparatus, by means of which we could record freely damped longitudinal and transverse vibrations and use the data obtained to determine the damping and Young's modulus over a wide range of temperatures in the frequency interval from 0.5 to 75 kc [9]. The alloys Cu—Be, Cu—In, Cu—Al, Co—W were melted in a high-frequency furnace in an argon atmosphere and then remelted. As the initial materials we used Cu (99.99%), Be (99.9%), In (99.99%), Al (99.99%), Co (99.98%), W (99.85%). The chemical compositions of the investigated alloys were as follows: Cu + 1.8 wt.% Be; Cu + 15 wt.% In; Cu + 11.7 wt.% Al; and Co + 31.89 wt.% W. To obtain the necessary structure and to achieve homogenization, each of the alloys was subjected to a suitable heat treatment. Samples of the copper—beryllium alloy were annealed in vacuum at 820°C for 10 h and were quenched in water from this temperature. The copper—indium alloy was annealed at 600°C for 15 h and also quenched. Samples of the copper—aluminum alloy were heated, after homogenization and quenching, to 900°C and cooled in the furnace to 600°C. The samples were kept at this temperature for 2 h and quenched in water. Samples made from the copper—tungsten alloy ingots were homogenized in vacuum for 15 h at 1100°C and quenched in water.

The experimental data on the damping of elastic vibrations, Young's modulus, and on electrical resistivity in the case of phase transitions in the alloys Cu—Be, Cu—In, and Cu—Al are given in Figs. 1-5.

The investigation of the logarithmic decrement of the Cu—Be alloy (Fig. 1) showed that the isothermal treatment at 300°C increased the damping, which reached a maximum after about 2 h. Comparison of the data on the change in the logarithmic decrement at frequencies

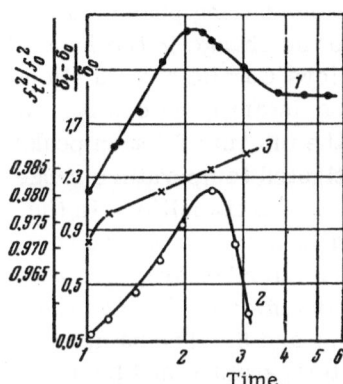

Fig. 1. Relative change in the damping at 1 kc (1) and 21 kc (2), and in Young's modulus (3) during the precipitation in Cu−Be.

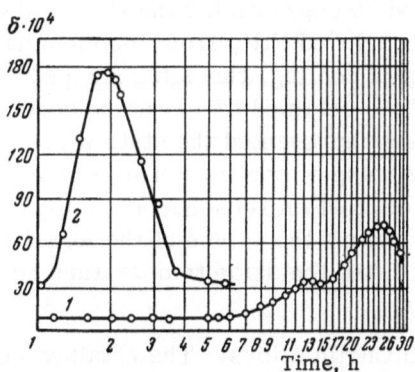

Fig. 2. Change in the damping in the Cu−Zn alloy at 1.5 kc at 245°C (1) and 420°C (2).

of 1 and 21 kc showed that the position of the decrement maximum, considered as a function of the duration of treatment, was the same for all these frequencies. However, the decrement maximum at 21 kc was twice as large that at 1 kc. The metallographic investigations, the measurements of the hardness, as well as comparison of the data obtained on the damping with the available x-ray diffraction data [3], showed that the damping peaks were due to the decomposition of the supersaturated solid solution of beryllium in copper, which gave rise to the γ-phase. The rise of the damping, shown in Fig. 1, was accompanied by an increase in Young's modulus, which was associated with the precipitation of the γ-phase. We investigated the damping in the Cu−In alloy during the isothermal treatment at various temperatures (Fig. 2). For this alloy the damping during the isothermal treatment increased with time. At 245°C the damping maximum was observed after 23-24 h. Comparison of the experimental data on the damping with the available values of the hardness [4-5] showed that the damping of elastic vibrations could be used to detect the various stages of the process of precipitation. A considerable rise of the logarithmic decrement was observed on isothermal annealing at 245°C even as soon as 7-8 h. Increase of the annealing temperature to 420°C caused a sharp rise of the damping after 1 h (Fig. 2). Exactly as in the case of the Cu−Be alloy, the position of the decrement maximum on the time axis, was constant when the frequency of elastic vibration was increased. Comparison of the data on the damping with the data on the hardness, as well as with the x-ray diffraction results, showed that the increase in the damping was caused by the precipitation of the supersaturated solid solution Cu−In.

In the Cu−Al alloy of the eutectoid composition (11.7 wt.% Al) the increase in the logarithmic decrement during the isothermal annealing was much stronger than in the Cu−Be and Cu−In alloys [10]. The decrement of the Cu−Al alloy, subjected to one hour at 425 and 460°C, increased by a factor of about 50-60, compared with its value at the beginning of annealing (Fig.3). Increase of the annealing temperature from 425 to 460°C shifted the damping maximum toward shorter treatment durations. On increase of the frequency from 1.5 to 21 kc, no shift of the decrement maximum along the time axis was observed, but its value increased strongly. Simultaneously with the rise of the decrement of the copper−aluminum alloy, Young's modulus of this alloy increased as well. Comparison of the obtained data on the damping and on Young's modulus with the investigations of the phase transitions in the Cu−Al alloy [7] indicated that the change in the damping was caused by the eutectoid precipitation and paralleled the kinetics of this precipitation.

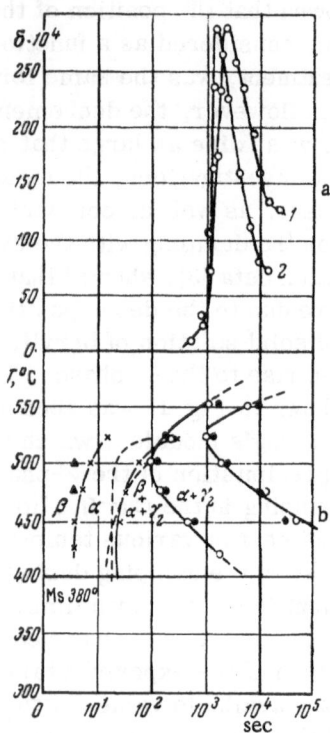

Fig. 3. Change in the damping of elastic
vibrations in the Cu−Al alloy (a) at 425°C
(1) and 460°C (2) (f = 1.5 kc) and the
phase diagram of this alloy of the eutectoid composition (b).

To determine the mechanism of the damping of elastic vibrations during phase transitions,
we investigated also the change in the logarithmic
decrement of the alloys Cu−Be and Cu−In during
isothermal-annealing-interrupted cooling. In
order to show that the internal friction peaks, observed in the investigated temperature range of
aging, were caused by the transitions in the
Cu−Be and Cu−In alloys and not by the thermal
conductivity or some other factors, we carried
out additional experiments and obtained the results shown in Figs. 4 and 5. The sample was
heated to the aging temperature and held at this
temperature for a time necessary to reach the
value of the logarithmic decrement corresponding to a point on the left-branch of the curve
(Figs. 4 and 5). Then the sample temperature
was reduced to 50°C, and the logarithmic decrement was measured during cooling. The cooling
was carried out at the rate of 8 deg/min for the
Cu−Be alloy and 16 deg/min for the Cu−In alloy.
From this we concluded that, from the moment
of the beginning of cooling, the aging stopped
completely and the obtained temperature dependences of damping represented the state which applied at the beginning of cooling. This was confirmed by the fact that on subsequent heating to
the aging temperature the values of the logarithmic decrement fitted the same temperature dependence curve.

Such measurements were carried out on the Cu−Be and Cu−In alloys. The obtained results on the damping in the Cu−In alloy during aging at 420°C at 1.5 kc and the temperature dependence of damping are given in Fig. 4. The same figure (curve 2) includes the temperature
dependence of the decrement for a sample of the Cu−In alloy, obtained in an 80-min period
while cooling from 420°C. After cooling to 50°C, the sample was again rapidly heated to the
aging temperature and the isothermal investigation was continued. Comparison of the logarithmic decrement with the data of Fig. 2 (curve 1) showed that there were no changes in the
form of the damping curve. From the obtained experimental data for the Cu−In alloy, it followed that the nature of the change in the logarithmic decrement, considered as a function of
the duration of treatment, was related to a phase transition.

Investigation of the logarithmic decrement of the Cu−Be alloy showed (Fig. 5) that
for this alloy the change in the decrement was also related to a phase transition. From the obtained data it was evident that, both the Cu−In and the Cu−Be alloys had logarithmic decrement curves which changed in the same way during straightforward isothermal measurements
and during isothermal measurements interrupted by cooling. This indicated that internal friction peaks, observed in the investigated temperature range of aging, were due to phase transitions in the Cu−Be and Cu−In alloys and not due to any other factor.

From the temperature dependence of the damping we determined the activation energy of
the process of aging of these alloys. For the Cu−Be alloy we found the value H = 18 kcal/mole,

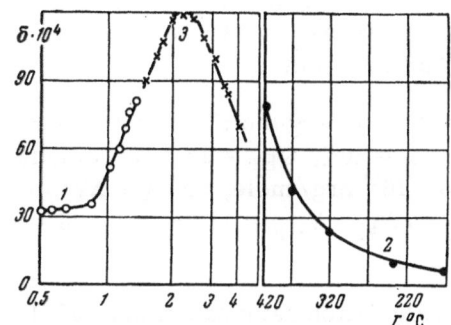

Fig. 4. Changes in the damping in the Cu—Zn alloy: 1) after treatment at 420°C; 2) temperature dependence during cooling; 3) after continued annealing at 420°C.

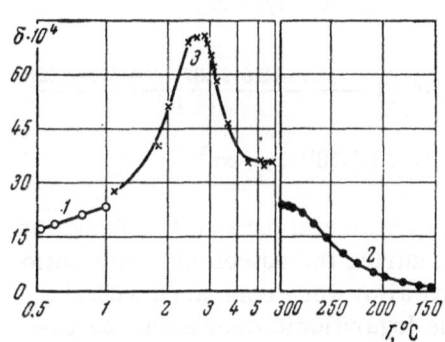

Fig. 5. Changes in the damping in the Cu—Be alloy: 1) after treatment at 300°C; 2) temperature dependence during cooling; 3) after annealing continued at 300°C.

and for the Cu—In alloy we obtained H = 22 kcal per mole. These results were in good agreement with the values of the activation energy of the aging process determined by other methods.

We attempted to compare the obtained experimental data on the changes in the damping during phase transitions in the Cu—Be, Cu—In, and Cu—Al alloys with the values of the logarithmic decrement calculated from the theory of Krivoglaz [1, 2]. According to this theory, the damping of low-frequency vibrations (when $R^2\omega \ll D$, where R is the distance between particles of the second phase, ω is the angular frequency, and D is the effective diffusion coefficient) could be calculated from the formula

$$\tan\ \delta = \frac{C_0^2 - C_\infty^2}{C_0^2}\ \frac{\omega\tau}{1 + \frac{C_\infty^2}{C_0^2}(\omega\tau)^2}\ ,$$

(1)

where C_0 and C_∞ are the velocities of propagation of elastic vibrations at low and high frequencies, respectively; τ is the relaxation time which, for an adiabatic process, can be calculated from the expression

$$\tau = \tau_0 \frac{K_0}{K_\infty},\quad \tau_0 = \frac{r_0^2}{3Dx_2},$$

(2)

where r_0 is the dimension of precipitated particles; x_2 is the volume concentration of the second phase; K_0 and K_∞ are the bulk moduli at low and high frequencies, respectively, which, for a two-component system, can be given by the following equations:

$$K_0 = K\left\{1 + K\frac{\alpha T\Delta V_\sigma}{q_\sigma} - K\left(\alpha + \frac{C_p\Delta V_\sigma}{Vq_\sigma}\right)\frac{c_2\Delta V_A + (1 - c_2)\Delta V_B}{c_2 Q_A + (1 - c_2)Q_B}\right\}^{-1},$$

(3)

$$K_\infty = K\left(1 - \frac{K\alpha^2 TV}{C_p}\right)^{-1},$$

(4)

where K is the average bulk modulus; α is the volume expansion coefficient; T is the transition temperature; ΔV_σ is the change in volume at the phase transition; q_σ is the heat of transitions; C_p is the specific heat at constant pressure; V is the molar volume of the alloy; c_2 is the concentration of the second component in the precipitated phase; ΔV_A and ΔV_B are the changes in the volume upon the introduction of one A atom or one B atom, respectively; Q_A and Q_B are the corresponding heats of solution.

To calculate $(K_\infty - K_0)/K_0$ for Cu—Be from Eqs. (3), (4), we used the data obtained by us as well as the published values of the following quantities:

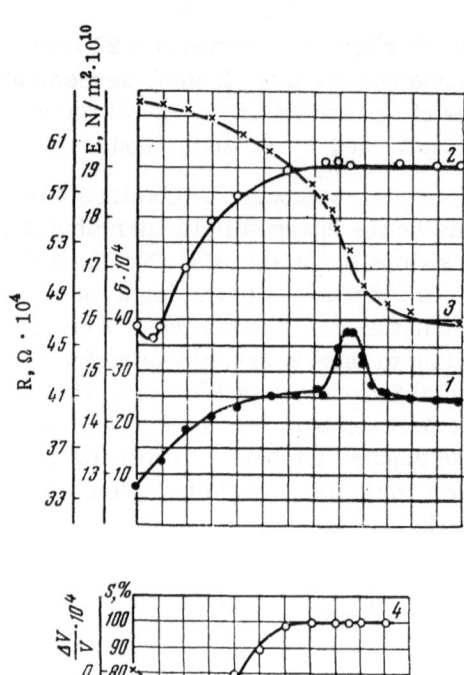

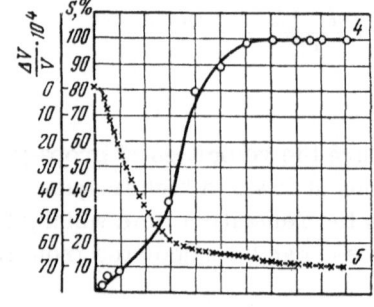

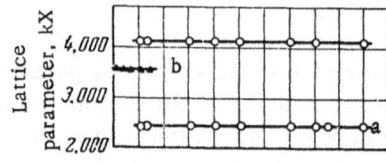

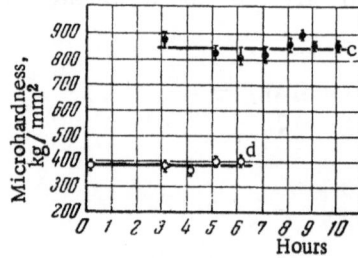

Fig. 6. Changes in the logarithmic decrement (1), Young's modulus (2), electrical resistivity (3), degree of conversion to a new phase (4), volume (5), and in the lattice parameter and microhardness during isothermal annealing (700°C) of the Co−W alloy: a) hexagonal Co; b) cubic Co; c) dark zones; d) light zones.

K = 12.7 · 10^{11} dyn/cm², α = 52.8 · 10^{-6} deg^{-1}, T = 573°K, ΔV_σ = 39 · 10^{-2} cm³/mole, q = 17.6 · 10^{10} erg/mole.

C_p = 25 · 10^7 erg·mole^{-1}·deg^{-1}; V = 6.89 cm³/mole, c_2 = 50%, Q_{Be} = 15 · 10^{10} erg/mole, Q_{Cu} = 5.36 · 10^{10} erg/mole, $(\Delta V_A + \Delta V_B)$ = $V_{A_2} + V_{B_2} - (V_{A_1} - V_{B_1})$ = 13.984 − 9.536 − 4.428.

The calculation carried out showed that the third term in Eq. (3), which defines K_0, was at least two orders of magnitude smaller than the second term. Therefore, K_0 was governed mainly by the first two terms of Eq. (3):

$$K_0 = 12.73 \cdot 10^{11} \left(1 + \frac{12.73 \cdot 10^{11} \cdot 52.8 \cdot 10^{-6} \cdot 573 \cdot 0.39}{17.59 \cdot 10^{10}} \right)^{-1}$$

$$= 11.6 \cdot 10^{11} \text{ dyn/cm}^2,$$

$$K_\infty = 12.73 \cdot 10^{11} \left(1 - \frac{12.73 \cdot 52.8 \cdot 27.87 \cdot 10^{-10} \cdot 573 \cdot 6.89}{25 \cdot 10^7} \right)^{-1}$$

$$= 13.4 \cdot 10^{11} \text{ dyn/cm}^2.$$

Our experimental data on the change in the damping during the eutectoid precipitation in the Cu−Al alloy were compared with the values of the logarithmic decrement calculated from the theory. We calculated the bulk moduli at low and high frequencies for the Cu−Al alloy. The quantities required in this calculation were as follows:

K = 12 · 10^{11} dyn/cm², α = 52.5 · 10^{-6} deg^{-1}, T = 700°K, ΔV_σ = 24 · 10^{-2} cm³/mole, q_σ = 22.86 · 10^9 erg/mole, V_σ = 12.2 cm³ per mole, Q_{Cu} = 11.63 · 10² cal/mole, C_p = 41.8 · 10^7 erg/mole.

Substituting these quantities into Eqs. (3) and (4) for this alloy, we obtained

$$K_0 = 12 \cdot 10^{11} \left(1 + \frac{12 \cdot 52.5 \cdot 700 \cdot 24 \cdot 10^5 \cdot 10^{-6}}{22.86 \cdot 10^9} \right)^{-1}$$

$$= 10.9 \cdot 10^{11} \text{ dyn/cm}^2,$$

$$K_0 = 12 \cdot 10^{11} \left(1 - \frac{12 \cdot 10^{11} \cdot 27.56 \cdot 10^{-10} \cdot 7 \cdot 10^2 \cdot 12.2}{41.8 \cdot 10^7} \right)^{-1}$$

$$= 13 \cdot 10^{11} \text{ dyn/cm}^2.$$

Using Turnbull's formula [11], we could calculate the effective diffusion coefficient, which was needed to determine the relaxation time:

$$D = \frac{G \cdot l^2}{2\lambda \left(\dfrac{x_0 - x_l}{x_l} \right)},$$

where G is the rate of growth of the precipitated cells; l is the distance between precipitated platelets; λ is the width of grain boundaries; x_0 and x_l are the initial and final concentrations of the solid solution.

To calculate the effective diffusion coefficient for the Cu–Be, Cu–In, and Cu–Al alloys, we used the experimental data reported in [5, 12-14]. The calculations carried out gave the following relaxation times for the alloys Cu–Be, Cu–In, and Cu–Al: $\tau \simeq 10^{-4}$ sec for Cu–Be, $\tau \simeq 2 \cdot 10^{-3}$ sec for Cu–In, and $\tau \simeq 5.8 \cdot 10^{-3}$ sec for Cu–Al. It should be mentioned that these values were determined with an accuracy which gives only the order of magnitude. The relative changes in the bulk moduli and relaxation times, calculated for these alloys, made it possible to estimate the logarithmic decrement changes during aging of the alloys. Substituting the appropriate quantities into Eq.(1), we found that the change in the decrement of the Cu–Be alloy was $\tan \delta \sim 9 \cdot 10^{-3}$-$5 \cdot 10^{-2}$, while the experimentally determined values were $\tan \delta = 4.5 \cdot 10^{-3}$-$1.3 \cdot 10^{-2}$. Similar calculations were carried out for the Cu–In and Cu–Al alloys.

The obtained agreement between the experimental data on the change in the logarithmic decrement during the precipitation in the Cu–Be, Cu–In, and Cu–Al alloys with the values calculated from Krivoglaz's theory could be regarded as fully satisfactory, bearing in mind that the estimates of τ and of $(K_\infty - K_0)/K_0$ were approximate.

The investigation of the aging of the alloys based on cobalt is of considerable theoretical and practical interest. The aging of the alloys of cobalt with tungsten, molybdenum, and chromium has already been investigated by Sykes [15], Sykes and Graff [16], as well as Köster and Tonn [17].

We investigated also the damping, elastic properties, the electrical resistivity, and structural changes during the precipitation of a saturated solid solution of tungsten in cobalt in the initial phases of this transition and during prolonged isothermal treatment [18, 19]. The preliminary investigation showed that an isothermal treatment at 700°C made it possible to investigate the influence of the transition on the damping of elastic vibrations in the absence of the $\varepsilon \to \gamma$ transition and of the magnetic transition. The results of the investigation of the logarithmic decrement, Young's modulus, and the electrical resistivity of the Co–W alloy during the isothermal annealing are given in Fig.6. The isothermal treatment of the Co–W alloy at 700°C caused a rise in the logarithmic decrement and in Young's modulus. The simultaneous rise of these two quantities was due to the fact that the transformation in the alloy gave rise to a new phase, whose Young's modulus was higher than the modulus of the alloy in the quenched state. The rise in the damping and in Young's modulus was accompanied by a fall of the electrical resistivity and was associated with transformations in the cobalt–tungsten alloy. The data obtained indicated that at 700°C the process of precipitation of a supersaturated Co–W solid solution was completed after an isothermal treatment lasting 10 h. A maximum of the damping curve appeared after 7.5-9 h. The appearance of the peak in the time dependence of the logarithmic decrement was due to a departure from the phase equilibrium conditions when an elastic wave traveled through a mixture of two phases as predicted by Krivoglaz's theory [1, 2]. The irreversibility of the process led to the dissipation of elastic wave energy, which in turn caused a peak to appear in the elastic vibration damping during phase transitions in the cobalt–tungsten alloy.

We investigated also the structural changes during the precipitation of a supersaturated solid solution of tungsten in cobalt. We investigated the microstructure, the changes in the microhardness, and we carried out dilatometric and x-ray investigations of the Co−W alloy during aging [19]. The obtained microstructure data showed that the precipitation of the Co−W alloy was heterogeneous. Increase of the thickness of boundaries was observed even after 30-min aging at 700°C. The dark regions, when observed under large magnification, exhibited laminar eutectoid-like structure. During further annealing these regions increased rapidly in size and after aging for 7 h they occupied the whole surface of a section. The kinetics of this process was investigated by the quantitative metallography method. The transformed area of a section was determined statistically from photomicrographs of the whole surface of the section. The kinetic curve obtained had the usual form typical of such transformations (Fig. 6).

The high value of the microhardness of the dark regions in a section (Fig. 6) was probably associated with the properties of the precipitated equilibrium phase Co_3W, as well as with the high degree of dispersion of the structure produced during heterogeneous precipitation. The hardness of the lighter parts of the section was low during the whole period of the precipitation; its value was about the same as that for the quenched alloy. This indicated slower development of the processes of homogeneous precipitation under our experimental conditions. The results of the dilatometric investigations (Fig. 6) also showed that, in general, volume changes accompanied the process of heterogeneous precipitation of the supersaturated Co−W solid solution.

The obtained x-ray diffraction data for the Co−31.89 wt.% W alloy in the quenched state and for various durations of annealing indicated a very complex phase transition behavior in this alloy [19]. In the quenched state the Co−W alloy had the fcc lattice with the parameter 3.873 kX (cf. Fig. 6). In the initial transition stages during an isothermal annealing at 700°C for 30 min, the x-ray diffraction patterns, recorded at room temperature, showed lines which represented the formation of the α-phase. After one hour at 700°C, the diffraction patterns showed lines corresponding to the hexagonal modification of the cobalt solid solution, with a concentration close to the equilibrium value. The interference pattern observed in these experiments represented the precipitation of the intermetallic compound Co_3W, which had the hcp lattice. It was also established that the lattice parameter remained practically constant.

Literature Cited

1.   M. A. Krivoglaz, Fiz. Metal i Metalloved., 10(1) : 497 (1960).
2.   M. A. Krivoglaz, Fiz. Metal i Metalloved., 12(3) : 338 (1961).
3.   C. S. Barret, Metals Technol., p. 21 (1948).
4.   C. Guy and C. S. Barret, Trans. AIME, pp. 175, 216 (1948).
5.   H. Böhm, Z. Metallk., 50(2) : 87 (1959).
6.   F. Weibke and I. Pleger, Z. Anorg. Allgem. Chem., 231 : 199 (1937).
7.   E. P. Klier and S. M. Grymko, Metals Trans., 185 : 611 (1942).
8.   T. N. Bogacheva and V. D. Sadovskii, Tr. Inst. Fiz. Metal., No. 17 : 125 (1956).
9.   N. S. Mordyuk, Problems of Metal Physics and Metallurgy, Izd. AN UkrSSR (1962), No. 16, p. 190.
10.   N. S. Mordyuk, Problems of Metal Physics and Metallurgy, Izd. AN UkrSSR (1963), No. 17, p. 75.
11.   D. Turnbull and H. N. Treaftis, Acta Met., 3 : 43 (1955).
12.   P. D. Robertin and S. Bray, Collection: Aging of Alloys [Russian translation], IL (1962), p. 338.

13. H.K. Hardy and T. Dock-Hill, Collection: Progress in Metal Physics [Russian transla-
     tion] (1958), Vol. 2, p. 285. [English edition: B. Chalmers and R. King, eds., 8 vols.,
     Pergamon Press, Inc., New York.]
14. S.D. Gertsriken and I.Ya. Dekhtyar, Diffusion in Solid Metals and Alloys, Fizmatgiz,
     Moscow (1960).
15. W. Sykes, Trans. Am.Soc.Steel Treating, 21:385 (1933).
16. W. Sykes and H.F. Graff, Trans.Am.Soc.Metals, 23:249 (1935).
17. W. Koster and W. Tonn, Z. Metallk., 24:296 (1932).
18. N.S. Mordyuk, Ukr.Fiz.Zh., No. 1:87 (1964).
19. L.N. Larikov, I.G. Polotskii, N.S. Mordyuk, and O.A. Shmatko, Problems of Metal
     Physics and Metallurgy, Izd. "Naukova dumka" (1964), No. 20.

INVESTIGATION OF THE HEAT RESISTANCE
AND THERMOCYCLIC FATIGUE OF REFRACTORY
METALS BY THE INTERNAL FRICTION METHOD

L. N. Aleksandrov and V. S. Mordyuk

Changes in the defect state of a metal affect the internal friction, while changes in the nature of the interaction between defects lead to a temperature shift of the relaxation spectrum or some of its components. The determination of the nature of defects affected most by the investigated change and the influence of this change on the behavior of other defects requires the knowledge of the nature of the internal friction maxima. Comparison of the internal friction data with the results of other physical methods makes it possible to draw conclusions [1, 2, 3] on the nature of the internal friction peaks of tungsten, alloyed with admixtures of Al_2O_3, Fe_2O_3, Mo, Ni, SiO_2 in the temperature range 800-2500°C. X-ray investigations have shown that the peak A (Fig. 1a) is due to the activated displacement of dislocations, giving rise to the polygonization effect in the deformed metal;* the peak B is due to the diffusion of the admixtures, present in the tungsten lattice, along the boundaries of grains grown during the recrystallization process, preceding collective recrystallization; the peak C was of the grain-boundary type. No special investigations have yet been carried out on the nature of the peak D.

Grain boundaries have a complex structure. Therefore, the determination of the nature of the grain-boundary peak is undoubtedly of interest. On the basis of recent investigations [4], it has been suggested that, in the case of aluminum and cadmium single crystals subjected to thermal cycling, the peak appearing in the same temperature range as the grain-boundary peak in polycrystalline samples of these metals, is due to the activation of dislocation loops formed when vacancy "disks" collapse. We investigated in detail the nature of the "grain-boundary" maximum of the internal friction of tungsten using zone-purified polycrystalline and single-crystal samples. The total content of Al, Fe, Mo, K, Cu, Mn, Cl impurities was less than $2 \cdot 10^{-4}\%$.

An investigation of the $Q^{-1}(T)$ curve of deformed and annealed zone-purified samples (curves 1 and 2 in Fig. 1b) showed that the grain-boundary maximum was shifted to the left and observed at 1500°C, while in the case of industrial tungsten of the VA grade the same maximum was found at 1900-2000°C (maximum C in Fig. 1a). A fall in the recrystallization temperature

---

*This subject was discussed at the Eighth All-Union Conference on the Application of X-Rays in the Investigation of Materials (1964).

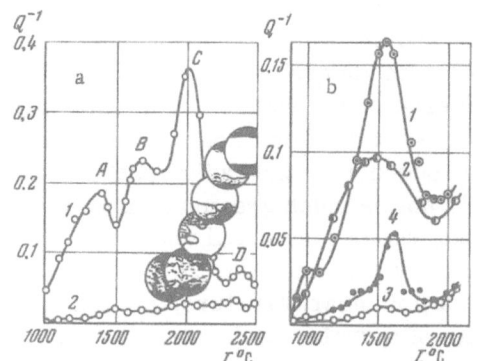

Fig. 1. Temperature dependence of the internal friction of tungsten. a) Industrial tungsten of the VA grade: 1) deformed sample; 2) recrystallized sample. The microstructure shows the temperature range in which grain growth occurred. b) Zone-purified polycrystalline tungsten: 1) in the deformed state; 2) in the recrystallized state and single-crystal state; 3) in the initial state; 4) after thermal cycling.

in the case of the zone-purified tungsten was observed earlier [5] within the same limits. Single-crystal samples exhibited the internal friction maximum in the same temperature range as polycrystalline samples. Evidently, the presence of a weak but well-defined inflection in the internal friction curve of the initial single crystal (curve 3 in Fig. 1b), not subjected to thermal cycling, was due to the special history of this sample. Assuming the mechanism proposed in [4] for the origin of this maximum in the case of single crystals, its presence in the initial tungsten single crystal, prepared by zone melting, could be explained by the formation of dislocation rings due to the cooling of the sample after zone melting. Thermal cycling of the sample increased markedly the amplitude of this peak (curve 4 in Fig. 1b).

The conclusion that the temperature of the "grain-boundary" peak is the temperature of activation of ring dislocations, which are thermally more stable than edge dislocations, makes it possible to explain, first of all, the considerable amplitude of this peak in the case of deformed polycrystalline samples; and, secondly, it may yield additional information on the secondary recrystallization process, taking place in deformed metals in this range of temperatures. While at low temperatures the loss of strength (softening) during polygonization and recrystallization treatments is mainly due to the reduction in the total number of edge dislocations, the secondary recrystallization stage (anomalous growth of grains in the case of tungsten) is responsible for further loss of strength (softening) due to additional annihilation of the remaining thermally-more-stable dislocations. Obviously, the strong drop in the internal friction beyond the "grain-boundary" maximum (Fig. 1a) during grain growth is not only due to a reduction in the total boundary area, but also due to the reduction in the density of such defects. Defects may be annihilated in different ways, including, as suggested in [4, 6], by the absorption in grain boundaries. The fracture of metals at high temperatures was investigated by the internal friction method. Figure 2 shows the results of an investigation of the change in peaks and background of the internal friction during the process of fracture of tungsten at an elevated temperature. The fracture took place at 1500°C under a stress $\sigma = 6.45$ kg/mm^2; the sample was subjected to a preliminary recrystallization at 2600°C and its structure was of the "single-crystal" type.* Bearing in mind that point defects played the dominant role in high-temperature investigations, all necessary measures were taken to retain the maximum number of vacancies and their clusters, generated during fracture, by quenching the samples. The experiments were carried out on thin tungsten samples (100-μ diameter), for which the rate of quenching, calculated from van Bueren's data [6] and bearing in mind the experimental conditions, amounted to ~5 · 10^4 deg/sec. The rate of dispersal of vacancies during the quenching depended strongly on the saturation of the lattice with defects which could serve as sinks for vacancies: these were grain boundaries and block boundaries, as well as single dislocations. According to [7], the surface layer of the sample was also important. Assuming a large num-

---

*We shall use the term "single-crystal" structure to describe samples in which, after recrystallization, the ratio of the grain length to the sample diameter was L/d $\approx$ 10^2-10^3.

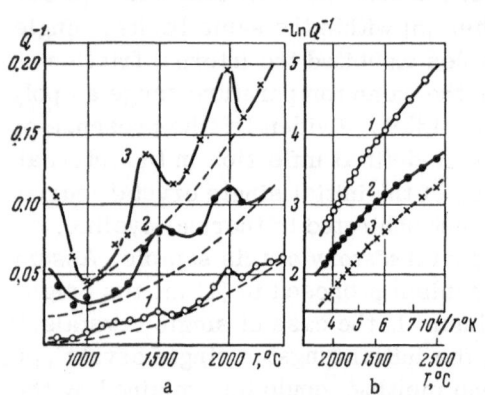

Fig. 2. Change in the relaxations spectrum (a) and in the high-temperature background (b) of the internal friction of tungsten during its fracture. 1) Initial (recrystallized) sample; 2, 3) the same sample fractured at 1500°C after three, and a further five minutes, respectively.

ber of vacancy jumps during the lifetime of one vacancy ($n \sim 10^9$), the thickness of the surface layer in which a considerable number of vacancies could emerge on the surface, was estimated to be of the order of 10^{-3} cm; consequently, to ensure effective quenching, the radius of the tested sample should not be much greater than this value. In this respect, the experimental conditions were favorable, namely: we used coarse-grained samples with a columnar structure, for which the average length of a grain was $L = (10^2-10^3)$ d, where d = 100 μ was the wire diameter, and the dislocation density in the material, calculated by counting the number of etch pits, corresponded to the number of dislocations in annealed crystals, $N = 10^6-10^7$ cm^{-2}; finally, rapid quenching was ensured by the fact that the fracture was carried out in a directly heated high-temperature machine. Under these conditions, the rapid cooling of a sample was ensured chiefly by switching off the heating current. The quenching of a sample under load made it possible to retain a considerably larger number of point defects than would be obtained by quenching-in the equilibrium concentration of vacancies from the same temperature. This difference should grow with the increase in the duration of the process leading to fracture. The method employed involved the interruption of the process leading to fracture of a metal after certain time intervals and simultaneous quenching of the defect state at a given moment, followed by further investigation by the internal friction method. Curve 1 in Fig. 2a represents the high-temperature internal friction background of the initial sample, while curves 2 and 3 represent the same sample which fractured after three minutes and a further five minutes, respectively. Figure 2b also gives, on semilogarithmic scale, the curves representing the change in the high-temperature background for these two cases.

Analyzing the experimental results, we note that the high-temperature deformation increases both the overall high-temperature background, as well as the individual relaxation maxima in the temperature range 700-2500°C. Enhancement of the dislocation (1500°C for tungsten of the VA grade) and "grain-boundary" (2000°C) peaks is observed. Moreover, the behavior of the curves below 1000°C is characterized by a general lowering of the background on cooling, as well as an appearance of a peak at 600-700°C, which had not been observed previously.

It should be mentioned that all the experimentally observed changes in the internal friction spectrum background (Fig. 2) were solely due to the influence of point defects and their clusters. Thick samples (of 250-μ diameter), in which quenching did not preserve point defects, showed no changes in the spectrum and background up to about 2000°C and only above this temperature the background did increase considerably with increase of the duration of application of the load. It has therefore been suggested that the 600°C peak was due to point defects. The published values of the activation energies of formation and motion of vacancies {40-70 kcal $\cdot$ (g-atom)$^{-1}$ [1, 3, 8, 9]} suggested the possibility of an internal friction peak, due to the diffusion of single vacancies in this range of temperatures. It should be mentioned that there are conflicting views of the existence of a vacancy maximum in metals. Some authors [10, 11] are of the opinion that a nonequilibrium change in the vacancy concentration during periodic deformation of a sample should not lead to the appearance of an internal friction maxi-

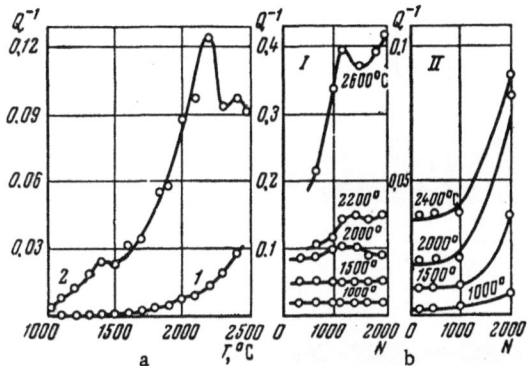

Fig. 3. Changes in the relaxation spectrum of "single-crystal" tungsten during thermocycling treatment (a) and changes in the high-temperature internal friction background of polycrystalline (average grain dimension 35 μ) I and "single-crystal" II samples as a function of the number of thermal cycles when temperature varied between 2500 and 30°C during one cycle (b). The temperatures at which the change in the background was observed are indicated alongside the corresponding curves: 1) initial sample; 2) the same sample after 2000 cycles.

mum in a certain range of temperatures, but it should give rise to a background increasing exponentially with temperature. According to [11], the absence of this maximum is due to the weak temperature dependence of the relaxation time of the vacancy concentration at high temperatures. On the other hand, like any other type of defect, vacancies have their own activation energy of motion, so that at a sufficiently high vacancy concentration, the peaks associated with the thermal activation of vacancies can be observed under certain conditions. There are published indications of the possibility of existence of peaks associated with the motion of vacancies [12] and their clusters [13], so that the assumption of the vacancy nature of the maximum observed in the temperature range near 700°C seems justified on the basis of an acceptable value of the activation energy ($\sim$50 kcal $\cdot$ (g-atom)$^{-1}$), calculated using Wert's formula [1]. Increase of the amplitude of this maximum during fracture confirms this conclusion. The peak associated with the diffusion of single vacancies in aluminum, detected by Pavlov [13], occurs at the same homologous temperature ($\sim$0.2 T_S).

The many vacancies nucleated during the fracture can, because of their high mobility, combine into large colonies whose collapse leads to the formation of ring dislocations. The applied load helps in the formation of clusters. We can thus explain the observed increase in the amplitude of the 2000°C peak, which indicates an increase in a number of such clusters and their growth during the process leading to fracture. Under a constant load, the average diameter of such clusters may increase to a dangerous level, when microscopic cracks begin to grow and one of them finally becomes the "critical" crack. The general increase in the background level, indicating, in accordance with the established correlations [2, 14], a reduction in the potential margin of the strength during the process leading to the fracture of tungsten provides a good confirmation of this conclusion. The increase in the amplitude of the 1500°C peak cannot be related to any increase in the dislocation density (we must bear in mind that the fracture occurs at temperatures at which intensive softening of tungsten takes place), but is most likely to be due to the interaction between dislocations and vacancies. For example, Cottrell has shown [7] that vacancies increase the mobility of dislocations in their slip planes at high temperatures. This is in good agreement with the observed tendency for this maximum to shoft toward lower temperatures. The conclusion that the observed increase in the background level and in the amplitudes of individual maxima is the consequence, not only of a quantitative accumulation of defects, but also of a change in the nature of their interactions, is supported by the results of plotting the background level in semilogarithmic coordinates (Fig. 2b). The decrease in the angle of the slope of each of the two parts of this dependence indicates a reduction in the energy of activation of the motion of divacancies and of single vacancies, respectively [1, 3], which can be treated as the consequence of a gradual reduction in the forces of interatomic binding with an increase in the duration of application of the load.

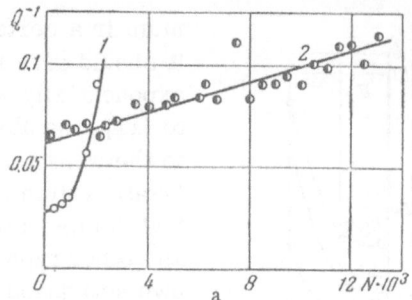

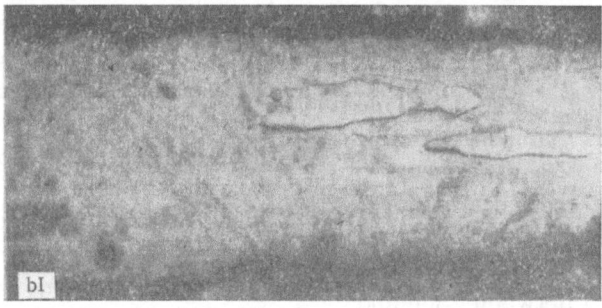

Fig. 4. Changes in the internal friction levels of two
tungsten samples (1, 2) at 2500°C during the process of
thermal fatigue (a); microstructure (I) and submicro-
structure (II) of tungsten subjected to 13,000 thermal
cycles (b).

An important characteristic of the behavior of metals is their cyclic (fatigue) strength.
The thermocyclic fatigue of refractory metals, which are frequently used under conditions of
large temperature drops, reaching 2000-2500°C, is particularly important. A typical example
is tungsten, which is used in incandescent lamps and in other sources of light, in which it is
subjected to thermal cycling. Consequently, an investigation of the change of the substructure
of tungsten during thermal fatigue is essential for the determination of the potentialities of this
metal in respect of the improvement of the quality of products made of it. Our investigation
was carried out using a directly heated relaxometer (torsional pendulum) with a special inter-
ruptor in the heating circuit of the sample. In order to determine the influence of grain boundaries on the thermocycling fatigue, the investigation was carried out on coarse-grained
samples (with the length of the crystal tens of times larger than its diameter), as well as on
fine-grain samples. Figure 3 shows the results of the investigation of the behavior of the spectrum and background of the internal friction of tungsten during thermal cycling. As in the case

considered earlier of fracture at high temperatures — and for the same reasons — it is found that the amplitudes of the 1500°C (dislocation) and the "grain-boundary" peaks increase (Fig. 3a) with increase of the number of thermal cycles applied to the coarse-grained metal. This is not observed in the case of the polycrystalline (fine-grained) samples. Figure 3b shows the results of the investigation of the behavior of various parts of the high-temperature internal-friction background during thermal cycling of the polycrystalline (I) and the "single-crystal" (II) tungsten samples. We note that:

1) the "single-crystal" samples exhibit an increase in the background level over the whole temperature range;

2) in the case of the polycrystalline samples, the parts of the background lying below the "grain-boundary" peak show practically no changes as N increases, but above this limit the changes are considerable. In earlier investigations [1, 3], it has been shown that the main contribution to the internal friction background in the temperature range 800-1700°C is made by double vacancies, while at higher temperatures single vacancies make the main contribution. Bearing this in mind, as well as estimates of the relative mobilities of these defects, we can explain qualitatively some of the results obtained. If we assume that the effective number of jumps executed by vacancies and divacancies is governed by the same expression: $n = A\nu zt \exp(-E/RT)$, where A is the activation entropy, ν is the frequency of vibrations of an atom, z is the coordination number of the lattice, t is the duration of annealing, and E is the activation energy of each defect, we find that, during the same annealing period, divacancies in tungsten may execute ~100 times more jumps than single vacancies, i.e., divacancies may move over considerably larger distances in a lattice. The presence of sinks of these defects will reduce their concentration in the surrounding volume. Thus, the appearance of an excess concentration of divacancies due to quenching from high temperatures is compensated, in the case of thermal cycling, by their rapid absorption into grain boundaries in the polycrystalline samples. Consequently, the part of the background to which these defects contribute most actively, remains unchanged in the case of thermal cycling. Single vacancies, having a considerably lower mobility, can gradually accumulate in the interiors of grains, and this process increases the high-temperature component of the background, corresponding to these defects. It is possible that the absence of R-dislocations near grain boundaries in the case of the thermal cycling of aluminum [4] is not only due to the ability of grain boundaries to absorb dislocations but, to a considerable extent, due to the aforementioned cause. It should be borne in mind that the formation of pairs is the first stage of the formation of clusters, so that the rapid loss of these defects at the boundaries may decelerate strongly the process of further clustering in the nearby regions. The formation of clusters without going through this stage (i.e., the direct formation of clusters consisting of three, four, and more vacancies) is not very likely.

We note that the microstructure investigations of the samples of tungsten subjected to thermocycling treatment do not exhibit the type of defects found in aluminum [4], possibly because of their small size. However, the investigation of the thermal fatigue of samples having various distributions of admixtures which are responsible for different laws of defect accumulation, showed some differences between the final microstructure. Figure 4a gives the results of an investigation of the influence of the thermocycling treatment (temperatures varying between 2500 and 30°C) on the internal friction level of two tungsten samples. A sample whose internal friction level increased rapidly with N (curve 1) withstood only 2000 thermal cycles, while a sample with a slowly rising dependence of $Q^{-1}(N)$ fractured after 13,000 cycles. After etching with an etchant which revealed grain boundaries [15], the first sample exhibited broadening of the grain boundaries and the characteristic "hilly" face of the section much more strongly than the second sample.

It is evident from Fig. 4b that, after etching the second sample for 15 h in a solution to reveal dislocations in tungsten [16], a very well developed network of sub-boundaries was observed.

Literature Cited

1. L.N. Aleksandrov and V.S. Mordyuk, Collection: Recrystallization Phenomena in Metals and Alloys, Metallurgizdat (1963), p.65.
2. V.S. Mordyuk, V.N. Orlov, and L.F. Savina, Collection: Sources of Light, No.2 VNIIÉM (1963), p. 41.
3. L.N. Aleksandrov and V.S. Mordyuk, Fiz.Metal.i Metalloved., 17(1) (1966).
4. V.S. Postnikov and N.V. Zolotukhin, Dokl.Akad.Nauk SSSR, 158(3): 590 (1964).
5. J.L. Orehotsky and R. Steinitz, Trans.Met. Soc.AIME, 224: 556 (June 1962).
6. H.G. van Bueren, Imperfections in Crystals [Russian translation], IL (1962). [English edition: 2nd ed. (Interscience) Wiley, New York.]
7. A.H. Cottrell, Vacancies and Point Defects [Russian translation], Metallurgizdat (1961), p.7.
8. A.A. Johnson, J.Less-Common Metals, 2(2-4): 241 (1960).
9. A.Ya. Kraftmakher and P.G. Strelkov, Fiz.Tverd.Tela, 4(8) (1962).
10. V.T. Shmatov and A.V. Grin', Fiz.Metal.i Metalloved., 12(4): 600 (1961).
11. M.A. Krishtal, Yu.V. Piguzov, and S.A. Golovin, Internal Friction in Metals and Alloys, Izd. "Metallurgiya" (1964).
12. R. Hasiguti, J.Phys.Soc.Japan, 8: 798 (1953).
13. V.A. Pavlov, Collection: Relaxation Phenomena in Metals and Alloys, Metallurgizdat (1960), p. 227. [English translation: B.N. Finkel'shtein, ed., Consultants Bureau, New York (1963), p.169.]
14. L.N. Aleksandrov and V.S. Mordyuk, Investigation of Steels and Alloys, Izd. "Nauka" (1964), p. 124.
15. Yu.S. Ezhova and S.M. Papin, Zavodsk. Lab., No. 8: 1044 (1961).
16. V.E. Wolf, Acta Met., 6(8): 559 (1958).

INVESTIGATION OF SOME RADIATION DEFECTS IN METALS BY THE METHOD OF MEASURING INTERNAL FRICTION AND YOUNG'S MODULUS

Yu. I. Pokrovskii, V. I. Vikhrov, and V. N. Perevezentsev

An investigation was made of the influence of small ($1.0 \cdot 10^{14}$ neutrons/cm^2) and large ($1.0 \cdot 10^{20}$ neutrons/cm^2) doses of neutron radiation on Young's modulus and internal friction, in the case of flexural vibrations at frequencies of 200-300 cps, of a number of metals having a range of melting points (copper, molybdenum, tungsten).

The results of the investigation, presented in this paper, are discussed from the point of view of the possibility of studying certain radiation defects. Moreover, the conditions are found under which the values of Young's modulus, obtained by different workers, could differ.

Method and Materials Investigated

The materials investigated were electrolytic copper, molybdenum, and tungsten of 99.98% purity. Before their irradiation, all the samples were subjected to high-temperature annealing: the copper samples were annealed at 1000°C, the molybdenum samples at 1500°C, and the tungsten samples at 1700°C.

The samples were irradiated in the working channel of the active zone of the RFT reactor using an integrated flux of 10^{20} neutrons/cm^2, and in a channel outside the active zone of the reactor IRT-1000 using a flux of 10^{14} neutrons/cm^2. The neutron fluxes were determined by means of the threshold reaction $Ni^{58}(n, p)Co^{58}$ using neutrons with E > 1 MeV.

The samples were irradiated outside the active zone in order to obtain simpler radiation defects, compared with the defects produced by irradiation in the active zone of a reactor.

The sample temperature during irradiation in the RFT reactor did not exceed 80-90°C, while in the IRT-1000 it was 25-30°C.

The apparatus used to obtain the results reported here, and the method of investigation, are described elsewhere in the present collection [5]. The value of Q^{-1} was measured in the stress amplitude (σ) range from 1 to 1000 g/mm^2. The maximum stress amplitude (σ)* was calculated from the vibration amplitude.

---

* This means the stress in the extreme "filaments" at the point where the sample was clamped.

The changes in Young's modulus (E), associated with the dependence of Q^{-1} on σ, were found from the changes in the square of the natural frequency of vibrations (f_0^2) of the sample, considered as a function of σ.

Influence of Irradiation on the Internal Friction

Figure 1 shows the dependence of Q^{-1} on σ applied to copper, molybdenum, and tungsten, before and after neutron irradiation in a flux of $1.0 \cdot 10^{20}$ neutrons/cm^2. Figure 2 shows the dependence of Q^{-1} on σ for copper before and after neutron irradiation with a flux of $1.0 \cdot 10^{14}$ neutrons/cm^2.

Considering Figs. 1 and 2, we conclude that the curves showing the dependence of Q^{-1} on σ of the investigated metals have a threshold (critical) stress. This critical stress (σ_{cr}) is shown in the figures and represents the value of the stress at which Q^{-1} begins to depend on the stress.

It has been reported [1] that the value of σ_{cr} of a zinc single crystal is related to the onset of plastic flow along the basal plane and is a measure of the critical shearing stress, while the value of σ_{cr} for a polycrystalline sample is related to the onset of plastic flow in the most favorably oriented grains, i.e., it is related to the motion of dislocations in a cyclic stress field along the slip planes of these grains.

It is evident from Figs. 1 and 2 that Q_{min}^{-1}* and σ_{cr} of the investigated metals was affected by irradiation.

Thus, for example, the value of σ_{cr} of copper increased after irradiation with a flux of 10^{20} neutrons/cm^2 by a factor of hundreds of times, while Q_{min}^{-1} decreased by a factor of 2.6, compared with the value before irradiation.

The changes in Q_{min}^{-1} of copper, accompanied by not too large changes in σ_{cr} (Fig. 1a) in the case of small radiation doses (10^{14} neutrons/cm^2), could be explained by the interaction† of dislocations with point defects, which were evidently vacancies; changes in Q_{min}^{-1} of copper accompanied by considerable changes in σ_{cr} in the case of large radiation doses (10^{20} neutrons per cm^2) could be explained by the interaction of dislocations with imperfections more complex than point defects. Such complex defects might be vacancy clusters.

Similar conclusions about the nature of defects produced by irradiation have been put forward for copper by Makin et al. [2]. They concluded that the defects were vacancy clusters. Makin et al. detected a relationship between radiation hardening and these clusters. In addition to vacancy clusters in copper, one may also observe dislocation loops, formed of interstitial atoms.

As is evident from Figs. 1b and 1c, the behavior of Q_{min}^{-1} of molybdenum and tungsten after irradiation with a flux of 10^{20} neutrons/cm^2 differed from the behavior of Q_{min}^{-1} of copper irradiated with the same flux.

The same was true of the effect of radiation on Young's modulus of molybdenum and tungsten (see next section). Q_{min}^{-1} of molybdenum and tungsten was increased by irradiation with a

---

*Q_{min}^{-1} is the minimum value of the internal friction, corresponding to the minimum stress amplitude.

† By the interaction of dislocations with point defects, we understand either the formation of a cloud consisting of defects around a dislocation, which retards the motion of the dislocation, or the precipitation of defects at a dislocation, pinning of a dislocation at certain points, and its splitting into shorter dislocations. It is probable that, under certain conditions, both mechanisms can act together.

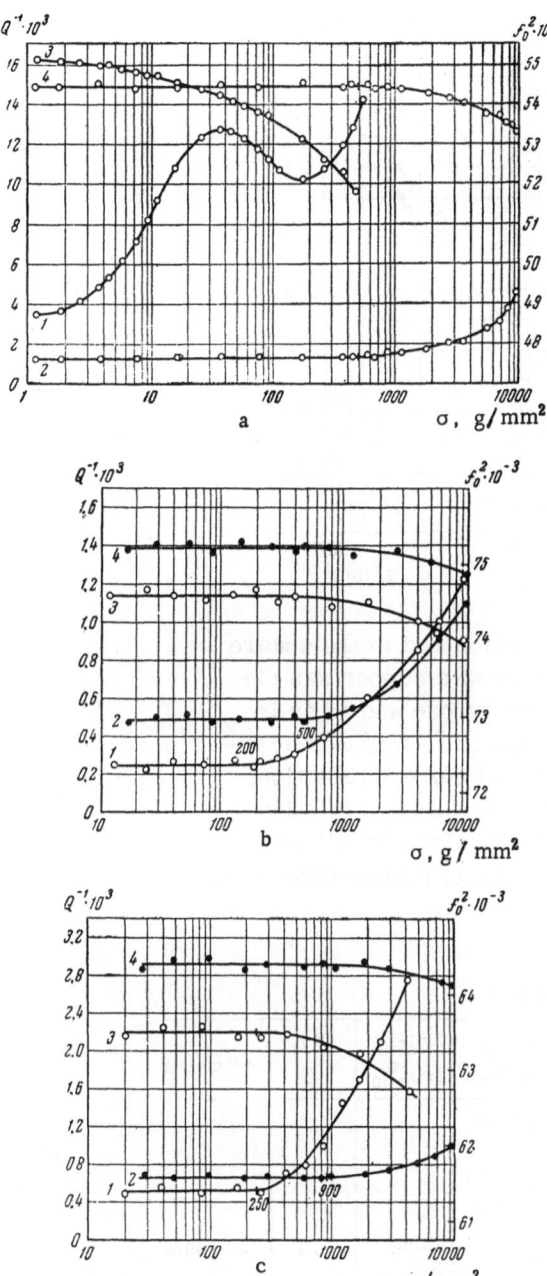

Fig. 1. Changes in the internal friction and the square of the natural frequency of vibrations of copper (a), molybdenum (b), and tungsten (c) samples as, a function of the stress amplitude, before and after irradiation with a flux of 10^{20} neutrons/cm². 1) Internal friction before irradiation; 2) internal friction after irradiation; 3) frequency squared before irradiation; 4) frequency squared after irradiation.

flux of 10^{20} neutrons/cm², while Q_{min}^{-1} of copper decreased after irradiation with the same neutron flux.

The difference between the behavior of Q^{-1} of molybdenum and tungsten and the behavior of Q^{-1} of copper after irradiation with a flux of 10^{20} neutrons/cm² could be due to the fact that molybdenum and tungsten contained "free" (i.e., not bound to dislocations) point defects, which were absent in copper. Judging by the increase in Young's modulus of molybdenum and tungsten, observed at the same time as the increase in Q^{-1} after irradiation (see next section), one could conclude that the main contribution to the change in Q^{-1} was made by interstitial atoms, and not by vacancies, which would also be present in irradiated molybdenum and tungsten.

The presence of free interstitial atoms, as well as vacancies, could have been due to the fact that they were not very mobile in the crystal lattices of molybdenum and tungsten and could retain their positions even during irradiation.

The increase in σ_{cr} of molybdenum and tungsten might indicate that, as in copper, dislocations were pinned by the interaction with radiation defects. The slight increase in σ_{cr} of these metals, compared with the large increase in σ_{cr} of copper, could be associated with the fact that the majority of the resultant radiation defects remained in the lattice because of their low mobility in molybdenum and tungsten compared with the mobility in copper.

To determine finally the nature of radiation defects in the investigated metals, and to study their properties, further experiments are necessary on the frequency and temperature dependence of Q^{-1} and on the various annealing stages (for the determination of the activation energy of the recovery process).

Figure 2 and Table 1* give the results of investigations of the influence of anneal-

_____

* Tables 1 and 2 give the values of Young's modulus calculated from the values of the square of the vibration frequency in the case $\sigma < \sigma_{cr}$.

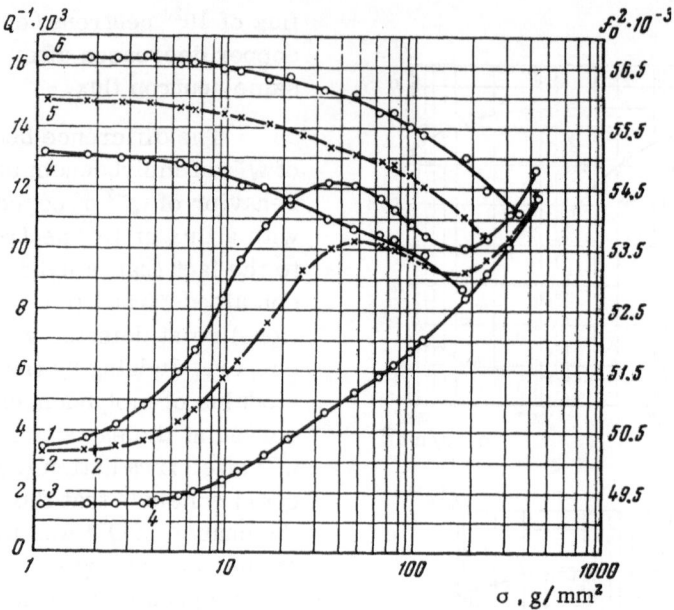

Fig. 2. Changes in the internal friction and in the square
of the natural frequency of vibrations of a copper sample
as a function of the stress amplitude before and after irradiation with a flux of 10^{14} neutrons/cm^2, followed by annealing at 100°C. 1) Internal friction before irradiation;
2) internal friction after irradiation; 3) internal friction
after annealing; 4) frequency squared before irradiation;
5) frequency squared after irradiation; 6) frequency squared
after annealing.

Table 1. Copper

	Q_{min}^{-1}	E, kg/mm^2	σ_{cr}, g/mm^2	ΔQ_{min}^{-1}, % Q_{min}^{-1}	$\Delta E/E$, %	$\Delta \sigma_{cr}/\sigma_{cr}$
Before irradiation	$3.4 \cdot 10^{-3}$	11430	<1.0			
After irradiation with 10^{14} neutrons/cm^2 . . .	$3.1 \cdot 10^{-3}$	11610	2.0			
After annealing at 100°C	$1.50 \cdot 10^{-3}$	11780	4.0			
Total change				−56	3.1	>4 times

ing at 100°C on Q_{min}^{-1}, σ_{cr}, and Young's modulus of copper irradiated with a neutron flux of
10^{14} neutrons/cm^2.

Examination of the data in Table 1 shows that the changes in Q_{min}^{-1} and in Young's modulus, caused by irradiation, became greater after annealing. The additional increase in Young's
modulus and in σ_{cr} and the additional reduction of Q_{min}^{-1} after annealing at 100°C could be associated with additional pinning of dislocations by defects. Such additional pinning could be due to defects liberated by annealing from impurities, which, during irradiation acted (like dislocations)
as "sinks" for defects.

Table 2. Copper, Molybdenum, and Tungsten

	E, kg/mm²			ΔE/E, %		
	Cu	Mo	W	Cu	Mo	W
Before irradiation	11430	33070	37910			
After irradiation in 10^{14} neutrons/cm² flux	11610	–	–			
The same, for 10^{20} neutrons/cm²	11210	33390	38470			
Changes due to 10^{14} neutrons/cm² flux				1.6	–	–
The same, for 10^{20} neutrons/cm²				−1.9	0.9	1.5

Further annealing of irradiated copper at 500°C for 2 h increased Q_{min}^{-1}, and reduced Young's modulus and σ_{cr} to the initial values before irradiation.

Influence of Irradiation on Young's Modulus

The upper curves in Figs. 1 and 2 represent the change in the square of the natural frequency of vibrations of a sample (f_0^2) as a function of the stress amplitude, obtained before and after irradiation of copper, molybdenum, and tungsten. Table 2 gives the changes, due to irradiation, in Young's modulus, calculated from the values of f_0^2.

Conclusions from curves similar to those given in Figs. 1 and 2 have been drawn in [3].

The main conclusion was that the magnitude of the change in Young's modulus due to irradiation depended on the value of σ. For $\sigma < \sigma_{cr}$, Young's modulus was reduced by irradiation with a large neutron dose.

From the data in Table 2, the following additional conclusions could be drawn: neutron irradiation could increase or decrease Young's modulus of copper, depending on the value of the integrated flux. Young's modulus of copper was increased by irradiation with a small neutron dose but was reduced by a large dose.

Increase of Young's modulus of copper after irradiation with a small neutron dose has been explained in [4] by an increase in the elasticity due to pinning of dislocations by point defects.

The reduction of Young's modulus of copper, due to irradiation with a large neutron dose in the case $\sigma < \sigma_{cr}$, was associated with a reduction of the elasticity of the lattice due to the formation of "free" vacancy clusters, not bound to dislocations.

The data in Table 2 show also that Young's modulus of molybdenum and tungsten increased after irradiation with a large neutron dose (a small dose caused no great change).

The increase in Young's modulus of molybdenum and tungsten, in the case $\sigma < \sigma_{cr}$, due to irradiation with a large neutron dose was ascribed to an increase in the elasticity of the lattices of these metals due to the formation of "free" interstitial atoms or complexes of such atoms.

The greater increase in Young's modulus of tungsten, compared with that in molybdenum, supported, to some extent, the conclusion that an increase in the melting point was accompanied by a reduction of defect mobility during irradiation, which led to a greater accumulation of interstitial atoms in the crystal lattice.

As in the case of copper, large stresses ($\sigma > \sigma_{cr}$) applied to molybdenum and tungsten affected the radiation-induced changes in Young's modulus, because of the dependence of the dislocation mobility on the stress because of interaction of dislocations with "bound" (i.e., interacting with dislocations) defects. This should be borne in mind in any comparison of the results of individual investigations of the influence of irradiation on Young's modulus.

Conclusions

The internal friction and Young's modulus are sensitive to neutron irradiation and can be used to investigate various radiation defects.

On the basis of the results of the investigation reported in the present paper, some hypotheses have been put forward about the nature of radiation defects produced by small and large neutron doses.

To check these hypotheses, additional experiments are needed on the dependence of the internal friction on the frequency, as well as experiments on the various annealing stages required to determine the activation energy of the recovery process.

In future investigations of point defects by the internal friction method, it would be advisable to use low temperatures, as well as nuclear radiation of other types (γ rays, electrons).

Literature Cited

1.   N. F. Pravdyuk, Yu. I. Pokrovskii, and V. I. Vikhrov, Soviet Journal of Atomic Energy, 10(4): 334 (1962) (Consultants Bureau, New York).
2.   M. J. Makin, A. D. Whapham, and E. J. Minter, Phil. Mag., 7: 285 (1962).
3.   S. T. Konobeevskii, N. F. Pravdyuk, Yu. I. Pokrovskii, and V. I. Vikhrov, Collection: Effects of Nuclear Radiations on Materials, Izd. AN SSSR (1962), p. 219.
4.   D. O. Thompson and D. K. Holmes, J. Appl. Phys., 27: 713 (1956).
5.   Yu. I. Pokrovskii, V. N. Perevezentsev, and V. I. Vikhrov, this volume, p. 195.

INTERNAL FRICTION IN AUSTENITIC STEELS

S. A. Golovin, K. N. Belkin, and B. M. Drapkin

The relaxation processes taking place during heating of fcc metals and alloys have been investigated most thoroughly by Finkel'shtein et al. [1-3]. The present paper reports the temperature dependence of the internal friction $Q^{-1}(T)$ of austenitic steels Kh18N9 and 45G17YuZ of industrial origin, at various vibration frequencies. The chemical compositions of the steels are given in the table.

The low-frequency measurements (~1 cps) were carried out on wire samples, 160 mm long and of 0.8 mm diameter. The dependence $Q^{-1}(T)$ was determined using a vacuum torsional pendulum of the RKF MIS type. At frequencies of 750-850 cps, we investigated samples 200 mm long and of 6-8 mm diameter, employing the method described in [4]. Annealing and heating of samples before quenching took place in 10^{-3} mm Hg vacuum; carbonization was carried out in benzene vapor at 1000°C for 1.5 h, and plastic deformation took the form of drawing. After mechanical or heat treatment, we allowed 15-20 min before the beginning of measurements of $Q^{-1}(T)$. The deformation amplitude during measurements did not exceed 10^{-6}.

Figure 1 shows the dependence $Q^{-1}(T)$ at $f = 1$ cps for the Kh18N9 steel, after austenization at temperatures of 925 and 1075°C, as well as the influence of carbonization. The increase in the carbon content in the solid solution due to increase of the quenching temperature or due to carbonization gave rise to the formation of a 300°C maximum of the internal friction. Plastic deformation of the Kh18N9 steel, quenched from 1075°C, led to a complex change in $Q^{-1}(T)$ (Fig. 2); increase of the degree of plastic deformation (cold working) led to an increase of $Q^{-1}(T)$ and two maxima were observed at 80-100°C and at 300°C. When the degree of plastic deformation was raised up to 27%, the amplitude of the 300°C peak and the area under this peak increased rapidly, but when the degree of plastic deformation (cold working) was increased still further, they both fell. The addition of titanium (0.51 wt.%) to the Kh18N9 steel destroyed completely the high-

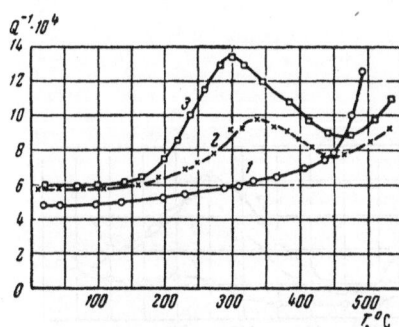

Fig. 1. Temperature dependence of internal friction of the quenched steel Kh18N9 ($f \approx 1$ cps). 1) Quenching from 925°C; 2) from 1075°C; 3) from 1075°C after carbonization.

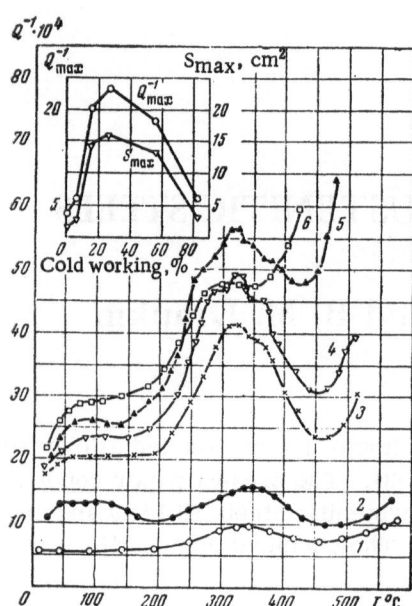

Fig. 2. Influence of plastic deformation on temperature dependence of internal friction of steel Kh18N9 quenched from 1075°C ($f \approx 1$ cps). 1) Without cold working; 2) 5% cold working; 3) 15% cold working; 4) 27% cold working; 5) 56% cold working; 6) 82% cold working.

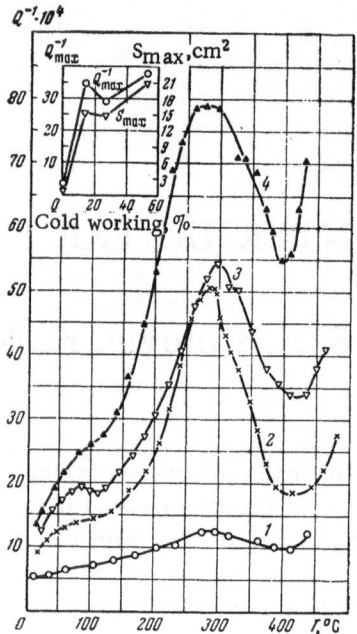

Fig. 3. Influence of the plastic deformation on the temperature dependence of the internal friction of the steel 45G17YuZ quenched from 1075°C ($f \approx 1$ cps). 1) Without cold working; 2) 15% cold working; 3) 27% cold working; 4) 55% cold working.

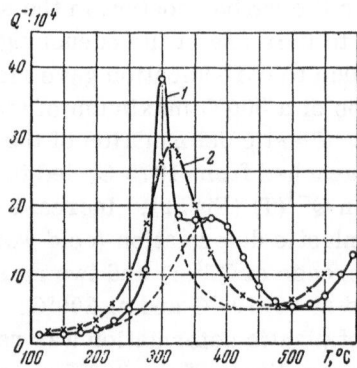

Fig. 4. Temperature dependence of the internal friction of the 45G17YuZ steel ($f \approx 780$ cps). 1) After forging; 2) after recrystallization at 780°C for 1 h.

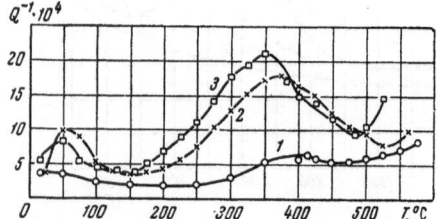

Fig. 5. Temperature dependence of the internal friction of steel Kh18N9 quenched from 1075°C ($f \approx 810$ cps). 1) Without cold working; 2) 14% cold working; 3) 22% cold working.

Steel designation	Chemical composition, wt.%						
	C	Cr	Ni	Mn	Si	Al	S ; P
Kh18N9	0.12	17.44	8.75	1.46	0.46		~0.01
45G17Yu3	0.45	0.24	—	17.45	0.18	2.76	~0.01

temperature maximum of $Q^{-1}(T)$ in the initial and deformed states. Plastic deformation of the 45G17YuZ steel (quenched from 1075°C) caused an increase in the parameters of the $Q^{-1}(T)$ maximum at 300°C (Fig. 3). The data in Figs. 1-3 show that the formation of the internal friction maximum at 300°C in plastically deformed austenitic steels of these two grades could be associated with the migration of carbon atoms in austenite and with the interaction of these atoms with dislocations in a solid solution subjected to alternating stress fields. It should be remembered that in the type 18-8 steel the motion of carbon atoms in a solid solution may take place along positions close to the carbide-forming alloying element (for example, chromium), and the activation energy of such a process is large.

To distinguish the contribution of various factors to $Q^{-1}(T)$, the decrement was measured also at frequencies of 750-850 cps. The investigated steels had $Q^{-1}(T)$ maxima at 300-375°C. The activation energy (H) was determined from the temperature of the peak, using the dependences given in [5, 6]:

$$H = RT_{max} \ln \frac{kT_{max}}{h \cdot f_{max}} + T_{max}\Delta S \tag{1}$$

and

$$H = RT_{max} \ln \frac{2}{\sqrt{\pi}} \frac{kT_{max}}{hf_{max}} + T_{max} \frac{R}{2} \ln \frac{H}{RT_{max}}, \tag{2}$$

where T_{max} is the temperature of the peak, f_{max} is the corresponding frequency; R is the gas constant; k is Boltzmann's constant; h is the Planck constant; ΔS is the entropy.

The activation energy was 34 ± 2 kcal · (g/atom)$^{-1}$, both for low-frequency and high-frequency dependences.

Figure 4* shows the dependence $Q^{-1}(T)$ for the 45G17YuZ steel after considerable deformation by forging (curve 1), as well as after annealing at 780°C for 1 h (curve 2). Two maxima at 300 and 375°C, were found for the deformed sample of the 45G17YuZ steel in the $Q^{-1}(T)$ curve. The activation energies, determined from the temperatures of the peaks using Eq. (2), were, respectively, 32 and 36 kcal · (g-atom)$^{-1}$. After recrystallization only one internal friction maximum remained near 320°C. In our opinion, the low-temperature maximum of curve 1 was associated with the migration of carbon in the solid solution subjected to a stress field, while the high-temperature maximum was due to the interaction of interstitial atoms with dislocations under the action of the periodic vibrations of the system. The decrease in the dislocation density in the annealed sample of the 45G17YuZ steel reduced considerably the fraction of the energy dissipated due to the interaction between dislocations and impurity atoms in Cottrell zones.

Figure 5 shows the dependence $Q^{-1}(T)$ at 800°C for the steel Kh18N9, quenched from 1075°C and subjected to various degrees of deformation by drawing. We found the same regularity in the changes of the peak parameters as the low-frequency measurements (Fig. 3),

*Along the abscissa one should read 150, 250, 350°C, etc. instead of 100, 200, 300°C, etc.

but there was a characteristic influence of plastic deformation. As the degree of compression was increased to 40%, the position of the high-temperature branch of the internal friction was not affected, but the increase of the area under the maximum took place mainly by displacement of its low-temperature branch. Evidently the relaxation in the high-temperature part of $Q^{-1}(T)$, in the region 375–400°C, was due to the migration of interstitial atoms in austenite along positions near carbide-forming elements. The low-temperature part of the maximum corresponded to the relaxation processes associated with the influence of plastic deformation. It was not clear why the amplitude of the internal friction peak at 300°C decreased in the case of Kh18N9 steel with high degrees of cold working. This was probably associated with phase transitions, formation of superequilibrium vacancies during deformation, and other causes.

Literature Cited

1.   K. M. Rozin and B. N. Finkel'shtein, Dokl. Akad. Nauk SSSR, 91(4): 811 (1953).
2.   V. D. Verner, B. N. Finkel'shtein, and A. V. Shalimova, Fiz. Tverd. Tela, 3(11): 3363 (1961).
3.   V. D. Verner, Investigation of the Behavior of Interstitial Atoms in Solid Solutions Based on the fcc Lattice by the Internal Friction Method, Dissertation for Candidate's Degree, Moscow (1963).
4.   M. A. Krishtal and B. M. Drapkin, Apparatus for Measuring Elastic Moduli and Damping, Zavodsk. Lab. (in press).
5.   C. Wert and J. Marx, Acta Met., 1: 113 (1953).
6.   B. M. Drapkin, Dissertation for Candidate's Degree, Tula (1965).

SOLUBILITY OF CARBON IN α-IRON

M. I. Bayazitov, I. N. Kidin,
and Yu. V. Piguzov

The solubility of carbon in α-iron is described by the QPG line in the phase diagram. It is assumed that the QP line represents the solubility limit of carbon in ferrite in equilibrium with cementite, and the PG line gives the solubility limit for ferrite in equilibrium with austenite.

The solubility of carbon in α-iron has been determined by many investigators who have used a great variety of metals. Comparison of the values of the solubility limit of carbon, obtained by different methods, shows that the metallographic and physicochemical methods give 0.03-0.04%, while the physical methods (internal friction, elastic aftereffect, electrical resistance) give 0.02% C.

In the internal-friction method the concentration of carbon in a solid solution is determined using the amplitude of the 40°C maximum (at a vibration frequency of about 1 cps). The amplitude of the 40°C internal friction maximum is proportional to the concentration of carbon in ferrite and, for a given carbon content, depends on the grain size. Seeman and Dickenscheid [1] used the results of several investigations to plot the dependence, on the grain size, of the coefficient of proportionality between the concentration of carbon in ferrite and the amplitude of the 40°C maximum (Fig. 1). Seeman and Dickenscheid were of the opinion that, other conditions being equal, the coefficient of proportionality depended also on the degree of contamination of iron with various elements. However, they did not explain the reasons for the variation of this coefficient.

Lagerberg and Josefsson [2] explained the reduction in the amplitude of the internal friction maximum with reduction of the grain size by the influence of grain boundaries. In the opinion of these authors, carbon atoms in a strongly distorted zone near grain boundaries (absorbed on boundaries) did not take part in the relaxation process in a stress field. Lagerberg and Josefsson concluded that the stresses applied to the sample during the internal friction measurements were far too small to displace these atoms from their equilibrium positions. They showed that the increase in the amplitude of the internal friction maximum of a sample saturated with carbon stopped at higher concentrations in the case of fine-grained samples.

It can now be regarded as proven [3-12] that many impurities, in particular carbon, are adsorbed on grain and block boundaries. Therefore, in the light of the cited investigations, the conclusions of Lagerberg and Josefsson seem to be fully acceptable. The values of the solubility of carbon in ferrite, obtained by various investigators, did not agree, probably because

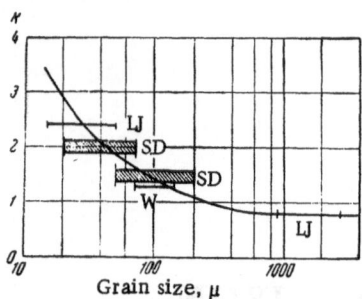

Fig. 1. Dependence, on the grain size, of the coefficient of proportionality between the amplitude of the 40°C maximum and concentration of carbon in α-iron. LJ) Lagerberg and Josefsson's data [2]; SD) Seeman and Dickenscheid's data [1]; W) Wepner's data [15].

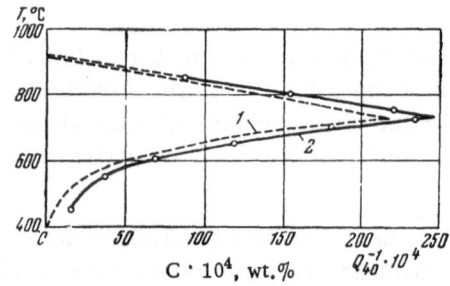

Fig. 2. Dependence of the concentration of carbon in ferrite and of the amplitude of the 40°C internal friction maximum on the quenching temperature. 1) Smith's data [12]; 2) data of the present authors (amplitude of the 40°C maximum).

the phenomenon of internal adsorption was not allowed for. If the carbon were distributed nonuniformly in ferrite, this would have affected the nature of the temperature dependence of the solubility and the phase transition in steels having structurally free ferrite. The structure of the core of a grain (block) would naturally differ from the structure of its periphery, which must have necessarily affected the mechanical and physical properties of ferrite.

Preparation of Samples and Experimental Method

The initial material was a powder of carbonyl iron of the V-3 grade, whose total impurity content was not more than $1 \cdot 10^{-2}$ wt.%. To prepare an ingot, brickettes of carbonyl iron were melted in a vacuum crucible, and then remelted in vacuum in a furnace with consumable electrodes.

The ingots were forged into rods of 8 mm diameter, from which a wire of 0.75 mm diameter was drawn. To remove nitrogen and carbon (the material was contaminated by the grease used in drawing), the wire was annealed in an atmosphere of moist hydrogen at 700°C for 50 h. After such a treatment the amplitude of the 40°C internal friction maximum (the measurements were carried out after quenching from 700°C) did not exceed $5 \cdot 10^{-4} Q^{-1}$.

Following hydrogen purification, the samples were carbonized in a mixture of benzene and hydrogen vapors at 700°C for various periods, necessary to obtain the required carbon content in the sample. Special measures were taken to ensure a uniform carbonization along the whole length of the sample. The carbonized samples were subjected to homogenizing annealing for 3 h at 700°C.

The internal friction measurements were carried out, using a vacuum torsional pendulum, on samples 100 mm long, subjected to vibrations at a frequency of about 1 cps. The maximum relative deformation on the sample surface did not exceed $2 \cdot 10^{-5}$. The measurements were carried out in a magnetic field of 120 Oe intensity, which was sufficient to suppress completely the magnetomechanical damping. The electrical resistance was measured by the potentiometric method at liquid nitrogen temperature.

All the measurements were begun not later than 30 min after quenching from the required temperature. The quenching was carried out in a vertical vacuum furnace and the samples were dropped into oil. It was established that samples of the 45-type steel quenched in this way contained martensite.

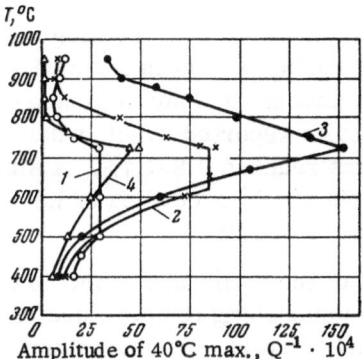

Fig. 3. Dependence of the amplitude of the 40°C internal friction maximum on the quenching temperature. 1) Carbon content 0.003%; 2) 0.01; 3) 0.04; 4) 0.4%.

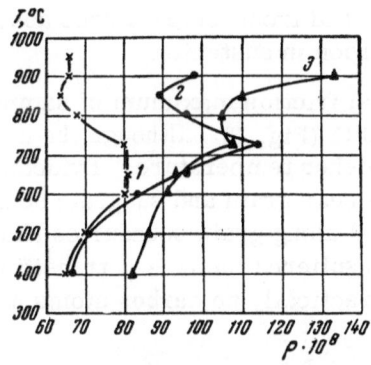

Fig. 4. Dependence of the electrical resistivity on the quenching temperature. 1) Carbon content 0.01%; 2) 0.04%; 3) 0.4%.

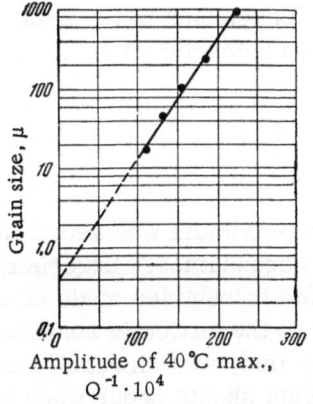

Fig. 5. Dependence of the amplitude of the 40°C internal friction maximum on the grain size.

Investigation of the Temperature Dependence of the Solubility of Carbon

It has been suggested above that the values of the solubility obtained by different investigators were not all equal because of the different grain size in the samples used. In the absence of any influence of grain boundaries, the values of the solubility should be the same. Figure 2 shows the dependences on the quenching temperature of the solubility of carbon in ferrite, obtained by Smith [12] by the decarbonization method, and of the amplitude of the 40°C internal friction maximum, obtained by the present authors for a sample containing 0.04% C and having the "columnar' structure with grain size greater than 1 mm (the sample diameter was |0.75 mm). In Smith's experiments, the role of grain boundaries could be neglected, because the maximum depth of the decarbonized layer did not exceed 30μ, while the grains in his samples were known to be large. Before his investigations, Smith annealed his samples for seven days at 700°C. Multiplication of the internal friction data (Fig. 2) by the factor 0.9 (which was the coefficient of proportionality, according to Seeman and Dickenscheid's curve) resulted in values of the solubility in good agreement with Smith's results.

We determined the temperature dependence of the solubility of carbon in α-iron from measurements of the internal friction and electrical resistivity using samples containing various amounts of carbon (Figs. 3 and 4). The internal friction and electrical resistivity curves were of the same nature. It was interesting to examine the internal friction and electrical resistivity curves for samples containing less carbon than the value at the solubility limit. For a uniform distribution of carbon over the whole volume of ferrite, the solubility curve, for samples containing carbon in amounts known to be less than the solubility limit, should be vertical up to intersection with the PG line at quenching temperatures above the QP line. However, when the quenching temperature was slightly higher than the eutectoid value, the values of the internal friction and electrical resistivity began to decrease rapidly. Since the PG line reflected the solubility limit of carbon in ferrite, in equilibrium with austenite, then, by analogy, we assumed that, at points of segregation on grain and subgrain boundaries, the concentration of carbon was close to the eutectoid value and at subcritical temperatures an austenitic film was formed. This austenitic film increased in thickness with increase of temperature, and, capturing an ever-increasing number of carbon atoms, tended to reduce rapidly the concentration of carbon atoms in the lattice of the remaining α-iron. Quenching transformed austenite into martensite, which was known not to have the 40° maximum. This was in agreement with the results reported in [6], where it was shown that the concentration of carbon at segregation points reached 0.8-1.7%. A similar effect has been observed in transformer steel [13], although this steel, according to its phase diagram, is of purely ferrite type up to temperatures of the order of 1200°C.

Figures 3 and 4 show, for comparison, the dependences of the amplitude of the 40°C internal friction maximum and of the electrical resistivity on the quenching temperature of samples containing 0.4% carbon. It is evident from these figures that at quenching temperatures above A_1, the amplitude of the 40°C maximum began to decrease. At these temperatures steel of this composition already contained austenite, and the concentration of carbon began to decrease because of the high solubility of carbon in austenite. On reaching the upper critical temperature (800°C for steel with 0.4% carbon) ferrite disappeared and austenite was transformed by quenching into martensite, which did not, as mentioned earlier, exhibit the 40°C internal friction maximum.

The structural changes were reflected also in the same way by the electrical resistivity. The sharp increase in ρ for a sample containing 0.4% C, quenched from temperatures above 800°C, was evidently due to a more uniform distribution of carbon in austenite.

It was worth noting that the amplitude of the 40°C internal friction maximum of samples containing 0.003% carbon was lowest after quenching from 850°C (Fig. 3) although the $\alpha \rightarrow \gamma$ transformation over the whole volume should take place at a higher temperature. Evidently, at these low concentrations, all carbon was in distorted zones near grain and subgrain boundaries. When heated before quenching, the austenite film formed along grain boundaries absorbed nearby carbon atoms. Although the bcc lattice, which had not suffered the $\alpha \rightarrow \gamma$ transition, was still retained in grain (subgrain) cores, this lattice had practically no carbon atoms and the 40°C maximum was not observed or was very small.

We may assume that at high concentrations carbon was distributed in ferrite very non-uniformly and, before the solubility limit was exceeded, carbon was found mainly near grain and subgrain boundaries, which represented only a small fraction of the grain volume. This possibility has been considered by Arkharov in one of his investigations [3] of aging processes. Only from this point of view can we explain the results of Kaminskii and Stelletskaya [14] who determined, by x-ray diffraction, the ferrite lattice parameter, and found no changes even after quenching from 700°C, which was carried out in order to "fix" carbon in the solid solution. Therefore, Kaminskii and Stelletskaya came to the erroneous conclusion that quenching could not "fix" carbon in the solid solution.

Influence of the Grain Size on the Amplitude
of the 40°C Internal Friction Maximum

The reduction of the amplitude of the 40°C internal friction maximum was ascribed by Lagerberg and Josefsson [2] to the adsorption of carbon on grain boundaries. Lagerberg and Josefsson were of the opinion that carbon atoms adsorbed on grain boundaries made no contribution to the internal friction because of the strong distortion of the lattice in zones near grain boundaries. We may assume that if grains are sufficiently fine, we can reach a situation when the 40°C internal friction maximum cannot be observed. This should occur when the adsorption zones (or zones where the lattice is strongly distorted, in accordance with Lagerberg and Joseffson's definition) converge within a grain. We made an attempt to check this hypothesis.

In this experiment we used samples containing 0.04% carbon and having the columnar structure. To obtain samples with a range of grain dimensions, a batch of samples (3 of them) was sealed in evacuated ampoules, heated for 15 min at 950°C, and then cooled in water inside the ampoule. Repetition of this treatment made it possible to reduce grain dimensions. The smallest grain dimensions obtained by us were 17 μ.

The results of the measurements were plotted on semilogarithmic scale (Fig. 5) and the obtained points fitted well a straight line. Assuming a linear dependence of the amplitude of the 40°C internal friction maximum on the grain size, we could, by extrapolating onto the grain-dimension axis, determine the critical grain dimension, for which the internal friction maximum would not be observed. According to our data, this occurred for the grain size of 0.4 μ. The critical grain size for samples of other compositions quenched from different temperatures might be somewhat different but, in principle, this critical effect should be observed in all cases whenever there was an internal friction maximum due to the diffusion of impurity atoms in a stress field. The method proposed here should make it possible to determine experimentally the width of the adsorption zone at grain boundaries. In our case this width was 0.2 μ, assuming that at critical grain size the zones of near-boundary absorption merged.

When viewing thin Fe + 0.015% C foils in an electron microscope, Phillips [10] observed a broad dark band near grain boundaries. The adsorption zone found in our experiments and this dark band (whose width was 0.1-0.2 μ) were probably of the same nature, because their dimensions were almost equal. Similar values for the widths of the adsorption zones have been reported elsewhere [6, 7].

Literature Cited

1. H. J. Seeman and W. Dickenscheid, Acta Met., 6(1): 62 (1958).
2. G. Lagerberg and A. Josefsson, Acta Met., 3:236 (1955).
3. V. I. Arkharov, Tr. Inst. Fiz. Metal., Akad. Nauk SSSR, Ural'sk. Filial, No. 8: 54 (1946).
4. N. G. Ainslie, R. E. Hoffman, and A. U. Seybolt, Acta Met., 8(8): 523 (1960).
5. V. N. Gridnev, Metalloved. i Term. Obrabotka Metal., No. 1: 19 (1959).
6. A. L. Tsou, J. Natting, and J. W. Menter, J. Iron Steel Inst. (London), 172: 163 (1952).
7. A. I. Gardin, Electron Microscopy of Steel [Russian translation], Metallurgizdat (1954), pp. 168-174.
8. P. Samuel and A. G. Quarrell, J. Iron Steel Inst. (London), 182: 20 (1956).
9. D. McLean, Grain Boundaries in Metals [Russian translation], Metallurgizdat (1960), pp. 109-140. [English edition: Oxford University Press, New York.]
10. V. A. Phillips, Acta Met., 11(10): 1139 (1963).
11. Yün P'ao-ts'ui and Ke T'ing-sui, Fiz. Metal. i Metalloved., 4(3): 407 (1957).
12. R. Smith, Trans. Met. Soc. AIME, 224(1): 105 (1962).
13. V. E. Kochnov and M. I. Bayazitov, Fiz. Metal. i Metalloved., 15(6): 937 (1963).
14. É. Z. Kaminskii and T. I. Stelletskaya, Collection: Problems of Metallurgy and Physics of Metals, Metallurgizdat (1951), Vol. 2, p. 176.
15. W. Wepner, Arch. Eisenhüttenw., 27:55 (1955).

"IMPURITY" INTERNAL FRICTION PEAK
OF IRON ALLOYS CONTAINING PHOSPHORUS

Yu. V. Grdina, E. É. Glikman, and Yu. V. Piguzov

It is known that polycrystalline samples of a pure metal, to which an impurity has been added, may exhibit, in addition to the high-temperature peak associated with the relaxation of stresses along grain boundaries, a so-called "impurity" internal friction peak with an activation energy close to the activation energy of the diffusion of the impurity in the solvent metal. Such peaks have been detected for a number of alloys based on copper and silver [1], as well as Fe−W alloys [2].

The general relationships established by investigating the "impurity" peak, have been analyzed by Shmatov and Grin' [3], who proposed a possible mechanism of the appearance of this peak.

On the basis of this mechanism, Shmatov and Grin' [3] obtained an expression for the degree of relaxation ΔE_T of the modulus in the form

$$\Delta E_T = AC\,(1-C),$$

(1)

where A is a constant, depending on the state of the sample, the nature of deformation, and temperature; C is the concentration of impurity atoms in grain boundaries.

It follows from Eq. (1) that the "impurity" peak may be fairly strong, provided the concentration of impurities in the grain boundaries is not too low; the amplitude of the peak reaches its maximum at C = 0.5.

Since it has been established experimentally that the "impurity" peak is observed even at such moderate impurity concentrations as 0.03 at.% (Cu + 0.03% Al [1]), its appearance evidently requires the concentration of impurity atoms in grain boundaries to be much higher than elsewhere. Among impurities forming substitutional solid solutions with iron, phosphorus has the strongest tendency to such aggregation in grain boundaries [4]. Therefore, iron-base alloys, containing phosphorus, should have an "impurity" internal friction peak, related to the migration of phosphorus atoms to grain boundaries (and conversely) under the action of cyclic deformation, and the investigation of this peak should be of interest, for example, in connection with the investigation of the reversible temper brittleness of steel, because the latter is associated with the aggregation of phosphorus atoms mainly in grain boundaries [5].

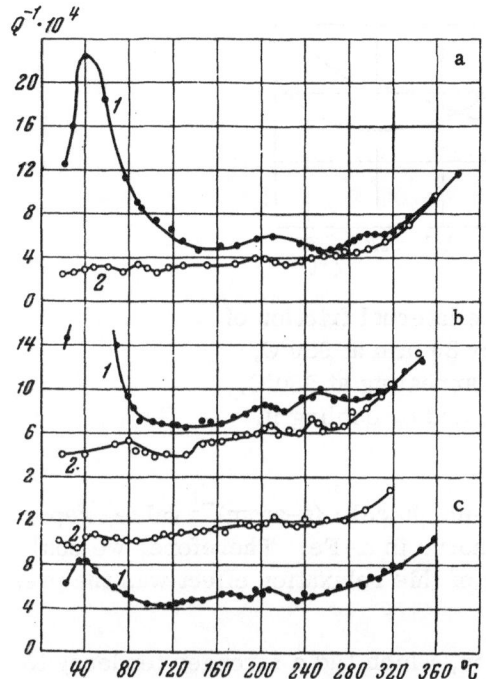

Fig. 1. Temperature dependence of internal friction of the investigated alloys subjected to heat treatment at 920°C for 30 min, followed by quenching, tempering for 2 h at 650°C, and cooling in water (1), and after additional tempering for 8 h at 530°C (2). a) Alloy 08G2R; b) alloy 08R; c) alloy 08G20R.

Alloy designation	Impurity content, at.%		
	C	Mn	P
08G2R	0.37	1.78	0,058
08R	0.37	<0.01	0,061
08G20R	0.42	1.88	< 0.002

We give below the results of an investigation of the internal friction of three alloys based on iron and containing admixtures of phosphorus, manganese, and carbon, * in which we detected this "impurity" peak and investigated some of the conditions for its appearance.

We investigated alloys prepared by melting carbonyl iron, refined in hydrogen. The chemical compositions of the alloys are given in the table.

The internal friction was measured using apparatus RKF MIS at the frequency $f = 1.1$ cps using samples of 0.8-mm diameter and 100-mm length in $8-9 \cdot 10^{-3}$ mm Hg vacuum.

Control measurements in a longitudinal magnetic field of 120 Oe intensity indicated the absence of the magnetomechanical contribution to the damping.

Heat treatment was carried out in vacuum furnaces and in evacuated quartz ampoules. The relative error in the determination of the value of the internal friction Q^{-1} amounted to about 3%. The results obtained are given in Figs. 1 and 2.

The temperature dependence of the internal friction of the alloy 08G2R with 0.058% phosphorus, which was quenched from the γ-region and then tempered at 650°C (Fig. 1), exhibited a wide peak with a maximum near 290°C (which was reproduced well in different samples), in addition to the 40°C peak and a small rise on its descending part (which was observed also in the 08R alloy and was evidently related to the transitions of some atoms of carbon, and possibly nitrogen, between the Fe$-$P positions in a stress field [6]), as well as a deformation peak in the region of 220°C.

The 290°C peak, shifted toward lower temperatures, was observed also in the 08R alloy, which did not contain manganese, but the peak disappeared when the phosphorus content was less than 0.002 at.% (the 08G20R alloy).

The activation energy of the relaxation process, causing the appearance of this peak, was found by the Wert$-$Marx method [8] and was 38,000 and 35,000 cal $\cdot$ (g-atom)$^{-1}$ for the alloys

_____
* The present investigation forms a part of a wider study of the nature of temper brittleness of steel by the internal friction method, and that has governed the selection of the material to be investigated.

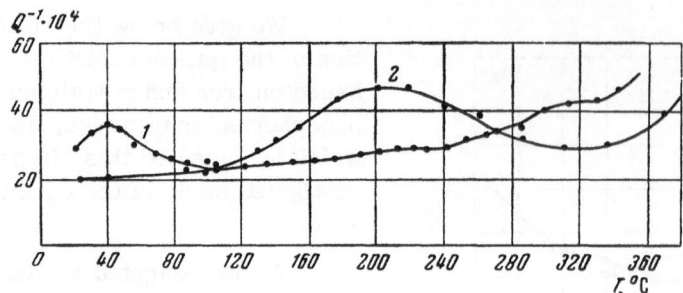

Fig. 2. Temperature dependence of the internal friction of
the 08G2R alloy after heat treatment for 30 min at 920°C,
followed by quenching (2) and after 1-h annealing at 950°C,
followed by tempering for 2 h at 650°C, and by cooling in
water (1).

08G2R and 08R, respectively, i.e., it was close to the 40,000 cal · (g-atom)$^{-1}$ value, reported
in [9] for the activation energy of the diffusion of phosphorus in α-Fe. Therefore, we could
assume that the necessary condition for the appearance of this relaxation effect was the presence of phosphorus.

In subsequent experiments we used the 08G2R alloy, which had a stronger tendency to
temper brittleness, and therefore the investigation of the behavior of phosphorus in this alloy
was of special interest.

Comparison of the results given in Fig. 1 (curve 1) and in Fig. 2 (curve 1) showed that
the peak in the region of 300°C, which was clearly observed after quenching followed by tempering at 650°C, was much weaker if tempering was preceded by slow cooling (annealing)*
from the γ-region.

The structure of the 08G2R alloy after quenching and tempering differed from the structure obtained after annealing and tempering because, first of all, in the former case, ferrite
retained the boundaries of the original austenitic grains considerably distorted by the quenching from the γ-region.

Electron-microscopic observations showed that in the 08G2R alloy annealed for 2 h at
650°C there was not a marked recrystallization of these boundaries. Only prolonged (tens of
hours) tempering at this temperature produced complete recrystallization of ferrite inside
former austenitic boundaries and destroyed these boundaries [5].

Slow cooling from the γ-region led to a more complete recrystallization of the boundaries of former austenitic grains, at which the concentration of phosphorus was higher [5, 10, 11];
the "phosphorus" peak discussed here was reduced by such a treatment. Thus, we could assume that the second condition for the appearance of a noticeable 300°C internal friction peak
was the presence of the boundaries of former austenitic grains at which the concentration of
phosphorus atoms was higher.†

To determine whether these boundaries were enriched with phosphorus in austenite and
then "fixed" by quenching, or whether this enrichment took place during tempering, we determined the temperature dependence of the internal friction after quenching from the γ-region.

---

* Cooling was carried out in air with the sample still inside a vacuum furnace.
† Phosphorus was adsorbed to a lesser degree on ferrite grain boundaries because these boundaries were less defective.

It was found that after such a treatment the 300°C internal friction peak disappeared (curve 2 in Fig. 2).

Evidently, the enrichment of the boundaries of former austenitic grains took place by the diffusion of phosphorus toward these boundaries in ferrite and tempering was therefore the third necessary condition for the appearance of the relaxation effect discussed here.

The necessary conditions for the appearance of the 300°C peak: the presence of relatively small amounts of the impurity (phosphorus), strong tendency to enrichment of boundaries, the presence of these boundaries and of conditions favoring enrichment (tempering), as well as similar values of the activation energy of the relaxation process, causing the appearance of a peak, and the activation energy of the diffusion of phosphorus in α-iron, led us to the conclusion that the observed internal friction peak was of the "impurity" type with a mechanism described in [3].

It was interesting to note that the internal friction peak in the region 300°C (at $f \approx 1$ cps) with the activation energy 40,000 cal $\cdot$ (g-atom)$^{-1}$ was detected also for some low-carbon steels containing phosphorus [12], and was ascribed to the presence of oxygen.

This peak was observed only after quenching from the intercritical region and disappeared when the sample was quenched from the γ-region.

In analyzing the results reported in [12], it should be remembered that, due to the large differences between the solubilities of phosphorus in α- and γ-iron, heating in the intercritical region caused a considerable enrichment of ferrite with phosphorus, which led to the appearance of the so-called "relief" phosphor ferrite [13]. The presence of a surface of separation between the "normal" and "phosphor" ferrite should lead to the appearance of the "impurity" peak described above.

Several workers established experimentally [1, 14, 15] that a considerable enrichment of grain boundaries with impurity atoms depressed and even destroyed the impurity peak of a number of alloys (this followed also from the extremal nature of the concentration dependence of ΔE_T reported in [3]), and simultaneously suppressed the internal friction maximum related to the relaxation along grain boundaries.

From this point of view, the disappearance of, or a drop in the amplitude of, the phosphor peak during the development of temper brittleness in the alloys 08G2R and 08R (curve 2 in Fig. 1a and in Fig. 1b), which was also accompanied by a reduction or total disappearance (in the 08GR alloy) of the grain-boundary maximum, indicated that the embrittlement was associated with the adsorption-enrichment of grain boundaries with phosphorus during tempering at 530°C.

Literature Cited

1. S. Weinig and E. Machin, J. Metals, 9:32 (1957).
2. Yu. G. Miller, Tr. Inst. Met. im A.A. Baikova, Akad. Nauk SSSR, No. 6:20 (1960).
3. V. G. Shmatov and A. V. Grin', Fiz. Metal. i Metalloved., 6:829 (1959).
4. M. C. Inman and H. R. Tipler, Acta Met., 6:73 (1958).
5. L. M. Utevskii, Reversible Temper Brittleness of Steel, Metallurgizdat (1961).
6. W. Dickenscheid and S. Brauner, Collection: Problems of Modern Metallurgy [Russian translation] (1959), pp. 3, 149.
6a. W. Dickenscheid and S. Brauner, Arch. Eisenhüttenw., 9:531 (1960).
7. D. P. Petarra and D. N. Beshers, J. Appl. Phys., 9:2739 (1963).
8. C. Wert and J. Marx, Acta Met., 1:113 (1953).
9. P. L. Gruzin and V. V. Mural', Fiz. Metal. i Metalloved., 4:551 (1963).

10. V.I. Arkharov, S.I. Ivanovskaya, et al., Fiz. Metal. i Metalloved., 1 : 57 (1956).
11. Yu. V. Gardina and E. É. Glikman, Izv. Vysshikh Uchebn. Zavedenii, Chernaya Met., No. 12 : 106 (1964).
12. L. F. Usova, Collection: Relaxation Phenomena in Metals and Alloys, Metallurgizdat (1960), p. 138. [English translation: B. N. Finkel'shtein, ed., Consultants Bureau, New York (1963), p. 100.]
13. V.I. Svechnikov and S.S. Golubev, Stal', No. 10 (1954).
14. L. Rotherham and S. Pearson, J. Metals, 8 : 881 (1956).
15. A.V. Grin', Fiz. Metal. i Metalloved., 4 : 561 (1957).

DETERMINATION OF THE TRUE DIFFUSION COEFFICIENTS AND OF THE THERMODYNAMIC ACTIVITY BY THE INTERNAL FRICTION METHOD

M. A. Krishtal

The true and effective values of the diffusion coefficient are related by the following expression:

$$D_{\text{eff}} \frac{\partial C}{\partial x} = D_{\text{true}} \frac{\partial a}{\partial x},$$

(1)

where C and a are, respectively, the concentration and the thermodynamic activity of the diffusing substance; x is the distance.

In the usual diffusion measurements, the diffusion coefficient is calculated from the dependence of the concentration on the distance, using the concentration gradient determined experimentally. Such calculations give the effective diffusion coefficient. To determine the true diffusion coefficient, additional measurements of the thermodynamic activity gradient $\partial a / \partial x$ are required. Such experiments are fairly complex, time-consuming, and require special care.

It follows from Eq. (1) that

$$D_{\text{true}} = D_{\text{eff}} \frac{\partial C}{\partial a},$$

(2)

i.e., the true diffusion coefficient can be calculated from the effective coefficient, knowing that the dependence dc/da, which also requires special measurements.

The effective diffusion coefficients are complex quantities, which reflect the influence not only of the kinetic but also of the thermodynamic characteristics of the diffusion process. They are not single-valued characteristics of the mobility of atoms and vacancies in a crystal lattice. The thermodynamic factor is frequently responsible for the dependence of the diffusion coefficient on the concentration of the diffusing substance, the anomalously low diffusion coefficients of intermetallic compounds, the influence of the third element on the diffusion coefficient of the principal impurity, and on the self-diffusion coefficient.

The true diffusion coefficients, in contrast to the effective coefficients, are simpler and reflect the kinetic characteristics of the elementary "jumps" of atoms in the process of diffusion and consequently they reflect the mobility, which in this case can be quite simply determined from

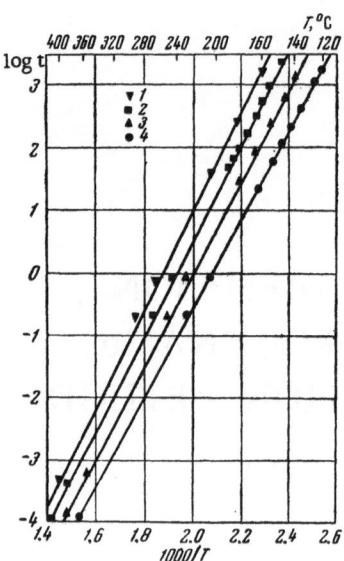

Fig. 1. Dependence of the logarithm of the relaxation time on the reciprocal of temperature of silver−zinc alloys [2]: 1) 15.8 at.% Zn; 2) 19.3; 3) 24.2; 4) 30.2.

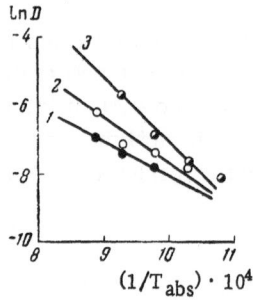

Fig. 2. Dependence of the diffusion coefficient of zinc in silver on the reciprocal of temperature [9]. 1) 7.99 at.% Zn; 2) 15.49; 3) 22.55.

$$U = \frac{D_{\text{true}}}{kT}. \tag{3}$$

In the case of the effective diffusion coefficients, the mobility is found from much more complex expressions, following from the phenomenological diffusion theory, which reflect very roughly the influence of the kinetic and thermodynamic factors. In Darken's approximation the influence of the thermodynamic factor is allowed for through the quantity $\partial \ln \gamma_i / \partial \ln C_i$, where γ_i is the thermodynamic activity coefficient, and the mobility is found from the following formula:

$$U_i = \frac{D_i}{kT \left[1 + \left(\frac{\partial \ln \gamma_i}{\partial \ln C_i} \right) \right]}, \tag{4}$$

where D_i is the effective diffusion coefficient of the i-th impurity.

The knowledge of the true diffusion coefficient is essential in the determination of the characteristics of the mobility and the determination of the influence of the various factors on the elementary "jumps" of atoms. The true diffusion coefficient makes it possible to draw conclusions about important characteristics of a solid, such as the magnitude and properties of active complexes. The value of the activation entropy, calculated from the pre-exponential factor occurring in the temperature dependence of the true diffusion coefficient, can be used to draw more reliable conclusions about the mechanism of the displacement of atoms during diffusion, than would be possible from $D_{0\text{eff}}$.

The relaxation time, found by the internal friction and elastic aftereffect methods, makes it possible to calculate the diffusion coefficient of impurity atoms, whose migration is responsible for the relaxation process. In a cubic lattice the diffusion coefficient of atoms is related to the average "residence" time of these atoms (τ) by the dependence deduced by Einstein:

$$D = \alpha \frac{\delta^2}{\tau}, \tag{5}$$

where δ is the interatomic distance, i.e., the magnitude of a random-walk step; α is a geometrical parameter. For the interstitial diffusion in a bcc lattice, $\alpha = \frac{1}{24}$, and in an fcc lattice, $\alpha = \frac{1}{12}$.

Equation (5) can be applied also to the vacancy mechanism of diffusion both in the self-diffusion and heterodiffusion cases. It is assumed that the successive "jumps" of an atom are

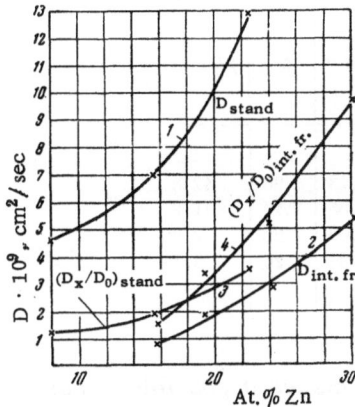

Fig. 3. Diffusion of zinc in silver as a function of zinc concentration. 1, 3) Concentration dependence of the diffusion coefficient measured by the standard method; 2, 4) the same dependence, measured by the internal friction method.

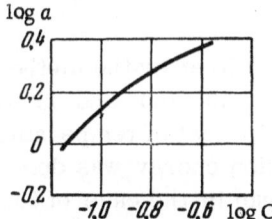

Fig. 4. Relationship between the logarithm of the thermodynamic activity of zinc and the logarithm of its concentration.

random and independent of one another, which is obviously incorrect. For the vacancy mechanism in an fcc lattice, $\alpha = 1/12$, and for a bcc lattice, $\alpha = 1/8$.

The formula (5) has been used by many authors to determine the diffusion coefficient [1-5] from the experimentally measured $1/\tau$

$$\tau = K \cdot \tau_R. \tag{6}$$

Here, K is a quantity of the order of unity.

A.D. Le Claire assumes that in the case of an interstitial solid solution, $K = 3/2$ for a bcc lattice. For a substitutional solid solution in an fcc lattice, and for polycrystalline samples, $K = 1$. The theoretical value of $1/\tau$ has been determined by Wert and Zener, from the theory of absolute rates of Eyring reactions.

The concentration does not appear explicitly in Eq. (5). The influence of the concentration is felt through the relaxation time, which depends on the kinetic characteristics of the atomic "jumps." Consequently, Eq. (5) does not contain thermodynamic parameters, and it can be used to calculate the true diffusion coefficient, bypassing the need for special measurements of the thermodynamic activity. The diffusion coefficient obtained in this way characterizes only the mobility of atoms in a field of weak stresses. If it is assumed that the activation processes taking place in a short-range-order region, surrounding a migrating atom, are the same for internal friction, elastic aftereffect, and diffusion; it follows also that the mechanisms of displacement of the migrating atoms are also identical. Such an identity is self-evident for low concentrations of the diffusing substance in solid solutions, when one relaxation mechanism predominates. At high concentrations the relaxation mechanism may be complex because of the formation of stable diffusing atom pairs and their interaction with point defects and dislocations. The high temperatures at which the diffusion is usually carried out can cause disordering. The activation energy of diffusion is therefore an average characteristic, related to disarrangement of various local bonds, formed in various interactions of the diffusing atom. The anelastic methods make it possible to determine the spectrum of the energies of interaction of the diffusing atom. For example, in alloys of iron with carbide-forming elements there are relaxation processes related to the interaction of impurity atoms with iron atoms, with atoms of the alloying element, with one another, with dislocations, and with point defects.

Still greater difficulties are encountered in the investigation of the diffusion of atoms forming substitutional solid solutions and of relaxation processes associated with their migration. Comparison of the true diffusion coefficients (determined by the anelastic methods for a relaxation process whose mechanism is basically the same as the diffusion mechanism) and the

Table 1

Zn content, at.%	Molar fract. of Zn, N_{Zn}	$D_{Zn}^{Ag} \cdot 10^9$ cm²/sec at 750°C determ. by		$\dfrac{\partial c}{\partial a} = \dfrac{D_{true}}{D_{eff}}$	$\partial\alpha/\partial c$	α	$\gamma_{Zn}^{Ag} = \dfrac{\alpha}{N_{Zn}}$
		evaporation method	anelastic method				
7.99	0.0799	4.6	—	—	—	—	—
15.5	0.155	7.0	0.864	0.123	8.13	1.96	12.6
19.3	0.193	9.4	1.36	0.198	5.05	2.16	11.2
22.55	0.226	13.0	2.60	0.200	5.00	2.22	9.84
24.2	0.242	15.0	3.00	0.200	5.00	2.26	9.34
30.2	0.302	—	5.36	—	—	—	—

effective diffusion coefficients, measured for the same alloys, makes it possible to calculate the thermodynamic activity [6]. The availability of such data for a series of alloys with a variable concentration of one element makes it possible to determine the influence of this element on the thermodynamic activity of the diffusing substance.

We shall consider several examples.

Nowick [7, 8] investigated anelastic phenomena in a silver−zinc alloy over a wide range of frequencies. Figure 1 shows the dependence of the logarithm of the relaxation time on the reciprocal of temperature over a wide range of temperatures for alloys containing 15.8, 19.3, 24.2, and 30.2 at.% Zn. The longer relaxation times were determined by the elastic aftereffect method, and the shorter times by the internal friction method. The mobility of the zinc atoms increased with the zinc concentration.

Bugakov and Sirotkin [9] investigated the diffusion of zinc in silver by the method of evaporation, using alloys containing 7.99, 15.49, 22.55 at.% Zn at 650, 750, and 850°C. The dependence of the logarithm of the diffusion coefficient on the reciprocal of temperatures is given in Fig. 2. The slopes of the curves, from which the activation energy was determined, depended on the concentration of the zinc to a greater extent than in the case of the internal friction and the elastic aftereffect. We shall now compare the data obtained by these methods.

The straight lines in Fig. 1 allow us to extrapolate reliably to a temperature lying in the range of measurements carried out by the standard method. We shall make this comparison at 750°C.

The diffusion coefficients were calculated using Eq. (5). For silver, $\alpha = 1/12$. A "jump" of an atom was assumed to be equal to the atomic diameter of silver, $\delta = 2.88 \cdot 10^{-8}$ cm, and $\tau = \tau_R$. The values of the diffusion coefficients, measured by different methods, were of the same order. However, the values obtained by the anelastic methods were 5-8 times smaller than the values measured by the vacuum evaporation method. The anelastic measurements gave a stronger dependence of the diffusion coefficient on the composition than would follow from Fig. 3. Curves 1 and 2 in Fig. 3 represent the dependences of the diffusion coefficients on the composition obtained by the standard and anelastic methods, respectively. Curves 3 and 4 give the concentration dependence of D_x/D_0 for both methods. As is known,

$$D_x = D_0 \frac{C_s}{C_s - C_x}, \qquad (7)$$

where D_x and D_0 are the diffusion coefficients for a solid solution at concentrations C_x and $C \to 0$; C_s is the saturation concentration at the test temperature. The agreement between the calculated and measured values was usually good up to $C_x \leq 0.9\,C_s$.

Table 2

$D_{eff}^* \cdot 10^7$, cm²/sec		$D_{true}^* \cdot 10^7$, cm²/sec		$\partial c / \partial a$	
Carbon content, %					
0.4	0.7	0.4	0.7	0.4	0.7
At 1200°C					
18.3	24.0	20.0	21.3	1.09	0.886
At 1100°C					
7.95	8.90	7.34	8.20	1.08	1.085
At 1000°C					
2.52	3.12	2.60	2.90	0.97	1.08

Table 3
(Temperature 1200°C)

$D_{eff} \cdot 10^7$, cm²/sec			$D_{true} \cdot 10^7$, cm²/sec			$\partial c / \partial a$		
Carbon content, %								
0.2	0.4	0.7	0.2	0.4	0.7	0.2	0.4	0.7
Pure alloy								
13.1	17.3	23.1	44.9	54.0	62.4	3.43	3.12	2.70
Alloy with 7% chromium								
2.16	3.20	4.34	20.8	30.0	38.8	9.66	9.39	8.96

Curves 3 and 4 in Fig. 3 show quite clearly that the diffusion coefficients determined by the anelastic methods increased with increase in the concentration much more strongly than the coefficients measured by the standard method.

Comparing the curves 1 and 3 with curves 2 and 4 in Fig. 3, we can determine the influence of the concentration of zinc on its thermodynamic activity. The results of the calculations are given in Table 1. It should be mentioned that the accuracy of the calculated values is the same as the accuracy of the data obtained in fairly old investigations of Bugakov and, therefore, not very high. Nevertheless, these data are of considerable interest, since they represent the first estimate of the thermodynamic activity of zinc in silver.

Some of the values of the diffusion coefficients given in Table 1 were determined from curves in Fig. 3. The value of α was determined from the values of N_{Zn} and $\delta a / \delta C$ by graphical integration. It is evident that the values of the thermodynamic activity were much higher than the values of the concentration. The activity coefficient f_{Zn}^{Ag} was found to be of the order of ten. Such a considerable positive deviation from Raoult's law is usually observed in dissociation. The activity coefficient decreased from 12.6 to 9.4 when the Zn concentration was increased from 15.5 to 24.2 at.%. The dependence of $\log a$ on $\log C$ is shown in Fig. 4.

An increase in the concentration of zinc atoms, which distorted the crystal lattice of silver, increased the mobility of zinc. Zinc atoms formed stronger bonds between one another than between themselves and silver atoms and, consequently, atomic clusters of zinc were formed in a solid solution, leading to an increase of the activity coefficient and of the diffusion flow.

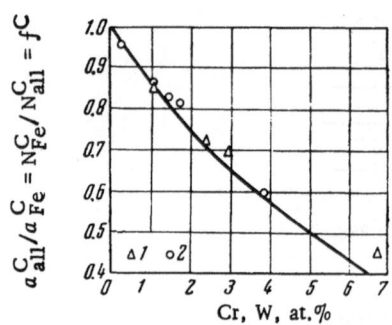

Fig. 5. Influence of chromium and tungsten on the thermodynamic activity of carbon in austenite. 1) Chromium; 2) tungsten; T = 950°C.

The state of a system with a lower mixing energy is energetically more favorable, because the free energy is less. In the silver−zinc system, at least in the solid phase, the energy of mixing was negative, i.e., bonds between like atoms were stronger than between unlike atoms.

To obtain more accurate values of the activity than those given in Table 1, measurements would be necessary at much lower zinc concentrations, because this would make graphical integration more accurate and the standard state could be selected more satisfactorily.

It seemed of interest to carry out a similar treatment for an interstitial solid solution. There are many data of the diffusion of carbon and nitrogen in iron [6]. The values given in the paper of Ke (K'o) and Tseng [3] and Blanter [11, 12] are suitable for such comparison. We used the data for the alloy with 18.5% Mn. The true diffusion coefficients at high temperatures were calculated from the data obtained by the internal friction method. It is evident from Table 2 that the calculated and measured values of the diffusion coefficients were in good agreement, indicating high accuracy of the measurements reported in the cited papers.

Carbon in the presence of manganese does not affect the thermodynamic activity strongly and its influence varies with temperature. Such a relatively weak influence of carbon is due to its fairly strong binding with manganese atoms, which affects the thermodynamic activity of carbon by lowering it.

Other alloying elements affect the thermodynamic activity and the diffusion of carbon in the same way as manganese. The coefficients of diffusion in the systems Fe−C−Cr and Fe−C−W were calculated from the dependence of the concentration on the composition in the diffusion layer. Such measurements give the effective values. To determine the true values of the diffusion coefficients, we calculated the thermodynamic activity for a series of alloys containing varying amounts of chromium and tungsten.

These additional measurements were carried out on diffusion pairs, consisting of the alloy Fe−C and of the alloy Fe−C−Cr (W), which differed only in the content of the alloying element. Annealing displaced carbon into the plates containing chromium or tungsten. The annealing was continued for a period necessary to establish equilibrium, for example, at 1150°C the duration was 16 h. Analysis of the carbon content was carried out chemically and by the local spectroscopic method. The thermodynamic activity was determined from the values of the equilibrium concentration of carbon in those parts of the sample which differed in the content of the alloying element, allowing for the known dependence of the influence of carbon on its activity in pure iron. In this case the activity coefficient was

$$f_{all}^C = \left(\frac{N_{Fe}^C}{N_{all}^C}\right)_{a=const} \simeq \left(\frac{a_{all}^C}{a_{Fe}^C}\right)_{N_{all}^c=const} \tag{8}$$

and

$$a_{all}^C = a_{Fe}^C \cdot f_{all}^C . \tag{9}$$

Figure 5 shows that tungsten and chromium, taken in equiatomic concentrations, reduced the activity coefficient of carbon to the same extent. Our data are compared with Schenk's measurements, represented by a continuous line (for alloys containing chromium).

Table 3 gives the true and effective coefficients of carbon in alloys containing chromium and in iron—carbon alloys at 1200°C; the carbon content in both alloys ranged from 0.2 to 0.7%. The true diffusion coefficients for the Fe—C alloys were several times greater than the effective values, and for the Fe—C—Cr alloys they were larger by an order of magnitude than the effective coefficients. The value of $\partial c / \partial a$ increased strongly when chromium was added. In alloys containing chromium, as well as in alloys containing manganese, carbon had little influence on the activity which was characteristic of alloys containing carbide-forming elements.

The results for tungsten were similar.

The reported results indicate that to find more accurate values of the thermodynamic activity, it is necessary to investigate anelastic characteristics and diffusion over the widest possible temperature range.

One further characteristic of the diffusion parameters measured by the anelastic methods should be pointed out. In substitutional solutions the diffusion activation energy, found from the relaxation peak, is usually less than the activation energy determined by the standard method. This characteristic is due to the different mechanisms of relaxation and diffusion. In the former case the participation of vacancies is essential. The activation energy of diffusion at high temperatures is a complex averaged-out characteristic.

Many examples of characteristics of the diffusion parameters determined by the anelastic methods are given in Darken and Gurry's book [14].

Literature Cited

1. C. A. Wert, Phys. Rev., 79(4):601 (1950).
2. A. S. Nowick, Progress in Metal Physics [Russian translation], Metallurgizdat (1956), Vol. 1. [English edition: B. Chalmers and R. King, eds., 8 vols., Pergamon Press, Inc., New York.]
3. Ke T'ing-hsü and Tseng Chih-chiang, Fiz. Metal. i Metalloved., 4(2):291 (1957).
4. M. A. Krishtal and S. A. Golovin, Fiz. Metal. i Metalloved., 8(2):295 (1959).
5. G. Bichter, Ann. Physik, 32:683 (1938).
6. M. A. Krishtal, Diffusion Processes in Iron Alloys, Metallurgizdat (1963).
7. A. S. Nowick, Phys. Rev., 88(9):925 (1952).
8. A. S. Nowick, Phys. Rev., 74(1):9 (1948).
9. V. Z. Bugakov and B. Sirotkin, Zh. Tekhn. Fiz., 7(11):1577 (1937).
10. Yu. V. Piguzov, Dokl. Akad. Nauk SSSR, 112(4):636 (1957).
11. M. E. Blanter, Zh. Tekhn. Fiz., 21(4):818 (1951).
12. M. E. Blanter, Zh. Tekhn. Fiz., 20(1):217 (1950).
13. M. A. Krishtal, Yu. V. Piguzov, and S. A. Golovin, Internal Friction in Metals and Alloys, Izd. "Metallurgiya" (1964).
14. L. S. Darken and R. W. Gurry, eds., Physical Chemistry of Metals [Russian translation], Metallurgizdat (1960). [English edition: McGraw-Hill, New York (1953).]

DETERMINATION OF DISLOCATION STRUCTURE PROPERTIES BY MEASURING THE AMPLITUDE DEPENDENCE OF THE INTERNAL FRICTION

M. A. Krishtal, S. A. Golovin, and S. I. Arkhangel'skii

The amplitude dependence of the internal friction is due to the interaction of dislocations with point defects. Under small alternating stresses, dislocation segments vibrate in accordance with the model of Granato and Lücke. The breakaway of dislocation fragments from impurity atoms gives rise to an amplitude dependence of the logarithmic decrement and takes place at a certain amplitude designated critical. A second critical amplitude represents the breakaway of dislocation loops from the mesh points of a dislocation grid in accordance with the model of a Frank–Read source. Consequently, the measured characteristics of the amplitude dependence of the internal friction are related directly to dislocation structure parameters: the binding energy between a dislocation line and point defects and mesh points in dislocation grid; the lengths of fragments; the dislocation density.

The present paper describes an attempt to estimate these dislocation structure parameters from the amplitude dependence of the damping in metals with the bcc lattice, by calculating the binding energy between dislocations and impurity atoms, allowing for an entropy correction.

Measurement Method

The measurements were carried out using a relaxator of the RKF MIS type at low amplitudes (visual methods) and at high amplitudes (using a nonvacuum optical recording system). The deformation amplitudes ranged from 10^{-6} to 10^{-3}. The magnitude of the decrement for wire samples was little affected by the energy dissipated in the apparatus, and this made it possible to carry out measurements over a wide range of amplitudes.

Figure 1 shows the apparatus schematically. The suspension system 3 was excited by means of electromagnets 4 in the usual way. Two mirrors 2 attached to the suspension of a torsional pendulum, a "relaxator," and two illumination systems 5 made it possible to record simultaneously the damped vibrations of a sample using a photographic attachment 1 and to measure the logarithmic decrement visually. The photographic attachment was used for large stress amplitudes. Visual measurements of the internal friction were carried out using a double-scale system, in which the two scales 6 and 7 were placed at distances of 5.5 and 0.7 m, respectively, from the relaxator mirror. Such a recording system made it possible to check additionally the amplitude dependence of the damping in those cases when the recording was automatic.

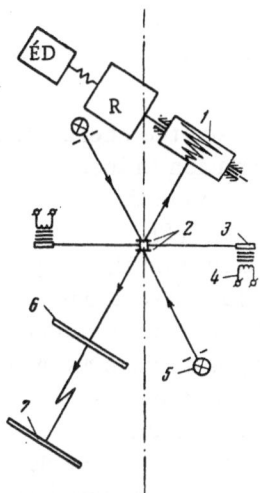

Fig. 1. Schematic diagram of apparatus. 1) Photographic attachment; 2) mirror; 3) suspension system; 4) magnets; 5) illuminating system; 6) scale at 0.7 m; 7) scale at 5.5 m.

Special attention was paid to the elimination of additional large-amplitude energy losses through the clamps. Control tests showed that, at any stress amplitude, we could neglect the energy losses in the apparatus compared with the value of the internal friction.

The damping of the vibrations was determined while increasing gradually the applied stress. After each cycle of vibrations, we determined the internal friction background at low deformations (the amplitude-independent internal friction δ_0). For example, having twisted a sample so that the deformation was $8 \cdot 10^{-5}$, we stopped the oscillations to measure the decrement for amplitudes from $8 \cdot 10^{-5}$ to $7 \cdot 10^{-5}$. Next, as the oscillations decreased in amplitude, we measured the internal friction background for deformations of 10^{-6}. Then the sample was deformed to larger amplitudes and the measurement process was repeated. This method made it possible to follow changes in the internal friction background when the amplitude was increased.

Critical Deformation Amplitudes and Plotting of Graphs $\ln(\delta_n\gamma) - 1/\gamma$

The amplitude dependence of the internal friction at low stresses is related to the vibrations of dislocation segments and at high stresses this dependence is related to the vibrations of dislocation fragments which have broken away from impurity atoms but are still rigidly locked at the mesh points of a dislocation grid. Under high stresses the internal dissipation of the energy in a material (hysteresis) is due to the breakaway of dislocations from the mesh points in the grid by displacement over large distances in grains of a metal or a solid solution. Further motion of dislocations leads to local deformations concentrated in microregions. As the density of dislocations in a material increases, its properties — among which is the damping effect — are altered substantially.

In view of the differences in the nature of the dissipation of energy at different stress amplitudes, it is very important to determine the values of the critical deformation amplitudes at which considerable changes in the mechanism of energy dissipation in a material take place. Figure 2b shows the amplitude dependence of the internal friction of annealed iron (δ) and the change in the corresponding internal friction background (δ_0). Parts of the dislocation damping curve are in good agreement with the suggested mechanisms of vibrational energy dissipation (Fig. 2a). Figure 2 shows two critical amplitudes: γ_{cr}' and γ_{cr}''.

In the first approximation, the dependence of the decrement on the amplitude can be divided into three regions. Region I is the amplitude-independent internal friction (background). The main contribution to the damping at low deformations is made by vibrations of dislocation segments (L_c) locked by impurity atoms (Fig. 2, k and l).

The critical deformation amplitude $\gamma = \gamma_{cr}'$ corresponds to the beginning of the amplitude dependence of the internal friction (region II). When the deformation $\gamma \geq \gamma_{cr}'$, dislocations break away from their impurity atom clouds and remain pinned only at the mesh points of a dislocation grid (L_N). The energy dissipation, due to the breakaway of dislocations from impurity

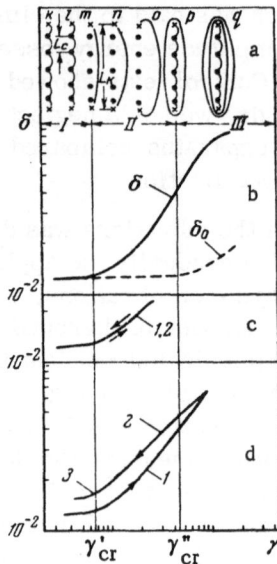

Fig. 2. Amplitude dependence of the internal friction of Armco iron. a) Energy dissipation models for various vibration amplitudes; b) amplitude dependence of the decrement (δ) and of the corresponding internal-friction background (δ_0); c, d) amplitude dependence of the decrement when the stress amplitude was being increased (1), after removal of a load (2), and during subsequent loading (3); c) $\gamma < \gamma''_{cr}$; d) $\gamma > \gamma''_{cr}$.

atoms and due to the vibrations of dislocations in a stress field, is the cause of the damping which depends on the amplitude (Fig. 2, mechanisms m, n, and o).

At higher stresses the energy dissipation is due to plastic deformation in local regions. The weakest structures provide the nuclei of such microplastic behavior [1, 2]. The value of the decrement in this case (region III) is governed by the long-range irreversible displacement of dislocations and by the behavior of the dislocation sources. There have been as yet no reports of the value of the critical stress amplitude above which such deformation loci appear in metals (hysteretic energy dissipation). In our opinion, the critical deformation amplitude (γ''_{cr}) is the amplitude corresponding to the beginning of the rise in the internal friction background (Fig. 2b), measured while the applied stress is increased, because the internal friction background is very sensitive to changes in the structure and increases even in the presence of small local residual deformations.

The irreversible motion of dislocations should lead to a hysteretic dissipation of the vibrational energy. Figures 2c and 2d show the changes in the decrement during gradual loading (curve 1), after removal of a load (curve 2), and during subsequent loading (3) of a sample of annealed iron. Up to deformation amplitudes $\gamma < \gamma''_{cr}$ (Fig. 2c) the amplitude dependence of the decrement is fully reproducible both during loading and unloading. When loading is increased up to $\gamma > \gamma''_{cr}$ (Fig. 2d) the vibration damping curve after removal of a load runs higher and, consequently, a hysteresis loop is formed. Similar behavior has been observed for iron and steel after hardening treatment. The value of the stress causing the generation of dislocations ($\approx 10^{-4}$ [3]) is in good agreement with the value of the stress corresponding to the critical deformation amplitude (γ''_{cr}) which also confirms this conclusion.

The critical amplitudes γ'_{cr} and γ''_{cr} may have similar values for dislocation damping in very pure single crystals and in polycrystalline samples at high temperatures. The behavior of these characteristics is more complex in alloys. We note that at the critical points γ'_{cr} and γ''_{cr} one energy dissipation mechanism cannot replace another one completely. At deformation amplitudes higher than the critical value, the damping is due to various sources of the internal energy dissipation, but one of them is the main one.

To obtain qualitative and quantitative information on the dislocation structure, it is necessary to plot the graphs $\ln(\delta_n \gamma)$ as a function of $1/\gamma$, in accordance with the model of Granato and Lücke (Fig. 3).

The graphs can be plotted only between γ'_{cr} and γ''_{cr}, because, at lower and higher deformations, the dissipation mechanism differs from the model assumed by Granato and Lücke [4]. Therefore, the data for other deformation amplitudes (for example the data on the appear-

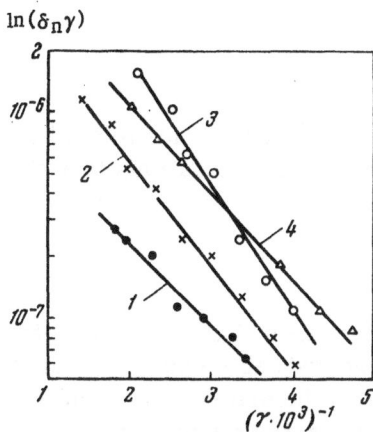

Fig. 3. Dependences of $\ln(\delta_n\gamma)$ on $1/\gamma$ for iron containing 0.03 wt.% C and N. 1) Annealing at 930°C, 1 h; 2) deformation by 2.8%; 3) deformation by 11.2%; 4) quenching from 950°C.

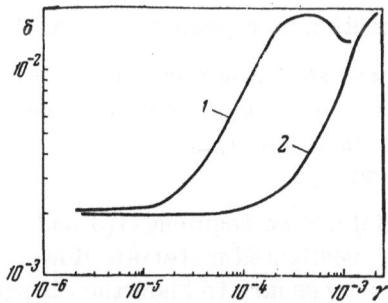

Fig. 4. Amplitude dependence of the internal friction of iron, containing 0.03 wt.% C and N in H = 0 (1) and H = 200 Oe (2).

ance of a maximum in the amplitude dependence of the decrement at high stresses) cannot be discussed on the basis of the Granato and Lücke theory.

In some cases the dependence of $\ln(\delta_n\gamma)$ on $1/\gamma$ deviates from linearity. Then, a quantitative estimate of the nature of the dislocation structure becomes difficult. In some cases it is necessary to analyze the data by the least-squares method, which makes it possible to distinguish clearly the various parts of the $\ln(\delta_n\gamma) - (1/\gamma)$ curve. To refine the theory of Granato and Lücke, an improved model has been proposed [5] for the damping of vibrations in the amplitude-dependent region (Fig. 2, II), allowing for an inhomogeneous distribution of impurities in the matrix field near dislocations. This model makes it possible to explain why the experimental data for metals with high-energy stacking faults differ from the amplitude dependence predicted by Granato and Lücke. The proposed improvement is completely inapplicable to metals with low-energy stacking faults, in which dislocations are split. The same paper [5] also gives the conditions for the appearance of the vibrational energy loss in metals, which is best described by the Granato and Lücke model, and criteria are given for the mechanism of damping in the range of vibration amplitudes considered here.

In investigations of ferromagnetic metals, special attention must be paid to the analysis of that part of the amplitude dependence of the damping which is due to magnetoelastic hysteresis losses. It has been shown [6, 7] that the low-frequency internal friction (≈1 cps) of ferromagnets, considered as a function of the intensity of applied constant magnetic field, has either a maximum or is represented by a monotonic curve. Under conditions of strong magnetic saturation the damping of vibrations in a magnetic material is due to causes of nonmagnetic origin.

For this reason, the amplitude dependence of the internal friction of ferromagnets and of the critical deformation amplitudes should be determined in magnetic fields of high intensity. The value of a suitable magnetic field intensity for each material should be determined experimentally.

Figure 4 shows the amplitude dependence of the internal friction of annealed iron, containing 0.03 wt.% C and N, in the absence of a magnetic field (H = 0) and in a field H = 200 Oe. The nature of the damping in the latter case is governed only by the vibrations and displacements of dislocation segments and fragments. Plastic deformation and heating increase considerably the contribution of the dislocation damping to the dissipation of the energy of elastic vibrations in this material. It should be mentioned that the application of a magnetic field has a strong influence on the position of the critical deformation amplitude γ'_{cr} but has little effect on the position of γ''_{cr}.

Calculation Method

According to the Granato and Lücke theory [4], the energy dissipation due to the breakaway of dislocations from impurity atoms and the vibrations of these dislocations in a stress field, is the cause of the amplitude-dependent damping (Fig. 2).

$$\delta_n = \frac{C_1}{\delta}\exp\left(-\frac{C_2}{\gamma}\right),\tag{1}$$

where γ is the instantaneous amplitude of the shear deformation; C_1 and C_2 are constants of the metal. The dependence of $\ln(\delta_n\gamma)$ on $1/\gamma$ is a straight line, whose slope is C_2, and the intercept on the ordinate axis is C_1. The constants C_1 and C_2 can be determined by an analysis of the data on the amplitude dependence of the internal friction. On the other hand, it has been shown in [8] that the quantity C_1 is related to the dislocation density (Λ)

$$C_1 = \frac{A_1\Lambda L_N^3}{L_c^2},\tag{2}$$

and C_2 depends on the dimensions of a dislocation segment

$$C_2 = k\eta a/L_c.\tag{3}$$

Here, $A_1 = \frac{\Omega}{\pi Q}\cdot\frac{P_m}{4a\,M}$, where P_m is a force exerted by a dislocation on a pinning point at the moment of breakaway; Ω is an orientation factor; Q is the shear stress factor in the slip plane; a is the lattice parameter; M is the elasticity modulus. The value of P_m can be estimated from the value of the stress corresponding to the breakaway amplitude γ'_{cr}, and η is the difference between atomic radii of the solvent metal and the impurity.

Equation (3) can be used to determine the length of a dislocation segment (L_c) and then, Eq. (2) can be used to find the density of dislocations in the investigated material. However, to obtain a quantitative value for the dislocation density, it is necessary to know the change in the length of a dislocation fragment (L_N), which depends strongly on the previous treatment of the sample.

The critical deformation γ''_{cr} corresponds to the stress which generates dislocations. Such a stress ($\tau''_{cr} = \gamma''_{cr}M$) therefore corresponds to the value of the stress needed to break away dislocation fragments from the mesh points of a dislocation grid and to cause irreversible displacement; this stress is

$$\tau''_{cr} = \frac{Mb}{L_N},$$

where b is the Burgers vector.

Comparing these two expressions for τ''_{cr}, we find that

$$L_N = \frac{b}{\gamma''_{cr}},\tag{4}$$

where γ''_{cr} is found by measuring the amplitude dependence of the internal friction from changes in the damping background. Thus, from this analysis of the amplitude dependence of the internal friction, we find all the necessary data for a quantitative determination of the dislocation density, using Eq. (2).

Material and treatment	t_{meas}, °C	γ'_{cr}	γ''_{cr}	L_N, cm	L_c, cm	β	Q, eV	$\Delta M/M$, % [Λ from (2)]	Λ, cm⁻² from (2)	Λ, cm⁻² from (5)
Armco iron (0.03 wt.% C and N)										
Annealing at 930°C, 1 h	100	$8 \cdot 10^{-5}$	$3.2 \cdot 10^{-4}$	$\sim 10^{-4}$	$9.8 \cdot 10^{-6}$	0.4014	0.18		$2 \cdot 10^{7}$	$3 \cdot 10^{8}$
Plastic deformation, %										
2.8	100	$\sim 10^{-4}$	$6.1 \cdot 10^{-4}$	$4.5 \cdot 10^{-5}$	$6.7 \cdot 10^{-6}$	0.4263	0.196	~ 1	10^{8}	$\sim 10^{9}$
7.0	100	$2 \cdot 10^{-4}$	$8 \cdot 10^{-4}$	$3.5 \cdot 10^{-5}$	$6.0 \cdot 10^{-6}$	0.4511	0.21		$7 \cdot 10^{8}$	
11.2	100	$2.4 \cdot 10^{-4}$	$\sim 10^{-3}$	$2.8 \cdot 10^{-5}$	$5.4 \cdot 10^{-6}$	0.4712	0.226	6-8	$3 \cdot 10^{9}$	$4 \cdot 10^{8}$
Quenching from 950°C	100	$\sim 10^{-4}$	$5.9 \cdot 10^{-4}$	$4.7 \cdot 10^{-5}$	$8 \cdot 10^{-6}$	0.436	0.2	~ 6	$4 \cdot 10^{8}$	
Molybdenum (99.8%)										
Plastic deformation, ~60%	350	$1.1 \cdot 10^{-4}$	$2 \cdot 10^{-3}$	$1.8 \cdot 10^{-5}$	$1.7 \cdot 10^{-5}$	0.5-0.6	0.32-0.35		10^{11}	
Recrystallization at 1200°C, 4 h	350	$7.8 \cdot 10^{-5}$	$6 \cdot 10^{-4}$	$5 \cdot 10^{-5}$	$2.8 \cdot 10^{-5}$				$6 \cdot 10^{9}$	$4 \cdot 10^{9}$

Using the data on the length of a dislocation fragment, held rigidly at the mesh points of a dislocation grid, we can employ Eq. (4) in a simplified calculation of the density of dislocations in deformed metals subjected to a small degree of cold working [5]

$$L_N^3 = \frac{3}{\Lambda} . \tag{5}$$

Comparative calculations of Λ using Eqs. (2) and (5) are given below.

Weinig and Machlin [9] have shown that the binding energy of an atom to a dislocation (Q) can be found from the critical amplitude γ'_{cr}:

$$\gamma'_{cr} = \frac{Q \cdot c}{M \cdot b^3} , \tag{6}$$

where c is the concentration of impurity atoms near dislocations.

We must bear in mind that, in the form given above, Eq. (6) applies at absolute zero. At any other temperature, Q should be replaced by the free energy of breakaway, i.e.,

$$\Delta G = Q - T \cdot \Delta S, \tag{7}$$

where ΔS is the change in the entropy and T is the absolute temperature.

Allowing for the temperature dependence of the impurity concentration in dislocation zones [10]

$$c = c_0 \exp \left(\frac{\Delta G}{kT} \right), \tag{8}$$

we obtain

$$\gamma'_{cr} = \frac{Q - T \cdot \Delta S}{M b^3} \cdot c_0 e^{-\frac{\Delta S}{k}} e^{\frac{Q}{kT}} , \tag{9}$$

where c_0 is the average concentration of the alloy in at.%.

The entropy may be calculated using formulas given in the papers of Zener and Le Claire [11, 12]:

$$\Delta S = \beta \frac{\Delta H}{T_{mp}} \tag{10}$$

and

$$\Delta S = n \cdot \Delta H \frac{\partial (M/M_0)}{\partial T} , \tag{11}$$

where $\beta = \frac{\partial (M/M_0)}{\partial (T/T_{mp})}$, M and M_0 are the shear moduli at the test temperature and at absolute zero; ΔH is the self-diffusion activation energy. The values of $\partial (M/M_0)$ and of the coefficient β can be found knowing the temperature dependence of the square of the frequency (f^2) of torsional vibrations. The value of f_0^2 can be found by extrapolating the temperature dependence of f^2 to absolute zero. Then, having determined $\partial (M/M_0)$ and β, we can calculate graphically or analytically the activation entropy from Eqs. (10) or (11).

The coefficient n in Eq. (11) should, in each case, be estimated from the self-diffusion activation energy and from the heat of sublimation [12]. For bcc and fcc metals, n = 0.5-0.6.

Thus, from the measurements of the dislocation damping of vibrations and from the temperature dependence of the square of the frequency of these vibrations, we can find the data required in the calculation of the energy of binding of a dislocation to impurity atoms, using Eq. (9).

We should bear in mind also the possibility of determining the change in the modulus when Λ, L_C, and L_N are known, using the formula [8]:

$$\frac{\Delta M}{M} = \frac{6\Omega}{\pi^2} \Lambda L^2, \tag{12}$$

where L is the effective length found from

$$\frac{1}{L} = \frac{1}{L_N} + \frac{1}{L_c}. \tag{13}$$

Results Obtained and Discussion

We estimated the characteristics of the dislocation structure from the measurements of the amplitude dependence of the internal friction and from the temperature dependence of the square of the frequency of vibrations in Armco iron, containing 0.03 wt.% C and N, and in molybdenum of 99.8% purity. In the measurements of the amplitude dependence of the damping we determined the critical deformation amplitudes (γ'_{cr} and γ''_{cr}), plotted the dependences of $\ln(\delta_{n\gamma})$ on $1/\gamma$, from which we calculated the constants C_1 and C_2. Subsequent calculations were carried out as described above.

The annealing and heating before quenching of Armco iron were carried out in vacuum ($\approx 10^{-3}$ mm Hg). Samples of molybdenum wire were prepared by drawing in several stages between which the samples were annealed in vacuum. Next, the samples were annealed at 1200°C for 4 h (recrystallization). Plastic deformation took the form of elongation. The measurements were carried out on the samples after cold working and quenching in the aged condition. To avoid energy dissipation in iron due to the magnetoelastic hysteresis, the measurements of the amplitude dependence of the internal friction were carried out in a constant magnetic field H = 200 Oe. The results reported in [6] and our own data indicated that in this case the damping of vibrations was mainly due to causes associated with the vibrations and displacements of dislocations in the stress field.

According to the data tabulated in the present paper, the dislocation density in annealed iron [Eq. (2)] was about 10^7 cm^{-2}; after plastic deformation by 11.2% the density increased to $3 \cdot 10^9$ cm^{-2}, which was in good agreement with the increase of Λ in copper, reported in [3] on the basis of a calculation using a dislocation model with a variable number of barriers. The length of a dislocation fragment varied from 10^{-4} to $2.8 \cdot 10^{-5}$ cm, while the length of a segment ranged from $9.8 \cdot 10^{-6}$ to $5.4 \cdot 10^{-6}$ cm. The change in the modulus increased from 1 to 6-8% with increase in the degree of deformation. The energy of binding of dislocations to impurity atoms amounted to 0.18 eV after annealing and about 0.23 eV after deformation by 11%. The quenching of iron also increased somewhat the values of Q and Λ. The same regularities in the changes of the dislocation structure were found for molybdenum.

Several points are worth making. The values of the energy of binding of dislocations to impurity atoms given in the present paper should be regarded only as estimates, because of the inaccuracy of Eqs. (10) and (11) used to determine the value of the entropy. Moreover, the value of the coefficient β, found from the temperature dependence of the square of the frequency of torsional vibrations, was an average for an extended range of temperatures, which also reduced the accuracy of the calculations. To investigate the influence of the test temperature on the energy characteristics of the dislocation structure, the value of the coefficient β should be found by the tangent method. However, relative estimates of the binding energy under isothermal conditions are fully permissible.

The results obtained should be regarded as estimates because of the cumulative error in the analysis of the data on the amplitude dependence of the internal friction. For this reason the table gives the average values found in the calculations. Estimates of the density of dislocations using Eq. (5) give values which are too high because of the insufficiently exact determination of the critical amplitudes in a system not in vacuum.

Literature Cited

1. G.S. Pisarenko, Energy Dissipation in Mechanical Vibrations, Izd. AN UkrSSR (1962).
2. M.A. Krishtal, Yu. V. Piguzov, and S.A. Golovin, Internal Friction in Metals and Alloys, Izd. "Metallurgiya" (1964).
3. H.G. van Bueren, Imperfections in Crystals [Russian translation], IL (1962). [English edition: 2nd ed. (Interscience), Wiley, New York.]
4. A. Granato and K. Lücke, J. Appl. Phys., 27: 583, 789 (1956); Collection: Dislocation and Mechanical Properties of Crystals, J. R. C. Fisher, ed., Wiley, New York (1957).
5. J.C. Swartz and S. Weertman, Collection: Ultrasonic Methods of Investigating Dislocations [Russian translation], IL (1963), p. 58.
6. Yu. S. Avraamov, I.B. Kekalo, and V. Morgner, Collection: Relaxation Phenomena in Metals and Alloys, Metallurgizdat (1963), p. 184.
7. K. Mishek, Fiz. Metal. i Metalloved., 18(3) (1964).
8. B.B. Gordon, Acta Met., 10(4):339 (1962).
9. S. Weinig and E. Machlin, J. Appl. Phys., 27: 743 (1956).
10. A.H. Cottrell, Report of the Conference on Strength of Solids, Phys. Soc., London (1948), p.30.
11. C. Zener, Imperfections in Nearly Perfect Crystals, Wiley, New York (1952), Chap. 11.
12. A.D. Le Claire, Progress in Metal Physics [Russian translation], Metallurgizdat (1956), Vol. 1, p. 224. [English edition: B. Chalmers and R. King, eds., 8 vols., Pergamon Press, Inc., New York.]

CHANGES IN THE INTERNAL FRICTION OF CARBON STEEL UNDER THE ACTION OF PERIODIC LOADS

I. A. Oding,* L. K. Gordienko, and T. S. Mar'yanovskaya

The application of a periodic load causes irreversible changes in the fine structure of a metal, as indicated by the regular changes in its physicomechanical properties [1-2]. It has been reported [3] that the periodic loading of rails (under service conditions) first increases the strength and the finite fatigue limit of full-section rails, but after a finite number of loading cycles the strength properties of the rail steel cease to change, while the fatigue limit begins to decrease. On the other hand, the critical brittleness temperature increases monotonically right from the beginning of such periodic loading. The data obtained on the changes in the mechanical properties have been analyzed on the basis of the general relationships governing changes in the dislocation structure of a metal subjected to periodic loading [4], associated with the accumulation of fatigue damage. However, a direct check of the conclusions reported in [3] was not possible, because of the difficulty of visualization of the dislocation structure in rail steel, containing about 90% pearlite with thin laminar structure.

In our investigation we employed the internal friction method to study fine structure changes in this steel and to compare them with changes in the mechanical fatigue properties. As a measure of the internal friction we use the logarithmic decrement of vibrations δ, measured at room temperature in vacuum of the order of 10^{-5} mm Hg and reflecting the internal-friction background of the investigated steel. The logarithmic decrement was determined from the damping of free transverse vibrations (at 2300 cps) of cylindrical samples of 5-mm diameter and 100-mm working length, used in apparatus of the UIMD-2 type [5]. The investigated samples were cut from rails used in [3], which originated from the same melt, containing 0.65% C and 0.77% Mn, subjected during their service life to 100, 200, 300, 400, and 500 million tons of load. The samples were cut at a distance of 4 mm from the rolling surface (head) of a rail; for the sake of comparison, samples were cut also from as-prepared rails (not subjected to service loads).

The experimentally determined curve representing the change in the decrement as a function of the duration of periodic loading (this characteristic was represented by the number of tons of train loads, which was proportional to the number of loading cycles) is shown in Fig. 1.

---

*Deceased.

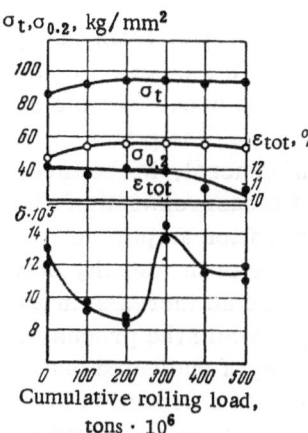

Fig. 1. Changes in the internal friction background (δ), the strength characteristics (σ_t and $\sigma_{0.2}$), and the relative elongation (ε_{tot}) of steel rails subjected to periodic loading.

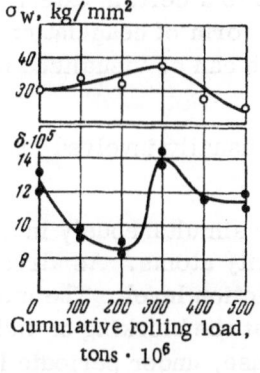

Fig. 2. Comparison of the changes in the internal friction with the changes in the fatigue limit (σ_W) of steel rails subjected to periodic loading.

The results of the measurements carried out showed that the changes in the internal-friction background were represented by a complex non-monotonic curve, with a minimum and a maximum. The same curve gave the limits of the scatter of the data obtained in measurements on two parallel samples (which had the same history). The experimental values of the internal friction of parallel samples were in good agreement, and the general features of the curve were far outside the limits of the scatter.

To compare the changes in the internal friction with the changes in other mechanical properties of the metal, the internal-friction curve was matched with the curves representing the changes in the strength (the tensile strength σ_t and the yield stress $\sigma_{0.2}$) and of the relative elongation ε_{tot} (Fig. 1), as well as the finite fatigue limit (Fig. 2) and the critical brittleness temperature (Fig. 3), plotted using the data given in [3].

The critical brittleness temperature was assumed to be the upper temperature limit of the scatter of the experimental results of the impact tests.

The results obtained showed that right from the beginning of the application of a periodic load, the decrement decreased, reaching its minimum value after $200 \cdot 10^6$ tons of load. At the same time the tensile strength, the yield stress, and the finite fatigue limit (Figs. 1 and 2) all increased. The greatest change in the strength and in the decrement was observed after the first $100 \cdot 10^6$ tons of load. Further periodic loading increased strongly the value of the decrement, which reached its maximum after $300 \cdot 10^6$ tons of load. The strength remained constant in this range, but the fatigue limit continued to increase. The maximum in the fatigue-limit curve coincided with the maximum in the decrement curve. Then, the decrement began to decrease again, and after $400 \cdot 10^6$ tons of load it remained constant within the limits of the investigated duration of periodic loading. At the same time, the fatigue limit decreased continuously, the greatest fall in its value σ_W (by 10 kg/mm^2) being observed after $400 \cdot 10^6$ tons of load, which was to some extent correlated with the large change in the decrement.

It was worth noting that, in contrast to the fatigue limit, the critical brittleness temperature increased monotonically throughout the duration of application of periodic loads; after $300 \cdot 10^6$ tons of load, the critical brittleness temperature increased more strongly than in the initial period, and this coincided with the beginning of the second fall in the decrement. At the same time a considerable reduction in the relative elongation was observed (Fig. 1).

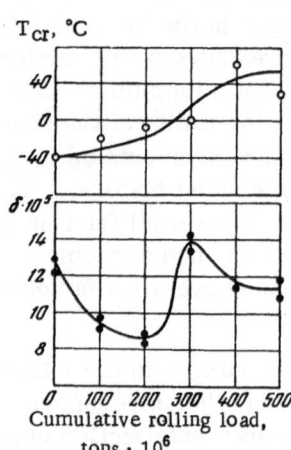

Fig. 3. Comparison of changes in the internal friction with the changes in the critical brittleness temperature T_{cr} of rail steel subjected to periodic loading.

To interpret these regularities in the changes of the mechanical properties from the point of view of structural changes in a metal, we shall use the well-known assumption about the role of crystal structure defects in the dissipation of mechanical energy [6, 7, 8]. In general, in the investigated frequency range of measurements of the decrement, at a fixed vibration amplitude, the processes leading to an increase of the density of mobile dislocations should cause an increase in the internal-friction background, while the processes associated with the blocking of nucleated and mobile dislocations should lead to a reduction of the internal friction. Thus, in analysis of structural changes caused by periodic loading, it is necessary to allow not only for the purely quantitative factors (increase of the density of defects), but also for the interaction of dislocations with impurity atoms and vacancies, for a redistribution of dislocations, and for the possibility of mutual blocking in the formation of clusters of sufficiently high density. The processes of mechanical-energy dissipation will be affected to a certain extent also by the processes of accumulation of irreversible fatigue damage in the form of coagulation of unhealable pores of vacancy origin and formation of microcracks which can be visualized metallographically.

However, since, in the case considered, periodic loading usually involved loads below the fatigue limit, these factors were unlikely to be important.

In the first stage of periodic loading, two processes occur simultaneously in a metal; dislocations are nucleated and blocked by interaction with impurity atoms. As already mentioned, these processes have opposite effects on the internal-friction level. The minimum in the curve representing the change in the decrement indicates that the blocking of dislocations is the dominant process at this stage. This is quite likely because, under periodic loading conditions at stresses below fatigue limit, the rate of rise of the density of mobile dislocations is small, but the concentration of dissolved impurities is relatively high. With increase in the number of loading cycles the concentration of atoms capable of interacting with the dislocations generated by loading increases gradually, while the density of line defects increases continuously. Therefore, finally, after a number of cycles, the internal-friction background should increase, and this was in fact observed after $200 \cdot 10^6$ tons of load.

The simultaneous occurrence of the processes of nucleation and blocking of dislocations is supported by the nature of the changes in the strength properties (Fig. 1) and in the fatigue limit (Fig. 2). Increase in the values of σ_t and $\sigma_{0.2}$ stops practically at the moment of reaching the internal-friction minimum; however, the value of σ_W continues to increase right up to the second fall of the decrement (beyond the maximum of the internal-friction curve). Evidently, the difference between the end points of the rise of these properties under fatigue conditions may be associated with their different sensitivities to the formation of dislocation clusters having the critical density [4], which is a necessary condition for the nucleation and subsequent accumulation of fatigue damage.

The formation of such dislocation clusters seems very likely under the conditions discussed here, in view of very small local volumes in which these structural processes take place. In the investigated case such "working" volumes are very small ferrite plates (whose dimensions do not exceed 2 μ) alternating with "rigid" cementite plates. Since the processes associated with the nucleation, motion, and interaction of dislocations and other defects take place in such tiny volumes, the formation in ferrite of sublocal dislocation clusters having the critical density becomes possible even at relatively low average dislocation densities (compared with the density over the whole volume of the metal) in microvolumes of the platelike component.

The reduction of the mobility of dislocation segments in such clusters, as well as the mutual blocking by elastic stress fields generated by the dislocation clusters themselves, should cause the internal-friction background to decrease from some moment during periodic loading, which is indeed observed experimentally after the passage of $300 \cdot 10^6$ tons of load (Fig.1). Consequently, the beginning of the second fall of the internal friction may be used as an indicator of the appearance, in ferrite grains, of dislocation clusters of the critical density, and this fall of the internal friction indicates a continuous rise in the number of such clusters.

Under these conditions there is one further structural process leading finally to a reduction in the internal friction: the formation of vacancies, at a rate which increases strongly with increase in the probability of intersection of dislocation lines in microvolumes having a high dislocation density. The interaction of these vacancies with dislocations and their precipitation at dislocations also favor the blocking of dislocations.

As shown in [9], the action of force fields existing around dislocation clusters leads, when they approach one another sufficiently, to the formation of nuclei of fatigue cracks. At the same time, the process of coagulation of vacancies generated by fatigue, and their precipitation at micropores may take place, "loosening" the crystal structure [10]. Both these processes, representing the beginning of accumulation of fatigue damage, will increase the internal-friction background, i.e., they will compete with the process of growth in the number of dislocation clusters of the critical density. The consequence of this should be at least to arrest the fall of the internal-friction curve — which is confirmed by the results of measurements carried out: after the passage of 400-500 million tons of load, the decrement remains practically constant.

Thus, the nature of the changes in the internal-friction background during periodic loading of rail steel is not accidental and is in full agreement with the current ideas on the regularities and the sequence of structural changes in a metal under fatigue conditions. However, this is not sufficient for the solution of the inverse problem: determination of the changes taking place in the dislocation structure of a metal from the nature of the changes in the internal-friction level. For this purpose it is necessary to continue our comparison of the internal-friction curve with the behavior of a wide range of mechanical properties, in order to find a more complete correlation with structural changes.

Analysis of the curves representing the mechanical properties (Figs. 1 and 2) shows that the increase, during the fatigue processes, of the value of σ_W ends later than the increase in the tensile strength and the yield stress and, in contrast to the latter two properties which remain constant when the duration of loading is extended, the fatigue limit decreases considerably after $300 \cdot 10^6$ tons of load. This decrease is to be expected if the assumption that, from this load onwards, the number of dislocation clusters of the critical density begins to increase and each of these clusters is a potential nucleus of a fatigue crack. The decrease of the fatigue limit should be particularly interesting at the stage when the drop in the internal-friction level stops because of accumulation of fatigue damage. At this stage, σ_W not only decreases to its initial value (30 kg/mm^2), but even becomes 6 kg/mm^2 smaller, which may be regarded as a convincing proof of the development of damage in the metal. At the same time, the strength characteristics of the material (σ_t and $\sigma_{0.2}$) can and should remain constant under the described experimental conditions because, as shown in [9], the formation of nuclei of fatigue cracks should not affect greatly these strength characteristics, since the growth of such nuclei and their opening out into cracks is only possible under periodic and not static loading. Consequently, the behavior of the characteristics of the static strength cannot be regarded as indicating the beginning of damage of the metal under fatigue conditions; the onset of this process can be found only from the parallel behavior of the internal friction and the fatigue limit curves (Fig. 2). On the other hand, such a characteristic as the relative elongation is more sensitive to structural changes which give rise to damage loci. In particular, a considerable change in the plasticity begins from the moment of the second fall in the internal-friction background and it may be related directly to the formation of dislocation clusters of the critical density in the metal.

Thus, the nature of the nonmonotonic internal-friction curve and the behavior of a range of mechanical properties, allow us to establish a correlation with the following structural processes, taking place in succession in a metal subjected to cyclic loading:

1) the increase in the dislocation density and the blocking of mobile dislocations by interacting with point defects;

2) the formation of dislocation clusters of the critical density;

3) the formation of nuclei of fatigue cracks by the interaction of stress fields around dislocation clusters, as well as the coagulation and precipitation of vacancies, generated at discontinuities by the fatigue processes.

It is particularly important for the strength of structural members that all these changes have the same effect (as far as their sign is concerned) on the critical brittleness temperature T_{cr}: in all cases the value of T_{cr} is raised by the application of periodic loads (Fig. 3). The embrittling effect of the blocking of dislocations in the first fatigue stage is confirmed indirectly by the results reported in [11], in which the values of the low-frequency internal-friction parameters suggest that the initial stage of deformation aging of technical-grade iron is associated with the blocking of dislocations by carbon atoms. At the same time, it is known [12] that, right from the beginning of the process of deformation aging, one should expect an increase in the critical brittleness temperature.

The temperature T_{cr} rises particularly strongly from the moment of formation of dislocation clusters having the critical density (Fig. 3). This confirms the suggestion made earlier [13] that an increase in the critical brittleness temperature of carbon steel with increase in the number of cycles of preliminary loading is associated with the formation of dislocation clusters, which are unable to disperse before the next impact test and which cause brittle fracture at higher temperatures. In the investigated case, intense embrittlement of steel is indicated by the second fall in the internal friction, caused by a rise in the number of dislocation clusters and formation of damage nuclei.

In conclusion, it must be mentioned that the nonmonotonic nature of the changes in the internal friction is to be expected for materials which have a tendency to exhibit the first stage of the deformation aging, in fatigue tests at stresses close to the fatigue limit or little below it. Earlier investigations [14-16] on the changes in the internal friction during fatigue tests under stresses lower than the fatigue limit have not yielded an unambiguous dependence of the damping decrement on the number of cycles. Thus, it has been shown in [14, 15] that an increase in the number of cycles causes the internal friction to increase somewhat at first, but this is followed by a practically constant value of the internal friction. When stresses considerably below σ_w are applied, the internal friction is practically unaffected, within the limits of the experimental error. It has been reported [16] that under stresses lower than σ_w there is some initial rise in the internal friction, followed by a drop to a value lower than the initial value. These data agree, in our opinion, with the hypothesis of two processes (growth of the density of mobile dislocations and blocking of such dislocations) in the initial fatigue stage, the two processes having opposite effects on the internal friction level. Depending on the efficiency of each of these processes, the internal friction may either increase or decrease, or it may remain unchanged. Evidently, the nature of changes in the internal friction in the initial fatigue stage will be governed in each actual case by the previous history of the samples, the test conditions, and other extraneous factors. All this points to the need for a parallel investigation of other structure-sensitive properties to give an unambiguous interpretation of the internal-friction results.

Finally, we must mention that the last remark applies to the internal-friction level measured after the second fall of the decrement (Fig. 3). Under the periodic loading conditions described here, and in spite of the nucleation of the fatigue damage, the final value of the internal friction is in practice not greater than the initial value, although it should be greater in the case of explicitly observed fatigue damage. We may conclude that in our case the accumulated damage was, to a considerable extent reversible, and could be eliminated by subsequent annealing, so that the investigated mechanical properties, including the fatigue limit and T_{cr}, should recover their initial properties. The results of an experimental check of this hypothesis are published in the next paper.

Literature Cited

1. V.S. Ivanova and L.K. Gordienko, Metallurgy, metallography, and physicochemical investigation methods, Tr. Inst. Met. im. A.A. Baikova, Akad. Nauk SSSR, No. 13 (1962).
2. T.S. Mar'yanovskaya and A.G. Nikonov, Collection: Strength of Metals Subjected to Alternating Loads, Izd. Akad. Nauk SSSR (1963).
3. I.A. Oding, A.G. Nikonov, and T.S. Mar'yanovskaya, Izv. Akad. Nauk SSSR, Otd. Tekhn. Nauk, Met. i Gorn. Delo, No. 5:101 (1964).
4. V.S. Ivanova, Fatigue Fracture of Metals, Metallurgizdat (1963).
5. I.G. Polotskii and V.F. Taborov, Zavodsk. Lab., 23(8):986 (1957).
6. I.A. Oding, M.G. Lozinskii, and L.K. Gordienko, Metalloved. i Term. Obrabotka Metal., No. 12 (1959).
7. A. Nowick, J. Appl. Phys., 22:1182 (1951).
8. V.S. Postnikov and G.A. Gorshkov, Collection: Strength of Metals under Alternating Loads, Izd. Akad. Nauk SSSR (1963), p. 283.
9. I.A. Oding, Permissible Stresses in Machine Construction and Periodic Loading of Metals, Mashgiz (1962).
10. I.A. Oding, Metalloved. i Term. Obrabotka Metal., No. 2 (1955).
11. S.O. Suvorova, V.I. Sarrak, and R.I. Éntin, Fiz. Metal. i Metalloved., 17(1):105 (1964).
12. W. Wepner, Arch. Eisenhüttenw., 2:71 (1955).
13. I.A. Oding, A.G. Nikonov, and T.S. Mar'yanovskaya, Dokl. Akad. Nauk SSSR, 143(6):1332 (1962).
14. V.A. Zhuravlev, Zavodsk. Lab., 14(5):614 (1948).
15. L.A. Glikman, V.A. Zhuravlev, and T.N. Snezhkova, Zh. Tekhn. Fiz., 14(4):448 (1949).
16. V.I. Shashlov, Proceedings of Conference on Investigation of Energy Dissipation in Vibrations of Elastic Bodies, Izd. Akad. Nauk UkrSSR (1958), p. 174.

INFLUENCE OF TEMPERING ON CHANGES IN
THE INTERNAL FRICTION AND COLD BRITTLENESS
OF CARBON STEEL SUBJECTED TO PERIODIC LOADING

L. K. Gordienko and T. S. Mar'yanovskaya

In the preceding paper [1], an investigation was reported of the changes in the internal-friction background during prolonged periodic loading of carbon steel at stresses below the fatigue limit and a relationship was established between the nature of the behavior of the internal friction and the following structural processes taking place during fatigue tests:

1) an increase in the dislocation density and the blocking of mobile dislocations by impurity atoms and by vacancies;

2) the formation of dislocation clusters having the critical density;

3) the formation of submicroscopic nuclei of fatigue cracks.

The existence of such a relationship was confirmed by the corresponding changes in the following mechanical properties: the tensile strength, the yield stress, the relative elongation, and the finite (limited) fatigue limit. A comparison of the nature of the changes in the internal friction, in the mechanical properties, and in the critical brittleness temperature during fatigue tests made it possible to relate the strong rise of the critical brittleness temperature to the onset of the formation of dislocation clusters with a density close to the critical value, corresponding to a maximum in the internal-friction curve

As predicted in the preceding paper [1], the finite value of the internal-friction background observed in spite of the nucleation of damage during fatigue tests, was in practice not greater than the initial value, although an increase was expected because of the appearance of microscopic cracks. Hence, we concluded that the damage accumulated in this process was, to a considerable extent, reversible, and could be eliminated by heat treatment so that the mechanical properties and the critical brittleness temperature returned to their initial values.

An investigation of the influence of tempering at 650°C for 3 h on the critical brittleness temperature of rail steel, which was loaded periodically for varying periods under service conditions, showed that this temperature was reduced greatly by the tempering [2].

Consequently, it was of interest to investigate the influence of tempering on structure-sensitive properties (the logarithmic decrement of vibrations and the electrical conductivity) of rail carbon steel, subjected to periodic deformation. At the same time we determined also the changes in the criteria of the static strength and plasticity.

120

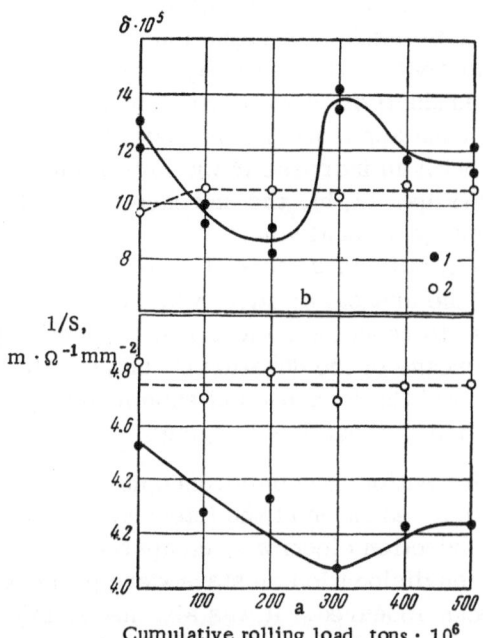

Fig. 1. Changes in the electrical conductivity (a) and in the logarithmic damping decrement (b) of rail steel, subjected to periodic loading under service conditions, plotted as a function of the cumulative rolling load: 1) before tempering; 2) after tempering.

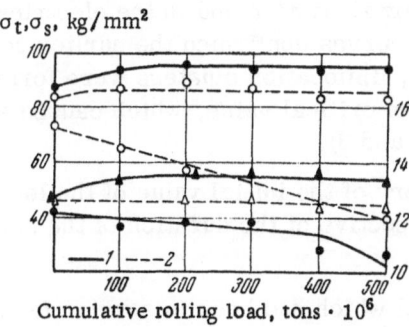

Fig. 2. Changes in the mechanical properties of rail steel subjected to periodic loading under service conditions, plotted as a function of the cumulative rolling load: 1) before tempering; 2) after tempering.

The method of measuring the logarithmic decrement has been described in the preceding paper [1]. The electrical conductivity was determined using apparatus constructed in the Strength of Materials Laboratory of the A.A. Baikov Metallurgy Institute [3]. The measurements of the internal friction and of the electrical conductivity were carried out on the same samples which had been used earlier in the investigation of the influence of the duration of the periodic loading on these characteristics [1]. Before tests the samples were subjected to tempering for 3 h at 650°C in vacuum.

After the measurements of the logarithmic decrement and the electrical conductivity of these samples, microsamples were cut from them for tensile tests.

The changes in the logarithmic decrement (δ), the electrical conductivity ($1/\rho$), the mechanical properties, and the critical brittleness temperature (T_{cr}) of rail steel, subjected first to periodic loading, are given in Figs. 1, 2, and 3 as a function of the cumulative total load on the rails, before and after tempering. The curves representing the changes in the logarithmic decrement, the mechanical properties, and the critical brittleness temperature of rail steel before tempering were plotted using the results reported in [1, 4]; the curve representing the changes in the critical brittleness temperature after tempering was plotted on the basis of data given in [2].

The curve representing the variation of the electrical conductivity of carbon steel subjected to periodic loading under service conditions was a mirror image of the curves representing the variation of the logarithmic decrement (Fig. 1a). Right from the beginning of the application of periodic loads, the electrical conductivity decreased, reaching its minimum value at 300 million tons of load. Further increase of the cumulative load passed over the rails led to some increase in the electrical conductivity, whose value reached saturation after 400 million tons of load.

The behavior of the electrical conductivity curve was in agreement with the structural processes (listed above) taking place in the metal during the fatigue process.

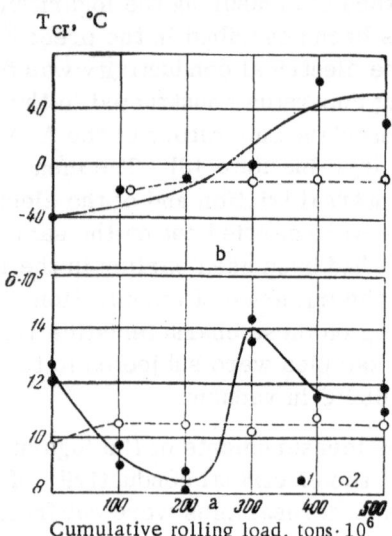

Fig. 3. Changes in the logarithmic damping decrement (a) and the critical brittleness temperature (b) of rail steel, subjected to periodic loading under service conditions, plotted as a function of the cumulative rolling load: 1) before tempering; 2) after tempering.

It is known [5-7] that an increase in the dislocation density leads to a reduction of the electrical conductivity. Consequently, since the electrical conductivity decreased right from the beginning of the application of periodic loads, in agreement with an increase in the logarithmic decrement (Figs. 1a, b), the increase in the dislocation density in local volumes was accompanied by a simultaneous formation of dislocation clusters of the critical density, which, in their turn, tended to reduce the electrical conductivity. Further increase of the duration of the periodic loading favored the formation of submicroscopic discontinuities in these local regions.

The formation of submicroscopic crack nuclei led to a reduction of the intensity of scattering of conduction electrons, compared with the case when dislocation clusters were present before cracks appeared. It was also necessary to allow for the absorption of vacancies by submicroscopic fracture loci. Therefore, as the process of nucleation of submicroscopic discontinuities developed, the fall of the electrical conductivity slowed down.

The termination of the fall of the internal friction, caused by the development of submicroscopic crack nuclei and by the precipitation of vacancies at them, was observed in the same range of the periodic loading durations as the increase in the electrical conductivity (Figs. 1a, 1b).

Thus, the curves representing the changes in the internal friction and in the electrical conductivity were in good agreement. Comparison of these curves confirmed the earlier conclusion [1] that, after a cumulative load of 300 million tons, dislocation clusters were formed in the metal and the density of such clusters was close to the critical value, which caused a strong rise of the critical brittleness temperature (Figs. 1 and 3).

The tempering treatment produced a complete recovery of the initial value of the logarithmic decrement and of the electrical conductivity, irrespective of the duration of the periodic loading, within the limits investigated (Figs. 1a, 1b).

The changes in the mechanical properties of rail steel which had been in service for a time are shown in Fig. 2 as a function of the cumulative rolling load, before and after the tempering of the samples. We can see that the tempering produced a complete recovery of the initial values of the strength characteristics of the steel, irrespective of the duration of the periodic loading. The relative elongation was raised by the tempering, but this rise was reduced by an increase in the cumulative total rolling load, which had passed over the rail. At the same time the absolute values of the relative elongation remained higher than in the case of untempered steel; this was true of all cases (Fig. 2).

The curves showing the changes in the logarithmic decrement of rail steel (subjected to rolling loads under service conditions) before and after tempering are matched in Fig. 3 with the corresponding curves showing the changes in the critical brittleness temperature. After tempering, the critical brittleness temperature of rails subjected to from 300 to 500 million tons of rolling load, decreased sharply to the same level irrespective of the previous total rolling load, while the critical brittleness temperature of rails subjected to 300 million tons of load remained practically unchanged. It is worth noting that all rails subjected to rolling loads during their service life had, after tempering, approximately the same brittleness temperature, which was higher than the temperature of rails not subjected to service conditions.

Thus, tempering undoubtedly led to the recovery of all the initial properties. The results obtained could be explained by the "healing" of submicroscopic discontinuities, the dispersal of dislocation clusters having a density close to the critical value, and the redistribution of dislocations during tempering. Some of the fatigue damage accumulated in the metal during the service life of rails should be healed by tempering at 650°C. The fact that the critical brittleness temperature remained somewhat higher than the initial value even after tempering indicated that some of the fatigue damage still remained in the metal. However, the influence of this damage on the embrittlement of the metal was slight.

Conclusions

1. A study was made of the changes in the internal-friction background, in the electrical conductivity, and in the mechanical properties caused by the tempering of carbon steel, subjected first to a preliminary prolonged periodic loading at stresses below the fatigue limit.

2. It was established that tempering at 650°C led to a recovery of the initial values of the mechanical and physical properties and of the critical brittleness temperature.

3. Comparison of the nature of changes in the physical and mechanical properties and in the critical brittleness temperature after tempering made it possible to conclude that such tempering tended to heal submicroscopic discontinuities and to disperse the clusters of dislocations having a density close to the critical value.

Literature Cited

1. I.A. Oding, L.K. Gordienko, and T.S. Mar'yanovskaya, this volume, p. 114.
2. I.A. Oding, A.G. Nikonov, and T.S. Mar'yanovskaya, Dokl. Akad. Nauk SSSR, 61(3) (1965).
3. L.K. Gordienko, Advanced Scientific-Technical and Industrial Practice, Subject 32, VINITI Branch (1959).
4. I.A. Oding, A.G. Nikonov, and T.S. Mar'yanovskaya, Izv. Akad. Nauk SSSR, Otd. Tekhn. Nauk, Met. i Gorn. Delo, No.5: 101 (1964).
5. L.M. Clarebrough, M.E. Hargreaves, and M.H. Loretto, Acta Met., 8(11): 797 (1960).
6. W. Boas, Collection: Dislocation and Mechanical Properties of Crystals [Russian translation], IL (1960), p.272. [English edition: J.R.C. Fisher, ed., Wiley, New York.]
7. L.K. Gordienko and I.A. Oding, Izv. Akad. Nauk SSSR, Otd. Tekhn. Nauk, Met. i Toplivo, No.6: 52 (1959).

ANISOTROPY OF THE INTERNAL FRICTION
OF TRANSFORMER STEEL

I. B. Kekalo and V. K. Potemkin

An investigation was made of the internal friction Q^{-1} of samples cut from a single crystal of transformer sheet steel, along directions close to the crystallographic axes [100], [111], and [110] in the (110) plane. The deviation of the angle of cut from these crystallographic directions was not more than 2-3°. The samples corresponding approximately to the directions [100], [111], and [110] will be denoted as samples No. 1, 2, and 3, respectively. Sample No.1 was cut along the direction of rolling, sample No. 2 at 55° to the rolling direction, and sample No. 3 at right angles to rolling

The samples were cut from a single-crystal sheet by a hack saw with in-line teeth, held in clamps which directed the plane of the saw. The damaged layer of the metal was removed by etching. Before the tests, all the samples were annealed in vacuum at 1100°C.

The internal friction was measured with a relaxator (torsional pendulum) in the form of a low-frequency pendulum. The apparatus used made it possible to carry out measurements below and above 0°C in magnetic fields. The frequency of the elastic vibrations was about 1 cps and the frequency of the alternating field was 50 cps. The path of a ray from a mirror to a scale was about 1.5 m long. The sample dimensions were $70 \times 1 \times 0.4$ mm. The internal friction was calculated from the well-known formula:

$$Q^{-1} = \frac{1}{\pi n} \ln \frac{A_0}{A_n},$$

where A_0 is the initial amplitude, A_n is the final amplitude, n is the number of vibrations in the amplitude range from A_0 to A_n. The maximum deformation, approximately equal to 10^{-4}, corresponded to the amplitude $A_0 = 2$ cm.

Some samples had a very high level of Q^{-1}, which made it difficult to estimate the damping sufficiently accurately. In these cases the amplitude range, in which the number of vibrations was calculated, had to be increased by reducing A_n. Each point was determined as an average from a large number of measurements (at least 4-5 measurements). From now on we shall only mention the initial amplitude A_0, since in all experiments the amplitude decreased during tests by 0.5 cm (with the exception of the cases referred to above when the damping of vibrations was strong).

Results of Experiments and Discussion

1. Our investigation of the temperature dependence of Q^{-1} of single-crystal samples of transformer steel having different crystallographic orientations showed that, in the demag-

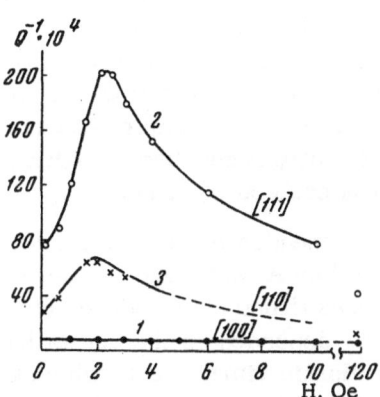

Fig. 1. Dependence of Q^{-1} on the intensity of an alternating magnetic field H for samples of different orientations:1) sample cut along the rolling direction (sample No. 1); 2) sample oriented at 55° to the rolling direction (sample No.2); 3) sample oriented at 90° to the rolling direction (sample No. 3). T = 20°C.

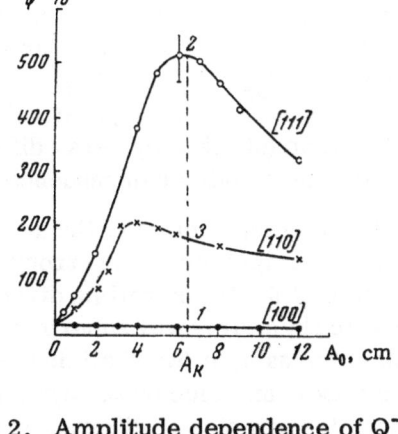

Fig. 2. Amplitude dependence of Q^{-1} of demagnetized samples: 1) sample cut along the rolling direction (sample No. 1); 2) sample oriented at 55° to the rolling direction (sample No. 2); 3) sample oriented at 90° to the rolling direction (sample No. 3). T = 20°C.

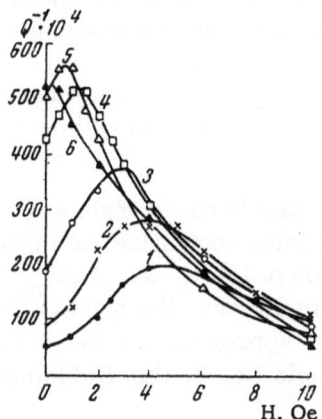

Fig. 3. Nature of the dependences of Q^{-1} on H for sample No. 2 (orientation angle 55°) for various elastic vibration amplitudes A_0: 1) 0.5 cm; 2) 1 cm; 3) 2 cm; 4) 4 cm; 5) 5 cm; 6) 7 cm. T = 20°C.

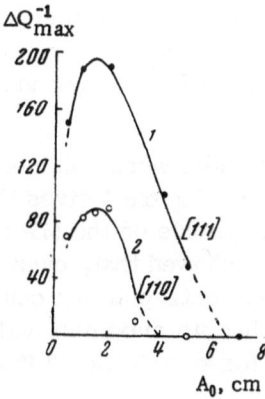

Fig. 4. Dependence of the MP amplitude (ΔQ^{-1}_{max}) on the vibration amplitude A_0. 1) Orientation angle 55° (sample No. 2); 2) orientation angle 90° (sample No.3). T = 20°C.

netized state, they had very different values of Q^{-1} in the investigated range of temperatures (from −196°C to above 0°C). The highest values of Q^{-1} were found for samples cut at 55° to the direction of rolling (that was the direction close to [111]), and the lowest values were found along the rolling direction (close to [100]). The differences at 20°C are given in Fig.1, from which it follows that Q^{-1} of sample No. 1 in H = 0 was almost an order of magnitude smaller than at Q^{-1} of sample No.2. The saturation magnetization of the samples reduced Q^{-1} and the difference between the values of Q^{-1} for different samples became less. The value of Q^{-1} of sample No. 2 in the magnetized state was greater than for samples Nos. 1 and 3.

2. Figure 1 shows that Q^{-1} did not vary monotonically with the magnetization, but the dependence of Q^{-1} on H had a magnetic peak. We shall consider the influence of the orientation on the parameters of the magnetic peak (MP) of the internal friction. The amplitude of the magnetic peak $\Delta Q_{max}^{-1} = Q_{max}^{-1} - Q_{H=0}^{-1}$ (where Q_{max}^{-1} is the maximum value of Q^{-1} in the dependence of Q^{-1} on H) reached its highest value for sample No. 2. For the sample cut exactly along the [100] direction (checked by x-ray diffraction) the MP was altogether absent. Such a sample did not exhibit any amplitude dependence of Q^{-1} or any dependence on H, either.

3. The internal friction of the demagnetized samples depended in various degrees on the elastic vibration amplitude. The strongest dependence of Q^{-1} on A was found for the sample cut at the angle of 55° to the rolling direction. Figure 2 shows that the dependence of Q^{-1} on A was represented by a curve with a maximum, whose position A_0 depended on the crystallographic orientation of the sample (we shall denote the amplitude of elastic vibrations at which Q^{-1} reaches its maximum value by the symbol A_k). Extrapolation of the amplitude dependences of Q^{-1}, obtained for samples of different orientation, showed that these curves converged to a single point when $A \rightarrow 0$. The value of Q^{-1}, corresponding to this point of convergence, was the same as that observed for the saturation-magnetized samples, i.e., as observed in the case when the losses due to magnetic processes were absent (Q^{-1} of the magnetized samples was independent of A). The presence of a maximum in the dependence of Q^{-1} on A is in agreement with the theoretical representations.* A similar dependence of Q^{-1} on A has been reported for some ferromagnets in [8] and elsewhere.

Figure 3 shows the dependence of Q^{-1} on H for sample No. 2, for various values of A_0. Figure 3 shows that first the amplitude of the MP (ΔQ_{max}^{-1}) increased with increase in A_0, but then it decreased, and at a certain amplitude of the elastic vibrations the MP disappeared altogether, and the dependence of Q^{-1} on H was described by a curve without a maximum (curve 6 in Fig. 3). Secondly, the field in which Q^{-1} had its maximum value decreased with increase of A_0. Thirdly, the amplitude of the elastic vibrations at which the MP in the dependence of Q^{-1} on H disappeared, corresponded to the amplitude A_k in the dependence of Q^{-1} on A_0 (see Fig. 2).

Similar results were obtained in the investigation of the sample cut at right angles to the rolling direction. Figure 4 gives the data for samples Nos. 2 and 3 indicating the influence of the vibration amplitude on the MP amplitude. The extrapolation of the values of ΔQ_{max}^{-1} to zero amplitudes showed that, even in the case of very small amplitudes, the MP existed in alternating fields. Extrapolation could be used also to determine approximately those fields in which Q^{-1} reached its maximum value, when $A_0 \rightarrow 0$. For sample No. 2, this field amounted to 5 Oe, while for sample No. 3 it was 2 Oe (Fig. 5).

The internal friction of the demagnetized samples depended on the preliminary elastic deformation. Thus, if the measurements of Q^{-1} were carried out under the same conditions (for example, when $A_0 = 1$ cm, $A_n = 0.5$ cm, as in Fig. 6) but the sample was deformed by twisting before measurements, then the value Q^{-1} at first increased with increase in the preliminary deformation, but became stabilized at some level (Fig. 6). Magnetization reversal reestablished Q^{-1} at the value found before deformation. This allowed us to conclude that the observed effect was not associated with plastic deformation. The effect has been discussed in detail for iron [1].

---

*$Q^{-1} \approx \Delta W/W$, where ΔW is the energy dissipation due to magnetic processes, per vibration period; $\Delta W \sim A^3$; W is the vibrational energy, $W \approx A^2$. Hence, at low vibration amplitudes $Q^{-1} \approx A$; at high amplitudes, when the magnetic saturation is reached and $\Delta W = $ const, $Q^{-1} \approx A^{-2}$ [6].

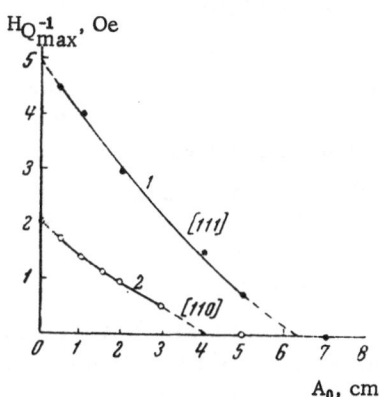

Fig. 5. Dependence of $H_{Q_{max}^{-1}}$ on A_0 for various orientations: 1) sample No. 2; 2) sample No. 3. T = 20°C.

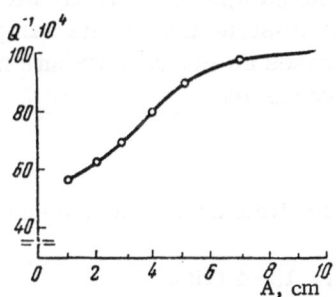

Fig. 6. Influence of preliminary elastic deformation by twisting (corresponding to an amplitude A) on the value of Q^{-1} measured over a constant range of amplitudes ($A_0 = 1$ cm; $A_n = 0.5$ cm). Sample No. 2.

The investigation of the frequency amplitude showed that, within the limits of the experimental error, the amplitude of the MP was independent of the elastic vibration frequency.

4. The damping of vibrations due to magnetic processes in demagnetized ferromagnets at low elastic-vibration frequencies is usually associated with irreversible displacements of domain boundaries. The internal friction caused by the magnetoelastic hysteresis (MEH) is stronger for ferromagnets having high permeability and high magnetostriction and mechanostriction.

Analysis of our results obtained showed that the damping of vibrations in the demagnetized samples of transformer steel was due to the MEH losses because of irreversible displacements of domain boundaries. The most convincing proof of this conclusion was provided by the detected dependence of Q^{-1} of the demagnetized samples on the amplitude of elastic vibrations, which was absent in the saturation-magnetized samples. Extrapolation of the dependence of Q^{-1} on A (Fig. 2) showed that the magnetic processes causing the damping of vibrations were absent when $A \rightarrow 0$, which also indicated a relationship between the observed losses and the MEH.

The large differences between the values of Q^{-1} for samples having different orientation were probably due to the fact that, first, they had different mechanostrictions; secondly, they could have differed from one another by the type of domain structure (for example, the relative volumes of 180° and 90° domains);* thirdly, they may have had different permeabilities in weak fields. A check of these suggestions would require an investigation of the domain structure and magnetic properties of the samples.

Of greatest interest were two observations made in the investigation of transformer steel: the disappearance of the magnetic peak, which took place at amplitudes A_k, corresponding to Q_{max}^{-1} in the dependences Q^{-1} on A, and the presence of the MP when $A_0 \rightarrow 0$, found by extrapolation of the dependence of ΔQ_{max}^{-1} on A_0 (Fig. 4). The latter observation has been made earlier in an investigation of nickel [2, 3] and iron [4]. The former observation had been made for the first time, although the model proposed earlier in [2, 4] to describe the dependence of the MP amplitude on the amplitude of elastic vibrations A assumed that the magnetic peak should be absent for A_k and higher vibration amplitudes. In fact, if the MP appeared only when some domain boundaries were displaced reversibly by elastic stresses in H = 0, then this condition was not satisfied when $A > A_k$, because A_k was the amplitude of elastic vibrations at which all do-

---

* Zaikov and Shur [Fiz. Metal. i Metalloved., 18 : 348 (1964)] proved this by direct observation.

main boundaries were displaced irreversibly and at which irreversible processes reach their saturation level [5, 6].

The results obtained in the investigation of the properties of the MP show that this peak may be associated with two processes. One of them depends on A and can therefore be identified with the processes causing the damping of vibrations because of the MEH losses. The second process, which may be separated from the first by extrapolating ΔQ_{max}^{-1} to zero amplitudes, is identified by some authors with the damping due to the appearance of micro-eddy currents, caused by reversible displacements of grain boundaries [8, 3]. This form of loss is independent of A but it does depend on the elastic vibration frequency.

The frequency dependence of the MP amplitude is very weak, within the limits of the frequency range used in this investigation (the weak frequency dependence of the MP amplitude over a relatively narrow range of frequencies has been reported earlier for iron and Invar [4]). Moreover, we are not yet sure whether the extrapolation of ΔQ_{max}^{-1} to zero amplitudes is fully justified; it is possible that at very low amplitudes the MP is absent altogether, that it appears at some amplitude and eventually the dependence of its amplitude on A obeys the law which was found by us. Consequently, we may draw the conclusion that the problem of the nature of the magnetoelastic damping in the case of very low amplitudes requires further investigation.

The absence of the dependence of Q^{-1} on H and on A for the sample cut strictly along the [100] direction is evidently due to the special nature of the domain structure of this sample (for example, 180° neighborhoods) because of which the elastic stresses do not redistribute magnetic phases and do not give rise to losses because of magnetic processes.

Literature Cited

1.   I.B. Kekalo, B.G. Livshits, and V. Morgner, Collection: Relaxation Phenomena in Metals and Alloys, Metallurgizdat (1963), p. 190.
2.   I.B. Kekalo and B.G. Livshits, Fiz. Metal. i Metalloved., 13:4 (1962).
3.   K. Mishek, Fiz. Metal. i Metalloved., 15:932 (1963).
4.   Yu.S. Avraamov, I.B. Kekalo, and V. Morgner, Collection: Relaxation Phenomena in Metals and Alloys, Metallurgizdat (1963), p. 184.
5.   A.W. Cochardt, Collection: Magnetic Properties of Metals and Alloys, [Russian translation], IL (1961).
6.   A.S. Nowick, Progress in Metal Physics [Russian translation], Metallurgizdat (1956). p. 7. [English edition: B. Chalmers and R. King, eds., 8 vols., Pergamon Press, Inc., New York.]
7.   R.M. Bozorth, Ferromagnetism [Russian translation], IL (1956). [English edition: D. Van Nostrand Co., Inc., Princeton, New Jersey.]
8.   K. Misek, Czech. J. Phys., Vol. 6 (1956); Vol. 7 (1957); Vol. 8 (1958).

INFLUENCE OF SOME ALLOYING ELEMENTS
ON THE ELASTICITY MODULI AND THE POSITION
OF THE HIGH-TEMPERATURE BRANCH OF THE
INTERNAL FRICTION OF IRON ALLOYS

M. A. Krishtal and B. M. Drapkin

The present communication reports the results of measurements of Young's modulus, the shear modulus, and the internal friction of binary alloys based on iron, whose chemical composition is given in Table 1.

The selection of this system for investigation was dictated by the following considerations. Vanadium, tungsten, and molybdenum are widely used as alloying elements in steel. They are introduced into steel usually because of their influence on the properties of ferrite. Thus, tungsten and molybdenum, which have similar chemical properties and atomic diameters (differing strongly from the corresponding properties of iron), affect the interatomic forces in iron in different ways [7]. Cobalt is very similar to iron both in its atomic dimensions and in other properties. Vanadium occupies an intermediate position. The elasticity moduli and the internal friction of the iron—cobalt system have been investigated very little (only one paper [1] is known to the present authors). The same can be said of the system

Table 1

Alloys	Chemical composition, %									
	C	Si	Mn	S	P	V	Mo	W	Co	Ni
1	0.040	0.04	0.13	0.026	0.006	—	—	—	—	0.27
2	0.051	0.16	0.47	0.027	0.006	0.62	—	—	—	0.27
3	0.046	0.16	0.45	0.025	0.006	2.01	—	—	—	0.27
4	0.036	0.14	0.43	0.027	0.010	2.60	—	—	—	0.27
5	0.033	0.36	0.39	0.020	0.003	—	0.98	—	—	0.27
6	0.048	0.15	0.35	0.023	0.005	—	0.86	—	—	0.27
7	0.053	0.22	0.46	0.020	0.006	—	1.67	—	—	0.27
8	0.05	0.13	0.37	0.023	0.008	—	—	—	0.90	0.27
9	0.05	0.12	0.29	0.023	0.008	—	—	—	2.86	0.27
10	0.05	0.11	0.21	0.023	0.008	—	—	—	4.53	0.27
11	0.043	0.16	0.46	0.023	0.008	—	—	0.50	—	0.27
12	0.044	0.11	0.37	0.023	0.008	—	—	0.76	—	0.27
13	0.040	0.12	0.26	0.030	0.006	—	—	4.40	—	0.27

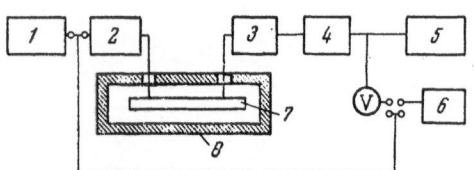

Fig. 1. Schematic diagram of the apparatus. 1) Audio-frequency oscillator "TESLA"; 2) generator of mechanical vibrations; 3) electromagnetic transducer; 4) UM-50A amplifier; 5) S1-19 oscillograph; 6) counter of the VSP type; 7) sample; 8) heating furnace (or cold chamber).

iron—tungsten. Investigations of the influence of cobalt on the diffusion of carbon in austenite [2, 3] suggest very complex behavior of cobalt in iron alloys.

The iron—vanadium system has been investigated most. However, even in the case of this system the experimental data on the influence of vanadium on the binding forces in iron are contradictory. Thus, according to the results reported in [4], Young's modulus at room temperature is reduced by the introduction of vanadium. The weakening of binding forces in iron by vanadium is supported also by the drop in the characteristic temperature [5]. On the other hand, the shear modulus of iron is increased by the introduction of vanadium and has a maximum at 2.0% vanadium [4]. The high-temperature branch of the internal friction curve of iron alloys shifts toward higher temperatures when the vanadium concentration is increased [6]. The enhancement of the binding forces in iron by vanadium is suggested by the reduction of the linear expansion coefficient of the alloy [7].

Our measurements were carried out using apparatus which is shown schematically in Fig. 1. Employing this apparatus we were able to measure simultaneously the natural frequency of flexural and torsional vibrations of the sample, as well as the decrement over a wide range of temperatures.

Young's modulus was found from the formula [8]:

$$E = 1.6388 \, (l/d)^4 P/l f_l^2, \tag{1}$$

where l is the sample length (cm), d is the sample diameter (cm), P is the sample weight (g), f_l is the frequency of the natural flexural vibrations (cps), and E is Young's modulus (kg/mm^2).

The shear modulus was found from the formula:

$$G = 5.1934 \cdot 10^{-8} l/d^2 p f_\tau^2, \tag{2}$$

where G is the shear modulus (kg/mm^2), and f_τ is the frequency of natural torsional vibrations (cps).

When the dimensions of the samples were accurate enough (length to within 0.1 mm and diameter to within 0.01 mm), then the main source of error was in the measurement of the natural vibration frequency. Therefore, it seemed important to use a precision audio-frequency oscillator of the "TESLA" type, by means of which the frequency could be measured with an accuracy of up to $5 \cdot 10^{-4}$, which contributed an error of 0.1%. The total maximum error in the determination of Young's and shear moduli was 0.8 and 0.5%, respectively.

Figure 2 shows the temperature dependence of Young's modulus of Fe—V, Fe—Mo, Fe—W, and Fe—Co alloys. Figure 2 shows that Young's modulus of these alloys and of pure iron decreased linearly with temperature and, beginning from a certain temperature, the rate of fall of the modulus increased. The slope of the linear portion and the temperature at which there was a departure from the linear dependence, were different for different alloys. The cause of the deviation of the temperature dependence of Young's modulus from its initial nature has been interpreted in different ways by different authors. Some [9, 10] have ascribed this effect to the viscous behavior of grain boundaries, while others [14, 15] have associated it with a

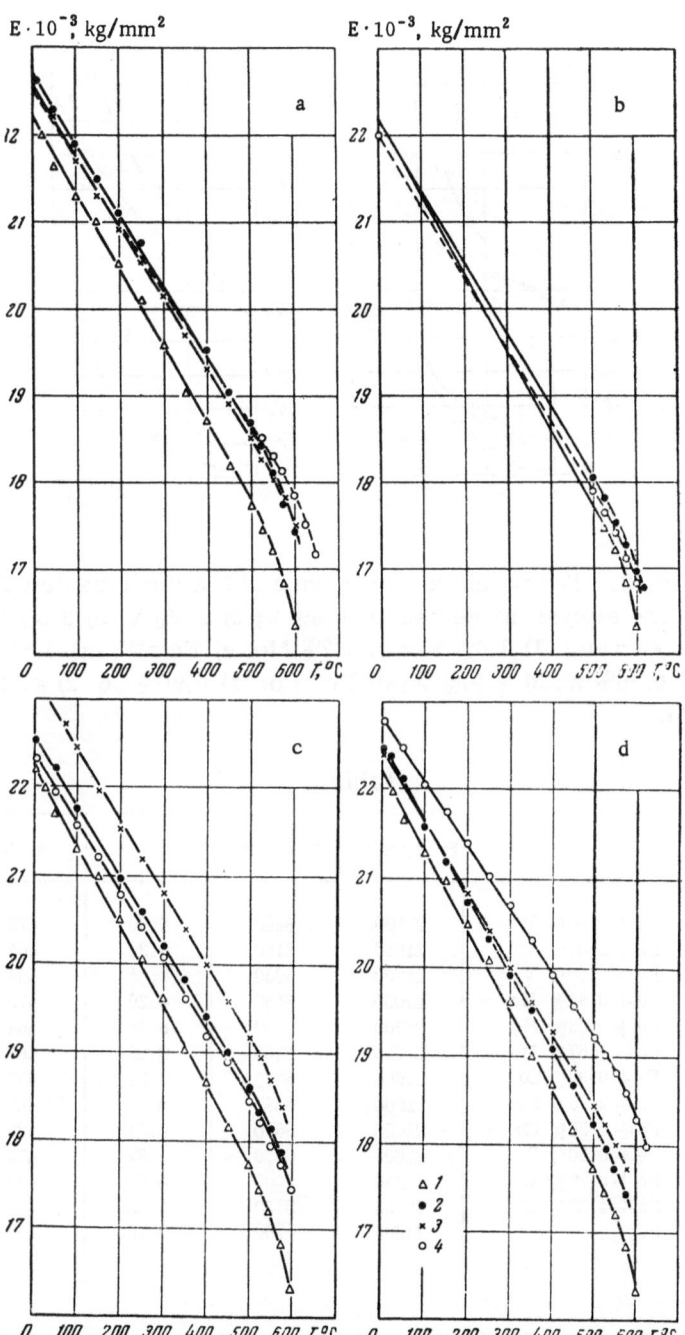

Fig. 2. Temperature dependence of Young's moduli of alloys:
a) Fe−V: 1) Fe; 2) Fe + 0.6 wt.% V; 3) Fe + 2.0 wt.% V;
4) Fe + 2.6 wt.% V; b) Fe−Mo: 1) Fe; 2) Fe + 0.86% Mo;
3) Fe + 1.67% Mo; c) Fe−W: 1) Fe; 2) Fe + 0.5% W; 3) Fe +
0.76% W; 4) Fe + 4.40% W; d) Fe−Co: 1) Fe; 2) Fe + 0.90%
Co; 3) Fe + 2.86% Co; 4) Fe + 4.53% Co.

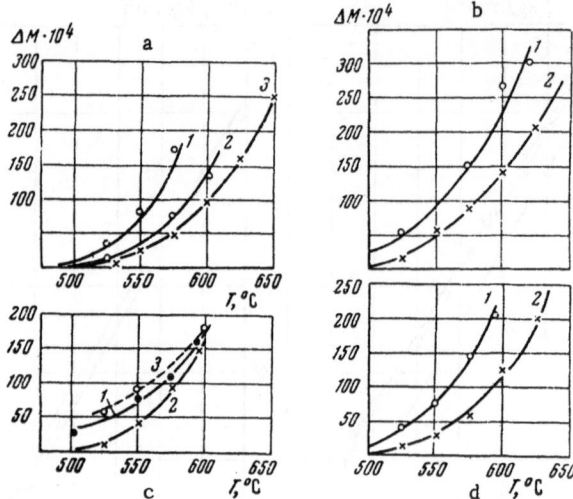

Fig. 3. Temperature dependence of the modulus defects
of the alloys: a) Fe−V: 1) 0.6% V; 2) 2.0% V; 3) 2.6% V;
b) Fe−Mo: 1) 1.0% Mo; 2) 1.7% Mo; c) Fe−W: 1) 0.50% W;
2) 0.76% W; 3) 4.40% W; d) Fe−Co: 1) 0.90% Co; 2) 4.53%
Co.

Table 2

Alloy	E, kg/mm²	G, kg/mm²	μ	θ, °K
Fe + 0.60% V	22100	8400	0.30	475
Fe + 2.01% V	21950	8150	0.34	465
Fe + 2.60% V	21700	8050	0.35	463
Fe + 0.86% Mo	22000	8550	0.29	475
Fe + 0.98% Mo	21700	8150	0.34	464
Fe + 1.67% Mo	21350	8400	0.28	467
Fe + 0.90% Co	21900	8375	0.30	469
Fe + 2.86% Co	22000	8880	0.24	482
Fe + 4.53% Co	23000	8600	0.34	480
Fe + 0.50% W	22600	8550	0.32	475
Fe + 0.76% W	23250	9000	0.29	486
Fe + 4.40% W	22150	8150	0.35	458
Fe	21900	8200	0.33	464

sharp increase in the number of defects, mainly vacancies. In the latter case the temperature
at which Young's modulus began to fall more rapidly could serve as a measure of the interatomic forces. On the basis of these considerations the derivative of the change in the modulus with respect to temperature could serve as a measure of the binding forces, or we could
use the position of the high-temperature branch of the internal friction for this purpose. Figure 3 shows the temperature dependence of the fall in the modulus (modulus defect), while
Fig. 4 gives the temperature dependence of the internal friction of these alloys. Table 2 lists
the values of the shear modulus, Poisson's ratio, and of the characteristic temperature of the
investigated alloys at room temperature.

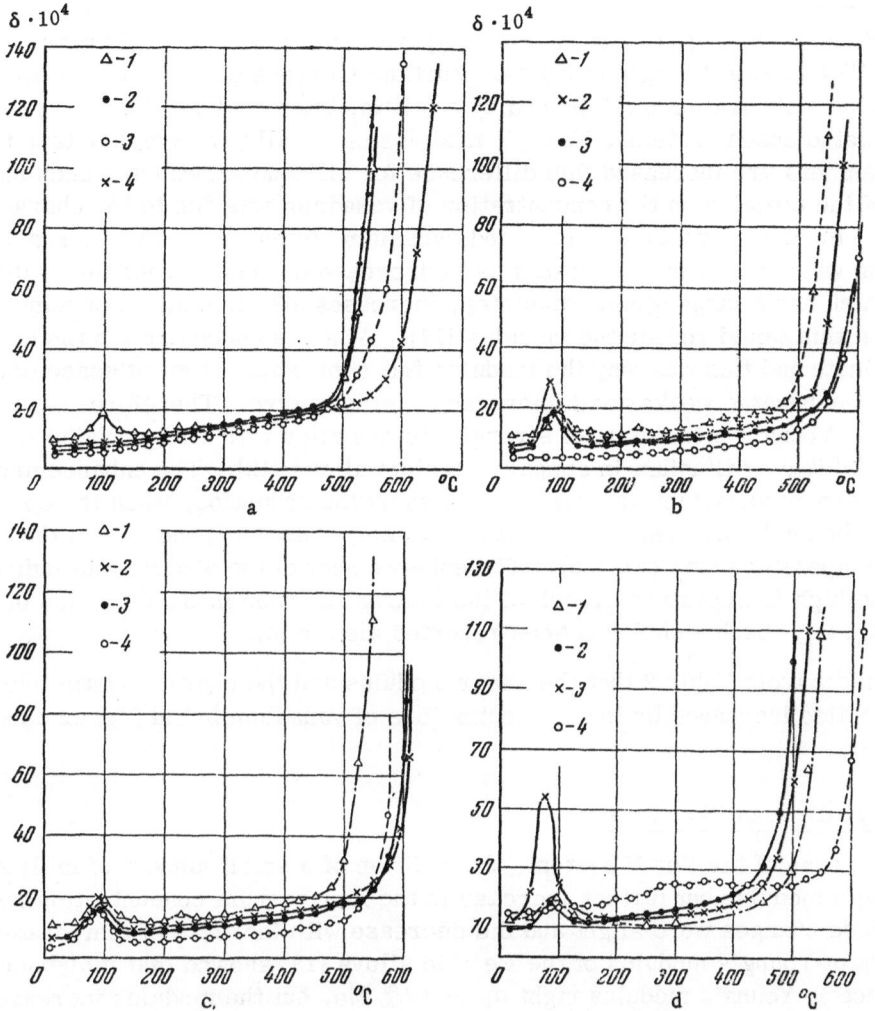

Fig. 4. Temperature dependence of the internal friction of the alloys.
a) Fe−V: 1) Fe; 2) Fe + 0.60% V; 3) Fe + 2.0% V; 4) Fe + 2.6% V; b)
Fe−Mo: 1) Fe; 2) Fe + 0.86% Mo; 3) Fe + 0.98% Mo; 4) Fe + 1.67% Mo;
c) Fe−W: 1) Fe; 2) Fe + 0.50% W; 3) Fe + 0.76% W; 4) Fe + 4.40% W;
d) Fe−Co: 1) Fe; 2) Fe + 0.90% Co; 3) Fe + 2.86% Co; 4) Fe + 4.53% Co.

Poisson's ratio μ and the characteristic temperature θ were calculated from the formulas

$$\mu = \frac{E}{2G} - 1 \tag{3}$$

and

$$\theta = \frac{h}{k}\left(\frac{3N}{4\pi A}\right)^{1/3} 2lf_\tau \sqrt[3]{\frac{3\rho}{2 + 5.6\left(\frac{f_\tau \cdot d}{f_e \cdot e}\right)^3}}, \tag{4}$$

where h is Planck's constant (erg · sec), k is Boltzmann's constant (erg/°K), N is Avogadro's number, A is the atomic weight (g), ρ is the density (g/cm³).

Iron — Vanadium System

It is evident from Fig. 2a that, at low temperatures, small concentrations of vanadium (up to 0.6 wt.%) increase Young's modulus. Further increase of the concentration of vanadium gives rise to the stabilization of, followed by a fall in, the modulus. However, within the limits of the investigated concentrations, Young's modulus of the alloy was higher than that of iron. Increase of temperature increased this difference for all concentrations. Such nature of the dependence of the modulus on the concentration of vanadium was due to the characteristic features of the electron structure of iron and vanadium atoms, the latter increasing the inter-atomic binding which was then manifested as an increase in Young's modulus. However, vanadium, which has a large atomic diameter, increases also the lattice parameter of the alloys [7], and this would reduce the modulus [11]. At some concentration the second factor began to dominate and this was why the modulus fell somewhat. The influence of the atomic dimension factor became weaker with increase of temperature. Therefore, when temperature was increased, Young's modulus rose strongly for the alloy with 2.6% vanadium. Similar behavior was exhibited by titanium in copper—titanium alloys [12]. The enhancement of the binding forces in iron by vanadium was supported also by the reduction, when the concentration was increased, of the angle between the tangent to the temperature dependence of the modulus defect and the temperature axis (Fig. 3a). The enhancement of the binding was indicated also by the shift of the high-temperature branch of the internal friction in the direction of higher temperatures (Fig. 4a). Such a shift had been reported also in [6].

It is evident from Table 2 that the shear modulus and the characteristic temperature of the alloy were also increased by low concentrations of vanadium but at higher concentrations they decreased.

Iron — Molybdenum System

As in the case of the Fe—V system, the addition of a small amount of molybdenum increased Young's modulus but further increase in the molybdenum concentration reduced this modulus. These changes were slight and did decrease with increase of temperature. The measurements of Young's modulus of the Fe—Mo alloys [13] showed that molybdenum had practically no effect on Young's modulus right up to 2.5% Mo, but the modulus increased slightly on further increase of the concentration of molybdenum. The position of the high-temperature branch of the internal-friction background of the Fe—Mo alloys indicated a complex nature of the influence of molybdenum on binding forces in iron. Thus, all curves for the alloys were to the right of the curve for iron. However, the curve for the alloy with 1.6% Mo was to the left of the curve for the 0.86% Mo alloy.

The values of the shear modulus and of the characteristic temperature (Table 2) also indicated a complex influence of molybdenum on the binding forces in iron, which could be explained by the opposite effects of the interatomic binding factors and of the difference between the atomic diameters of iron and molybdenum.

Iron — Tungsten System

The atomic diameters of tungsten and molybdenum are similar (d_W = 2.80 Å, d_{Mo} = 2.78 Å). It was expected that because of the characteristic features of the electron structure of the tungsten atom, the latter would enhance considerably the strength of binding in iron alloys, while the weakening effect of the atomic dimensions factor would be apparent only at considerable concentrations of tungsten. The weakening effect decreased with increase of temperature. Thus, at 0°C, Young's modulus of the alloy containing 4.5% W was close to the modulus of iron. At 500°C, Young's modulus of this alloy was larger than the modulus of iron by

800 kg/mm^2 (about 5%). The temperature dependence of the modulus defect (Fig. 3c) and the position of the high-temperature branch of the internal-friction background (Fig. 4c) also indicated that tungsten enhanced the binding forces in iron alloys.

This was further supported by the data on the behavior of the shear modulus and of the characteristic temperature of the iron—tungsten alloys (Table 2).

Iron — Cobalt System

Cobalt increased considerably Young's modulus of iron and this increase became greater at higher temperatures. The position of the high-temperature branch of the internal friction of the iron—cobalt alloys (Fig. 4d) indicated a complex influence of cobalt on the binding forces in these alloys.

Table 2 shows that the addition of cobalt to iron increased considerably the shear modulus and the characteristic temperature of the alloy. However, when the cobalt content was 4.53%, these quantities fell somewhat and this also indicated the complex influence of cobalt on the interatomic binding forces in iron.

Literature Cited

1. N.S. Rysina and B.N. Finkel'shtein, Dokl.Akad.Nauk SSSR, 98:2 (1954).
2. V.A. Yurkov and M.A. Krishtal, Dokl.Akad.Nauk SSSR, 92(6):1171 (1953).
3. R. Smoluchowski, Phys.Rev., 62:539 (1942).
4. G.M. Ashmarin and B.N. Finkel'shtein, Nauchn. Dokl. Vysshei Shkoly, Met., 8:1 (1958).
5. V.A. Il'ina and V.K. Kritskaya, Collection: Problems of Metallography and Metal Physics, No. 4, Moscow (1955), p. 412.
6. G.M. Ashmarin, Collection: Relaxation Phenomena in Metals and Alloys, Metallurgizdat (1960), p. 146. [English translation: B.N. Finkel'shtein, ed., Consultants Bureau, New York (1963), p. 106.]
7. A.E. Vol, Structure and Properties of Binary Metallic Systems [Russian translation], Izd. Akad. Nauk SSSR (1962), Vol. 2.
8. G. M. Lozinskii, Structure and Properties of Metals and Alloys at High Temperatures, Metallurgizdat (1963).
9. C. M. Zener, Collection: Elasticity and Anelasticity of Metals [Russian translation], IL (1954). [English edition: University of Chicago Press, Chicago, Illinois (1948).]
10. Ke T'ing-sui, Collection: Elasticity and Anelasticity of Metals [Russian translation], IL (1954). [English edition: University of Chicago Press, Chicago, Illinois (1948).]
11. Ya.B. Fridman, High-Quality Steel, No. 1 (1937).
12. P.I. Mel'nichuk, Investigation of the Elasticity Modulus and Hardness of Aging Alloys Based on Copper, Kiev (1957).
13. R.B. Gershman, V.E. Kochnov, S.M. Vasil'eva, and V.A. Zvereva, Collection: Theory and Practice of Metallurgy, No. 5, Chelyabinsk, NIIM (1963).
14. V.S. Postnikov, Collection: Cyclic Strength of Metals, Izd.Akad.Nauk SSSR (1962), p.207.
15. V.T. Shmatov and A.V. Grin', Fiz.Metal.i Metalloved., 12(4):600 (1961).

CHARACTERISTICS OF THE TEMPERATURE DEPENDENCE OF THE INTERNAL FRICTION OF HARDENED IRON AND ALLOYED STEEL

S. A. Golovin, V. D. Korvachev, and A. N. Titov

A study was made of the temperature dependence of the internal friction, $Q^{-1}(T)$, of iron containing 0.04 wt.% C and N, and of the 15GSKhMFR steel of the following compositi   .17 0.17% C, 1.39% Mn, 0.71% Si, 0.026% S, 0.024% P, 0.1% Cr, 0.25% Mo, 0.09% V, 0.022% B. After quenching and tempering this steel at 650°C, the tensile strength was 75.2 kg/mm^2 and the relative elongation was 18.8%.

The measurements of $Q^{-1}(T)$ were carried out using a vacuum relaxator (torsional pendulum) of the RKF MIS type at a frequency of about 1 cps. The sample length was 140 mm, and the diameter was 0.7-0.8 mm. The deformation amplitude did not exceed 10^{-5}. The tensile strength was determined using standard wire samples, treated in the same way, and was measured in the relaxator. Next, we found the increase in the strength of each sample, which was defined as the difference between the tensile strength of the investigated sample and an annealed sample.

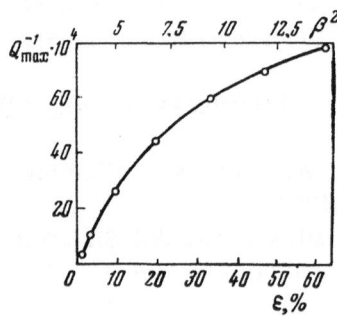

Fig. 1. Relationship between the square of the physical line broadening in plastically deformed iron, containing 0.04 wt.% C and N, and the corresponding amplitude of the 200°C internal friction maximum.

It has been shown [1-3] that the x-ray diffraction method makes it possible to estimate the density of dislocations from the change in the square of the physical line broadening β^2, where β is the physical broadening of the $(110)_\alpha$ line, expressed in radians ($\times 10^{-3}$) and determined by the method of Kurdyumov and Lysak [4]. The measurements were carried out using apparatus of the URS-50IM type with an ionization counter.

Armco iron was annealed and then quenched from 820°C in water. The annealing and heating before quenching were carried out in vacuum ($\sim 10^{-3}$ mm Hg). After this heat treatment, iron was deformed by drawing involving compression up to 62%. As the degree of cold working increased, a 200°C maximum,

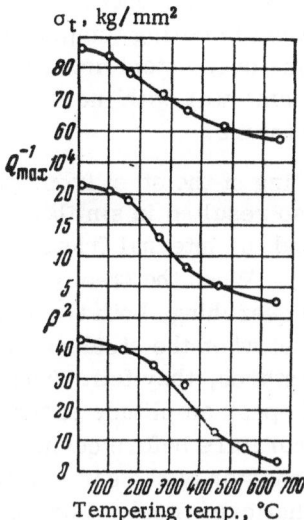

Fig. 2. Changes in tensile strength, amplitude of the 200°C damping maximum, and square of the physical line broadening during tempering of the steel 15GSKhMFR quenched from 930°C.

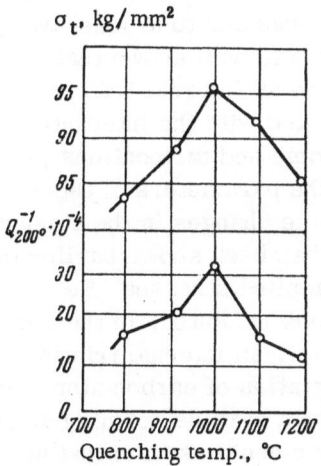

Fig. 3. Dependence of the tensile strength and the amplitude of the 200°C damping maximum on the temperature to which the steel 15GSKhMFR was heated before quenching.

reported earlier in [5, 9] appeared (and increased) in the temperature dependence of the internal friction. The measurements were carried out after prolonged storage at room temperature.

Figure 1 shows the variation of β^2 and Q_{max}^{-1} with the deformation of Armco iron. As the degree of cold working increased, the nature of the relationship between the square of the line broadening and the amplitude of the 200°C peak changed, and the relationship became nonlinear.

The variation of the amplitude of the 200°C maximum of the damping in iron was described by the formula

$$pQ_{max}^{-1} = \varepsilon^{1/2}, \qquad (1)$$

where ε is the relative value of the degree of plastic deformation; p is a coefficient equal to about 10^2.

Consideration of the simplest model of hardening in the process of plastic deformation, due to the generation of dislocation sources, gave the dependence

$$d \cong \varepsilon^{1/2}, \qquad (2)$$

where d is an increment of the dislocation density; ε is the plastic deformation introduced by Frank−Read sources.

Comparison of the dependences (1) and (2) made it possible to draw conclusions about the analogous variation of d and Q_{max}^{-1} during plastic deformation, which once again stressed the relationship between the parameters of the 200°C internal friction maximum on the one hand, and the density of dislocations and the strength of the steel on the other. In the case considered, the line broadening could be related to changes in the microstresses and dimensions of mosaic blocks, making different contributions to the hardening of iron. The amplitude of the 200°C peak was mainly due to the internal grain hardening. This explained the different nature of the variation of β^2 and Q_{max}^{-1} during plastic deformation.

We shall consider some results on the hardening treatment of the alloyed steel. The $Q^{-1}(T)$ curve of the 15GSKhMFR steel after quenching and tempering exhibited a broad maximum at temperatures of 170-220°C. The curves in Fig. 2 show changes in the tensile strength, the amplitude of the 200°C maximum of $Q^{-1}(T)$, and of the square of the physical line broadening,

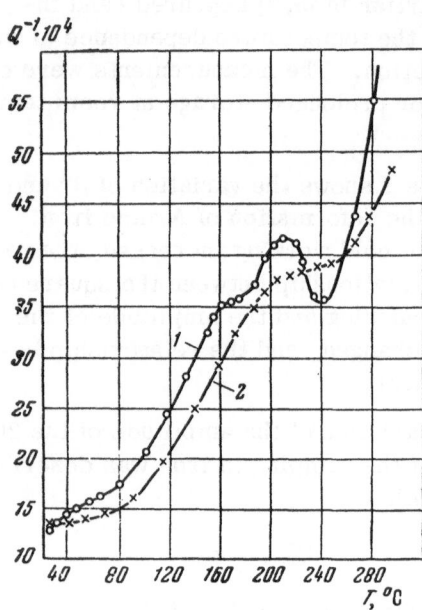

Fig. 4. Temperature dependence of the internal friction of the steel 15GSKhMFR plastically deformed by 92% (1) and of the steel containing 0.12 wt.% C deformed by 78% (2).

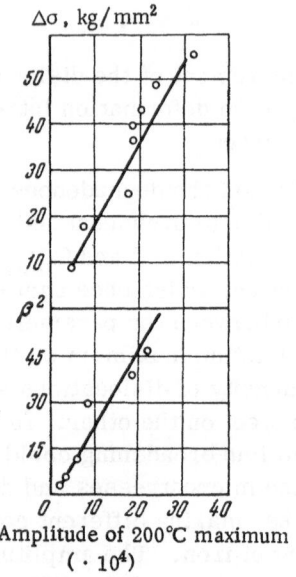

Fig. 5. Relationship between the increase in strength, the square of physical line broadening of hardened steel 15GSKhMFR, and the corresponding amplitude of the 200°C internal friction maximum.

for the steel 15GSKhMFR quenched from 930°C, as a function of the tempering temperature. All the characteristics of the steel decreased basically in the same way with increase in the tempering temperature.

The quenching of the steel from various temperatures also resulted in similar changes in the strength and the internal friction peak amplitude (Fig. 3). When the steel was heated to 1000°C before quenching, both these quantities increased, but when the steel was heated to higher temperatures, they fell. The dependence of the 200°C-peak amplitude of the alloyed steel on the temperature before quenching differed from the behavior of the same quantity in the case of the unalloyed steel. For the carbon steel, treated in the same way, the strength and the 200°C maximum were enhanced by heating to 1000-1100°C, and above this temperature stabilization was observed.

Analysis of the width of the relaxation maximum of the internal friction of the heat-treated alloyed steel 15GSKhMFR showed that this maximum was due to at least two relaxation times. It was well known that the hardened steel could have an internal friction peak at 200°C, associated with the interaction between interstitial atoms and dislocations [5]. For the carbon steel the parameters of this maximum matched well the changes in the strength [6]. Moreover, it has been shown earlier that, under the action of applied stresses, $Fe-C-Mo$ and $Fe-C-Cr$ alloys exhibited, in the temperature range 170-180°C, an internal friction maximum due to the migration of carbon atoms along positions close to a carbide-forming alloying element [7, 8]. We could assume that at these temperatures both relaxation processes would take place in the alloyed steel.

Figure 4 gives the temperature dependence of the internal friction of the 15GSKhMFR steel (curve 1), which was tempered at 650°C and plastically deformed by drawing involving 92% compression. For comparison, the same figure includes the analogous dependence for the carbon steel containing 0.12 wt.% C, after it was tempered at 650°C and deformed by drawing which involved 78% compression (curve 2). These experimental data showed that the alloyed steel had two internal friction maxima: at low temperatures (~160°C) and at high temperatures (~210°C). In our opinion, the former was associated with the migration of carbon atoms in ferrite near carbide-forming elements, and the latter was due to the interaction between interstitial atoms and dislocations in the stress fields.

Obviously the relaxation processes caused by the migration of carbon in ferrite near Cr, Mo, and V atoms under the action of stresses would affect the main (dislocation) part of the maximum, which was related to the strength of the steel. During heating before quenching the solubility of carbides increased and the number of carbon-alloying element positions in the solid solution increased, while the relative saturation of dislocation lines with carbon decreased. At high temperatures the strength and relaxation were also strongly affected by the formation of superequilibrium vacancies.

We analyzed, for the hardened steel 15GSKhMFR, the contributions of microstresses and of mosaic block dimensions to the broadening of the interference maximum line. The determination of the microstresses of type II was carried out using the $(110)_\alpha$ and $(220)_\alpha$ lines, employing the method described in [4]. It was found that the broadening was associated mainly with the microstresses and the contribution of the mosaic structure to the hardening was slight. This was responsible for the fact that, irrespective of the mode of hardening (heat treatment, plastic deformation), the alloyed steel 15GSKhMFR exhibited a satisfactory linear correspondence between the increase in the strength ($\Delta\sigma$) and β^2, on the one hand, and Q_{max}^{-1}, on the other (Fig. 5). The results reported in the present paper indicate that the main contribution to the maximum was made by the interaction between dislocations and impurity atoms.

Literature Cited

1. L.I. Mirkin and Ya.S. Umanskii, Izv. Vysshikh Uchebn. Zavedenii, Fiz., No.3 (1960).
2. G.K. Williamson and R.E. Smallman, Acta Cryst., 7:574 (1954).
3. P.B. Hirsch, Progress in Metal Physics [Russian translation] (1960), Vol. 3, p. 283. [English edition: B. Chalmers and R. King, eds., 8 vols., Pergamon Press, Inc., New York.]
4. G.V. Kurdyumov and L.I. Lysak, Zh. Tekhn. Fiz., 17:993 (1947).
5. W. Köster, L. Bangert, and R. Hahn, Arch. Eisenhüttenw., 25(3):569 (1954).
6. M.A. Krishtal, Yu.V. Piguzov, and S.A. Golovin, Internal Friction in Metals and Alloys, Izd. "Metallurgiya' (1964).
7. M.A. Krishtal, Fiz.Metal.i Metalloved., No. 5:10 (1960).
8. M.A. Krishtal and V.I. Baranova, Fiz.Metal.i Metalloved., No. 5:12 (1961).
9. M.L. Bernshtein and E.S. Tikhomirova, Collection: Relaxation Phenomena in Metals and Alloys, Metallurgizdat (1960), p. 279. [English edition: B.N. Finkel'shtein, ed., Consultants Bureau, New York (1963), p. 211.]

INFLUENCE OF REPEATED THERMO-MECHANICAL TREATMENT ON THE TEMPERATURE DEPENDENCE OF THE INTERNAL FRICTION OF IRON

L. K. Gordienko

The present author, together with Oding and Ivanova [1], has proposed a new method for increasing the strength of bcc metals by repeated thermomechanical treatment (RTMT). The increase of the strength is achieved by repeatedly loading the finished product up to the stage of permanent deformation, corresponding to the end of the yield plateau, alternating with aging (between deformation cycles) at temperatures below the recrystallization stage (Fig. 1).

Metallographic investigations carried out on iron of technical grade and on low-carbon steel [2] have shown that RTMT led to the establishment of uniformly distributed (in the volume being strengthened) planar dislocation clusters near grain boundaries, in which the density was very high even when the total deformation was still relatively small.

Increase of the number of RTMT cycles caused growth of "ring" clusters, which extended over an increasing fraction of a grain, and this was accompanied by a sharp increase in the strength (Fig. 2). The blocking of dislocations as a result of static deformation during aging [3] stabilized this high-strength state; the thermal stability of the imperfections so produced increased by this treatment, and the strength of the metal subjected to cyclic loading, as well as its resistance to creep, increased [1-2].

In the investigation reported here we measured temperature dependence of the internal friction to determine the structural changes, which were responsible for the increase in the strength due to RTMT. The investigated materials were iron of technical grade (0.045% C) and low-carbon steel (steel 3); they were annealed in vacuum before the treatment; the annealing and aging conditions (the latter used in the alternate stages of the RTMT) are given in Table 1.

Table 1

Material	Annealing conditions	Aging conditions in RTMT
Technical grade iron	Heating to 950°C, 3 h at this temperature, cooling with furnace	Heating to 150°C, 5 h at this temperature, cooling in air
Steel 3	Heating to 900°C, 6 h at this temperature, cooling with furnace	Heating to 100°C, 2 h at this temperature, cooling in air

140

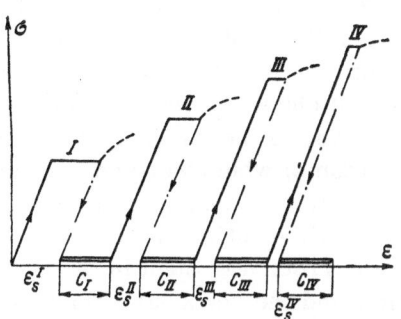

Fig. 1. Schematic representation of successive "deformation + aging" cycles during RTMT. ε_s^I, ε_s^{II}, ε_s^{III}, ε_s^{IV} — plastic deformation corresponding to the end of the yield plateau after I, II, III, IV cycles on RTMT; C_I, C_{II}, C_{III}, C_{IV} — successive stages of aging of samples in unloaded state between two consecutive deformation cycles.

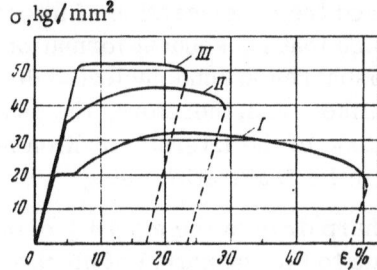

Fig. 2. Effect of the RTMT on the nature of the elongation curves of technical-grade iron samples. I) Aged state; II) after 4 RTMT cycles; III) after 5 RTMT cycles.

The deformation of samples in the RTMT and the subsequent elongation, in the determination of the mechanical properties of the strengthened control samples, were carried out using an IM4-R machine at room temperature.* The rate of deformation was 2.5 mm/min.

The measurements of the internal friction were carried out using cylindrical rods of 5-mm diameter and 100 mm long, prepared from the same samples which were used to measure the elongation, but the threaded ends were cut off. The temperature dependence of the internal friction between 20 and 450°C was determined using a modernized apparatus of the UIMD-2 type [4] in 10^{-5} mm Hg vacuum, employing a method described earlier [5].

The frequency of the natural transverse vibrations of the investigated samples at room temperature was 2200-2300 cps. The logarithmic decrement of vibrations was used as the measure of the internal friction.

Figure 3 shows the dependence, on the number of RTMT cycles, of the amplitude of the relaxation peak of the internal friction of iron, δ_{max}, found in the region of 150°C [5] and corresponding to the Snoek damping [6]. The same figure includes the dependence on the number of RTMT cycles of the internal-friction background, corresponding to the temperature of the Snoek peak. A sharp drop in the peak amplitude was observed after the first two cycles of the RTMT, and then the amplitude of the peak remained constant. The value of the internal-friction background continued to decrease with the number of RTMT cycles, although the peak amplitude remained constant.

Figure 4 shows the temperature dependences of the internal friction of steel 3 in the initial (annealed) state and after various numbers of the RTMT cycles. These curves show, in addition to the Snoek peak at 120°C, a second maximum in the region of 330°C, which appeared after the first RTMT cycle, and whose amplitude increased considerably with the degree of deformation (i.e., with the number of RTMT cycles). The presence of this 330°C peak was the consequence of the Köster damping [7], whose mechanism was related to the migration of carbon atoms in the stress fields of dislocations. The dependence of the amplitudes of the first ($\delta_{max\,1}$) and second ($\delta_{max\,2}$) relaxation peaks on the number of the RTMT cycles, as well as the amplitudes of the maxima from which the corresponding internal friction background was subtracted ($\Delta_1 = \delta_{max\,1} - \delta_b$ and $\Delta_2 = \delta_{max\,2} - \delta_b$) are shown in Fig. 5.

---

*The strengthening and the determining of the mechanical properties of steel 3 were carried out by P.V. Zubarev.

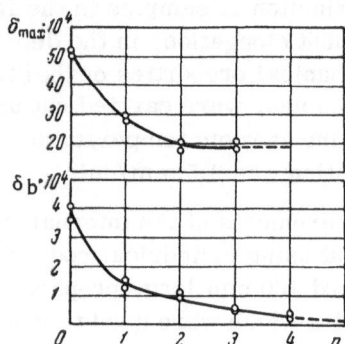

Fig. 3. Dependence of the Snoek peak amplitude (δ_{max}) and of the internal-friction background (δ_b) of iron on the number of the RTMT cycles (n).

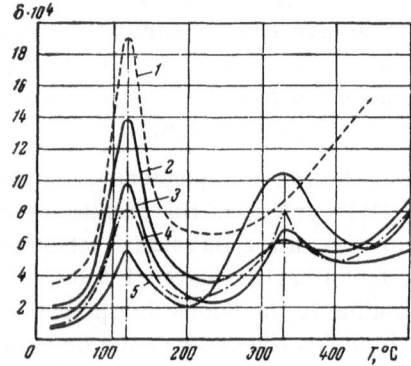

Fig. 4. Temperature dependence of the internal friction of steel 3. 1) Initial (annealed) state; 2, 3, 4, 5) after 1, 2, 3, and 4 RTMT cycles, respectively.

It was established earlier [5] that in the case of iron and steel 10 the reduction of the Snoek peak amplitude after a single deformation of the samples in the yield plateau region was caused by a change in the concentration, in the solid solution, of interstitial atoms (primarily carbon atoms) which were not associated with dislocations. The decisive factors were the unpinning of dislocations from the elastic atmospheres surrounding them in the first stage of plastic flow and the secondary blocking of dislocation loops after the plastic flow stage. Repetition of this process, alternated with the deformation aging between the loading stages, would cause a continuous reduction of the concentration of the interstitial solid solution and, therefore, a reduction of the amplitude of the Snoek peak with increase of the number of the RTMT cycles, as confirmed experimentally (Fig. 3).

It should be mentioned that in the case of technical-grade iron, the first two RTMT cycles reduced the concentration of interstitial atoms so much that subsequent formation of elastic atmospheres around nucleated dislocations was unlikely and, therefore, the changes in the amplitude of the internal-friction maximum reached the "saturation" stage.

The sharp drop in the mobility of dislocation loops due to the formation of blocked dislocation clusters and to a reduction in the number of mobile dislocations were indicated by the strong fall of the internal friction background after the first few RTMT cycles. However, the magnitude of the background, in contrast to δ_{max}, did not exhibit saturation when the number of cycles increased, and this could be regarded as an indirect proof of the continuous increase in the density of pinned dislocations even when the concentration of the solid solution reached its minimum value. The blocking of the nucleated dislocations in this case was not due to the formation of atmospheres, but due to the mutual superposition of elastic stress fields in the case of a high defect density and due to the interaction of dislocations with a dislocation "forest."

The same behavior of the Snoek peak amplitude and of the internal-friction background was found for steel 3 (Fig. 5), but in this case the fall of the value of $\delta_{max\,1}$ did not reach a constant value even after four RTMT cycles. This was expected because, in the initial state, the concentration of carbon in this steel was almost 4 times higher than in the investigated iron.

The variation of the value of $\delta_{max\,1}$ could be used for the qualitative estimate of the fraction of dislocation clusters near grain boundaries in the total volume of a grain. The amplitude of the Snoek peak, proportional to the concentration of carbon in ferrite, depended on the grain size, when carbon content was constant [8]. It has also been reported [9] that the reduction of

Table 2

Treatment	Total degree of deform. in RTMT	$\Delta_2 \cdot 10^4$	Mechanical properties of steel 3				
			σ_s, kg/mm^2	σ_b, kg/mm^2	ε_s, %	δ, %	ψ, %
Annealing	0	—	17	35	2.0	22	67
1 cycle RTMT	2.0	1.9	29	38	1.1	15	59
2 cycles »	3.5	3.4	35	39.5	1.3	13	62
3 » »	4.8	4.5	37.5	41	1.0	12	61
4 » »	6.1	6.6	43.5*	43.5	—	6.5	58

*The value of $\sigma_{0.2}$ is given.

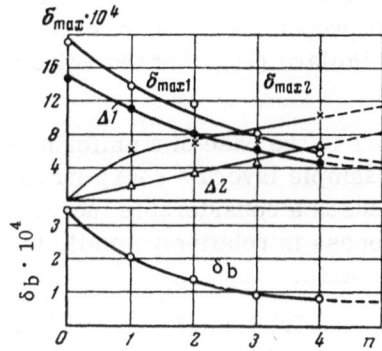

Fig. 5. Dependence of the Snoek peak amplitude ($\delta_{max\,1}$, Δ_1), the Köster peak amplitude ($\delta_{max\,2}$, Δ_2), and of the internal friction background (δ_b) of steel 3 on the number of RTMT cycles (n).

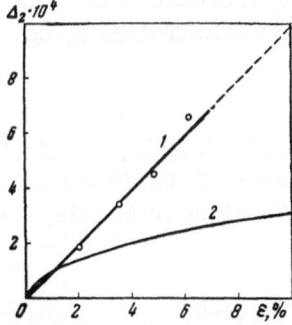

Fig. 6. Dependence of the amplitude of the Köster peak on the degree of plastic deformation. 1) Experimental dependence of Δ_2 on the total deformation during the RTMT process (steel 3); 2) curve representing the dependence $\Delta = A\varepsilon^{1/2}$.

the amplitude of this peak with decrease of the grain size was associated with the influence of grain boundaries, because carbon atoms in the strongly distorted zones near grain boundaries did not take part in the relaxation process. It has been suggested* that when the grain size was very small, the peak could disappear altogether because of the closure of boundary zones inside the grain. This was confirmed experimentally for iron with 0.04% C, in which the grain dimensions were reduced to 17 μ. A similar effect was also observed during the gradual growth of the dislocation clusters near grain boundaries generated by the RTMT. Obviously, the disappearance of the peak could correspond to the moment when, with the increase in the number of the RTMT cycles, the clusters at which carbon atoms were segregated spread over practically the whole grain volume. Simultaneously, the yield plateau should vanish [1].

Analysis of the variation of the Köster peak amplitude was of considerable interest in the determination of structural changes during the RTMT. With decrease of the number of RTMT cycles, the value of $\delta_{max\,2}$ decreased monotonically and the dependence of $\Delta_2 = \delta_{max\,2} - \delta_b$ on the number of cycles n was linear (Fig. 5). It is known that the Köster damping appears only in a plastically deformed metal and, although the elementary process causing this type of damping is not yet known in detail, we may assume that the internal-friction maximum is due to the interaction between interstitial dissolved atoms and defects produced by plastic deformation [10]. The increase in the amplitude of the maximum reflects directly the in-

*Bayazitov, Kidin, and Piguzov, Solubility of Carbon in α-Iron, present collection, p. 89.

crease in the dislocation density with increase of the degree of plastic deformation; the dependence of the Köster damping Δ on the degree of plastic deformation ε is given by the relationship

$$\Delta = A\varepsilon^{1/4}. \tag{1}$$

However, in the case investigated by us, the increase in the amplitude of the Köster peak with increase of the degree of the total plastic deformation was much more rapid (Fig. 6) and, when the increase of the peak amplitude was expressed as a function of the total plastic deformation, it was given by the equation

$$\Delta_2 = A\varepsilon, \tag{2}$$

where A was 10^{-2}.

These data indicate that when the number of RTMT cycles is increased, the total density of dislocations in the metal being strengthened increases at a higher rate than is the case when the same degree of plastic deformation is produced by a single deformation process. This is confirmed by the results of the determination of the mechanical properties of the steel 3 after the RTMT (Table 2) and is in agreement with the hypothesis of Pollard [11], that the aging intensifies the rise in the dislocation density.

The strong rise in the dislocation density during the RTMT in the case of a uniform distribution of the defects generated throughout the volume of the sample favors a sharp increase in the total energy content of the metal [12] and consequently causes a considerable increase in the strength. Since the degree of deformation in the RTMT process is relatively small, the metal retains a fully satisfactory plasticity in its high-strength state.

Thus, the measurement of the relaxation peaks of the internal friction of iron and low-carbon steel gives information on the following processes taking place in bcc metals during the RTMT:

a) the blocking of dislocations after the plastic flow stage and the formation of clusters of pinned dislocations near grain boundaries;

b) the growth of the dislocation clusters near grain boundaries which gradually extend over the major part of a grain as the number of the RTMT cycles increases;

c) the accelerated (compared with the normal cold working) increase of the total dislocation density due to alternation of the first stage of plastic flow with deformation aging.

Literature Cited

1. I.A. Oding, V.S. Ivanova, and L.K. Gordienko, Dokl. Akad. Nauk SSSR, 160(2):321 (1965).
2. V.S. Ivanova, L.K. Gordienko, V.N. Geminov, P.V. Zubarev, Z.G. Fridman, Yu.P. Liberov, et al., Role of Dislocations in the Hardening and Fracture of Metals, Izd. "Nauka" (1965).
3. J.D. Baird, Iron Steel (London), 36(10):450 (1963).
4. I.G. Polotskii and V.F. Taborov, Zavodsk. Lab., 23(8):986 (1957).
5. L.K. Gordienko, Collection: Relaxation Phenomena in Metals and Alloys, Metallurgizdat (1963), p. 263.
6. J. Snoek, Physica, 8:711 (1941).
7. W. Köster, L. Bangert, and R. Hahn, Arch. Eisenhüttenw., 25(11-12):569 (1954).
8. H.J. Seeman and W. Dickenscheid, Acta Met., 6(1):62 (1958).
9. G. Lagerberg and A. Josefsson, Acta Met., 3(3):236 (1955).
10. H.G. van Bueren, Imperfections in Crystals [Russian translation], IL (1962). [English edition: 2nd ed. (Interscience), Wiley, New York.]

11. P.M. Kelly and J. Natting, High-Strength Steels (Iron and Steel Institute, May 7, 1962), Special Report 76.
12. V.S. Ivanova and L.K. Gordienko, New Methods of Increasing the Strength of Metals, Izd. "Nauka" (1964).

INFLUENCE OF DEFORMATION AND TEMPERING ON THE INTERNAL FRICTION O PATENTED WIRE

S. V. Grachev and V. Ya. Zubov

An investigation was made of the influence of the carbon content and tempering temperature on the internal friction of patented cold-drawn wire containing from 0.20 to 0.80% carbon (see table).

The heat treatment and the drawing of the wire (to reduce its diameter to 0.7 mm) were carried out at the Beloretsk Steel Wire Factory.* The final patenting of the wire (1.4-mm diameter) was carried out at the austenization temperature of 910-930°C and at the cooling bath temperature, which depended on the carbon content: 470-480°C for steel 20; 500-510°C for steel 50; and 520-530°C for steel U8A. Thus, the degree of deformation of the wire of 0.7-mm diameter was 75%. The cold-drawn wire was subjected to tempering in the temperature range 150-550°C in steps of 50°C, 1 h at each step. The tempered wire was tested to find its tensile strength and the elastic limit in torsion.

The internal friction of the wire was determined by the method of low-frequency small-amplitude torsional vibrations [1]. The apparatus included photocells and counters, by means of which the number of vibrations of the sample could be recorded automatically. The vibration frequency of the sample was 2.08 cps.

Figure 1 shows the influence of the tempering temperature on the tensile strength σ_b and on the elastic limit σ_e of the cold-drawn wire. When the tempering temperature was increased from 200 to 300°C, a considerable increase of σ_b, and particularly of σ_e, was observed. The maximum value of the elastic limit was obtained at somewhat higher tempering temperatures than the maximum value of the tensile strength. The absolute rise in the tensile strength and

Chemical Composition of Wire (wt.%)

Steel designation	C	Mn	Si	Cr	Ni	Cu	S	P
Steel 20	0.19	0.47	0.22	0.09	0.11	0.12	0.022	0.020
Steel 50	0.53	0.48	0.26	0.09	0.08	0.08	0.029	0.019
U8A	0.76	0.22	0.23	0.07	0.09	0.08	0.020	0.014

---

*L. P. Makarova took part in the experimental investigation.

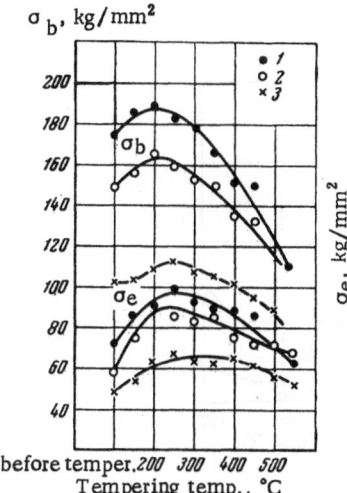

Fig. 1. Influence of the tempering temperature on the tensile strength σ_b and on the elastic limit σ_e of patented cold-drawn wire: 1) steel U8A; 2) steel 50; 3) steel 20.

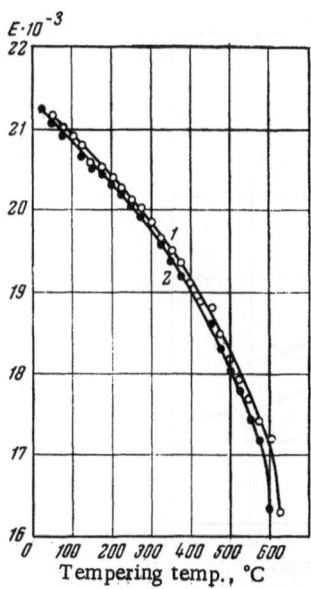

Fig. 2. Temperature dependence of Young's modulus of the U8A wire (4.0-mm diameter). 1) Patented wire; 2) cold-drawn wire (degree of deformation 55%).

in the elastic limit increased with increase of the carbon content in the wire. The increase in the strength and the elastic limit of the cold-drawn wire after tempering at 200-300°C was due to the development of deformation aging processes, consisting of the formation of impurity atom clouds around dislocations (Cottrell and Suzuki atmospheres) and partial precipitation of carbides, etc. (which caused the blocking of dislocations as well as an increase in the critical shearing stress). Higher tempering temperatures caused unpinning as well as partial annihilation of the dislocations, due to increase of the diffusion mobility of atoms. These processes reduced the strength of the cold-drawn wire.

It should be mentioned that Young's modulus of the cold-drawn wire was somewhat smaller than that of the patented wire (Fig. 2), which was explained, first of all, by the participation of pinned dislocation loops in the reversible deformation [2]. Young's modulus was in this case determined using samples of 4.0-mm diameter, employing the dynamic method and the formula

$$E = 1.6388 \cdot 10^{-8} \left(\frac{l}{d}\right)^4 \frac{P}{l} \nu^2, \qquad (1)$$

where E is Young's modulus (kg/mm²), l is the sample length (cm), d is the sample diameter (cm), P is the sample weight (g), ν is the vibration frequency (cps).

The influence of the carbon content of the cold-drawn wire on the value of the logarithmic decrement, corresponding to a peak in the region 40°C, is shown in Fig. 3a and where it is represented graphically by a straight line. It was interesting to investigate the influence of the tempering temperature on the internal friction peak in the region of 40°C (Fig. 3b). Up to tempering temperatures of 200-300°C, we observed a considerable fall in the value of the logarithmic decrement. Further increase of the tempering temperature increased the decrement again. The internal friction peak in the region of 40°C is usually ascribed to the presence of impurity atoms in the lattice of α-iron [3]. On the basis of this assumption, the fall in the decrement on tempering of the wire at 200-300°C was due to the formation of impurity atom clouds around dislocations and reduction of the concentrations of these atoms elsewhere. A strong rise of the tensile strength and of the elastic limit was observed in the same range of temperatures. Further increase of the tempering temperature could cause partial dispersal of the impurity atom clouds around dislocations due to such processes as, for example, annihilation or climb of dislocations.

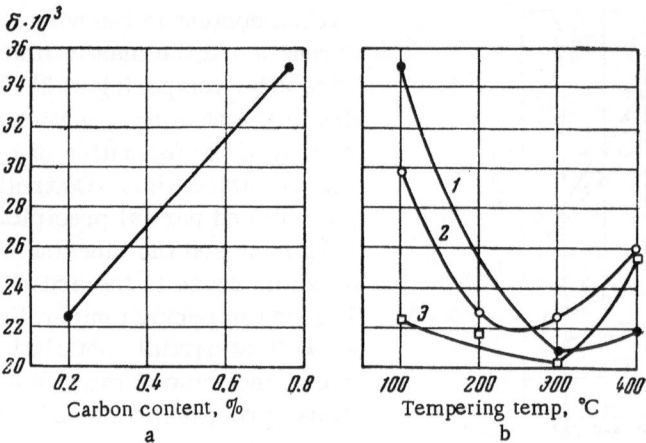

Fig. 3. Influence of the carbon content and of the tempering temperature on the amplitude of the internal friction peak in the region of 40°C for cold-drawn wire. a) Influence of carbon content; b) influence of tempering temperature. 1) Steel U8A; 2) steel 50; 3) steel 20.

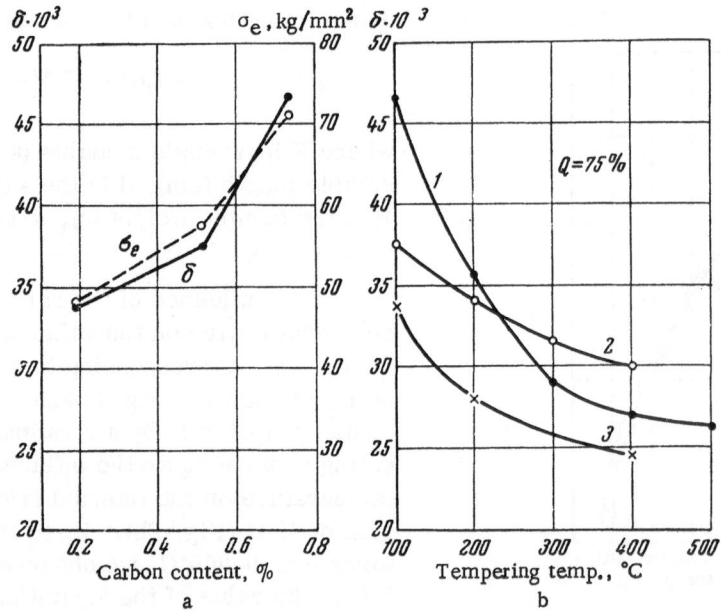

Fig. 4. Influence of the carbon content and of the tempering temperature on the amplitude of the internal friction peak in the region 200°C for cold-drawn wire. a) Influence of carbon content; b) influence of tempering temperature. 1) Steel U8A; 2) steel 50; 3) steel 20.

The influence of carbon in the cold-drawn wire on the position of the internal friction peak in the region of 200°C is shown in Fig. 4a. The same figure includes, for comparison, a dashed line representing the dependence, on the carbon content, of the elastic limit found in torsion of the cold-drawn wire. It should be mentioned that for the patented cold-drawn wire this peak was in the region of 190-220°C and was fairly broad. The internal friction peak in the region of 200°C has been ascribed by the majority of investigators to the presence of dislocations in an α-solution [3, 4, 5]. The good agreement between the influence of carbon on the elastic limit and on the value of the logarithmic decrement, corresponding to the internal friction maximum of the cold-drawn patented wire, also supported the dislocation nature of the internal friction peak in the region of 200°C. The influence of the tempering temperature on the value of the logarithmic decrement at the 200°C peak is shown in Fig. 4b. Increase of the tempering temperature produced a continuous fall of the amplitude of the 200°C internal friction peak, which was in good agreement with the hypothesis of the dislocation nature of this maximum and was due to the processes of recovery and polygonization.

We carried out an approximate estimate of the activation energy of the relaxation process near 200°C. The activation energy was determined from the formula

$$H = 2.3R \, \frac{\lg (f_2/f_1)}{\dfrac{1}{T_1} - \dfrac{1}{T_2}} , \tag{2}$$

where H is the activation energy (cal/mol), R is the universal gas constant (cal $\cdot$ mole$^{-1} \cdot$ deg^{-1}), f_1 and f_2 are the frequencies in two measurements; T_1 and T_2 are the corresponding absolute temperatures.

The measurements were carried out on samples of the steel U8A at the frequencies of 3.5 and 2.05 cps. The activation energy was found to be 25,000 cal/mole. This value of the activation energy was not very accurate because the internal-friction peaks were fairly broad. It should be mentioned that the published data also indicated a considerable scatter of the activation energy values of the relaxation process in the region of 200°C. Thus, Ke [1] mentioned Snoek's data on the activation energy on the 200°C peak, which had a scatter of the values from 24,000 to 40,000 cal/mole. Evidently, there were two closely spaced internal-friction peaks in the region of 200°C. For example, a wire which was deformed but not tempered showed a splitting of the peak into two in the temperature range 180-250°C.

Literature Cited

1.   Ke T'ing-sui, Collection: Elasticity and Anelasticity of Metals [Russian translation], IL (1954), p. 212. [English edition: University of Chicago Press, Chicago, Illinois (1948).]
2.   A.S. Nowick, Collection: Creep and Recovery [Russian translation], Metallurgizdat (1961), p. 166. [English edition: R. Maddin, ed., American Society of Metals, Novelty, Ohio.]
3.   M.L. Bernshtein and E.S. Tikhomirova, Collection: Relaxation Phenomena in Metals and Alloys, Metallurgizdat (1960), p. 279. [English edition: B.N. Finkel'shtein, ed., Consultants Bureau (1963), p. 211.]
4.   Yu.V. Piguzov and M.L. Bernshtein, Collection: Relaxation Phenomena in Metals and Alloys, Metallurgizdat (1963), p. 85.
5.   P.M. Robinson and R. Rawlings, Collection: Internal Friction in Metals [Russian translation], Metallurgizdat (1963), p. 89.

RELATIONSHIP BETWEEN COLD BRITTLENESS
AND THE BEHAVIOR OF CARBON IN IRON
IN THE PRESENCE OF MANGANESE

I. A. Azizov, E. S. Nosyreva, and K. V. Popov

Attempts have been made to relate the cold brittleness (the tendency of a metal to embrittlement at low temperatures) to the internal friction [1, 2]. Analyses of the tendency to embrittlement have been based on the assumption of the development and the dispersal of local regions in a volume-stressed state. The rate of relief of local stresses increases as the dislocation activation increases, i.e., with decrease of the tendency of dislocations to be pinned by impurity atoms.

Since the Snoek peak depends on the behavior of impurity atoms, changes in the cold brittleness of a number of alloys can be investigated by observing changes in this peak.

The reduction in the amplitude of the Snoek peak in the presence of manganese [3, 4] cannot be explained simply by a reduction in the solubility of carbon in the α-iron lattice. It has been shown [5] that the coefficient of proportionality (the "P" factor) between the value of the internal friction and the actual amount of carbon dissolved in a solid solution varies with the manganese content. Thus, for a manganese-free steel this coefficient is equal to 1.3, but for a steel containing about 2.8% manganese, this coefficient is equal to 6. Hence, we may conclude that the presence of manganese has a "paralyzing" effect on some interstitial atoms, preventing them from participating in the internal friction. If we explain this effect by the resultant local binding forces, then, in the presence of manganese, we would expect some of the carbon atoms located near manganese atoms to differ from other carbon atoms in the relaxation time associated with the jumps of these atoms from one lattice site to another. Such behavior

Alloy No.	Content of elements, %					
	C	Mn	Si	P	N	Cr
1	0.044	0.09	—	—	0.1	—
7	0.048	0.45	0.06	0.027	0.0064	0.09
13	0.05	1.12	—	—	0.086	—
19	0.05	2.3	—	—	0.01	—
6	0.26	0.15	—	—	0.0067	—
11	0.24	0.6	—	—	0.0084	—
18	0.25	1.21	0.19	0.027	0.0095	None
24	0.27	2.38	0.43	0.027	0.012	0.45

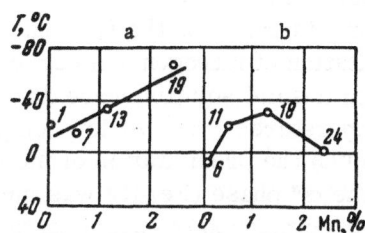

Fig. 1. Dependence of the cold brittleness threshold on the alloy composition.
a) 0.04–0.05% C; b) 0.25–0.27% C. Numbers by the curves indicate the alloy
designations.

should be reflected in the form of the Snoek peak.
This has not been observed in the investigations of
the "carbon" peak, but it has been found in the behavior of the "nitrogen" peak [6, 7].

Using a batch of specially prepared steels,
whose compositions are given in the table, we carried out impact strength tests at various temperatures. The steels were first forged into rods of
14 × 14 mm dimensions and annealed at a temperature 50°C higher than A_{C_3}. This was followed by
cooling in the critical point range at a rate of 100
to 130°/h. The impact strength tests were carried
out on samples of type 1; the cold-brittleness
threshold was defined as the temperature at which
the value of the impact strength fell to 2.5 kg·m·cm^{-2}.

Figure 1 shows the dependence of the cold brittleness threshold on the manganese content. In some steels, containing small amounts of carbon (0.04%), the threshold shifted toward
lower temperatures with increase of the manganese content. When the carbon content was
about 0.25%, manganese had a favorable influence on the cold brittleness only up to 1.2% Mn,
but a strongly unfavorable influence at contents above 2% Mn.

The internal friction was measured, with apparatus described in [8], using samples of
5-mm diameter at frequencies of about 1 and 3 cps. The large dimensions of the samples
made it possible to investigate the internal friction of a steel in the same state in which it was
subjected to the impact tests. The results of the internal friction investigations of a number
of low-carbon steels are given in Fig. 2. Similar curves were found for several alloys containing more carbon. It is evident from Fig. 2 that the Snoek peak was reduced by increase in
the manganese content up to 1.2%. The peak was broadened even by 0.6% manganese. At
1.2% Mn, the peak became asymmetric. It could be represented as a sum of two symmetric
peaks A and B, corresponding to simple relaxation processes (Fig. 3).

The peak B, measured at 3 cps, had a maximum at a temperature of about 52°C, which
corresponded to the temperature of the peak in a manganese-free alloy (Fig. 2, curve 1).
The peak A was found at about 17°C. When the vibration frequency was reduced to 1.3 cps, the
peak A shifted from 17°C to 9°C, and the peak B shifted from 52°C to 42°C. The activation
energy calculated from these data was 17,100 cal/mole for the peak A, and 17,700 cal/mole
for the peak B.

In addition to peaks A and B, the internal-friction curve of the alloy 13 (Fig. 3) had
also a peak C. Because this peak was small and strongly broadened, it was difficult to determine its temperature exactly. At a frequency of about 1 cps, this temperature was about
150°C and at 3 cps it was about 165°C.

The origin of the peak B could be ascribed with a fair certainty to the carbon in the solid
solution state. The temperature of this peak was in agreement with numerous published data.

Nothing definite could be said about the origin of peak A. The suggestion that this peak
was also of carbon origin is not supported by the published data, which indicate that, in the
presence of manganese, the carbon peak does not change its form. The explanation of this peak
by the motion of nitrogen atoms under the action of stresses is also insufficiently justified, because (according to the published data) manganese shifts the nitrogen peak toward high temperatures [10], while in our case peak A occurred at lower temperatures than the nitrogen peak.

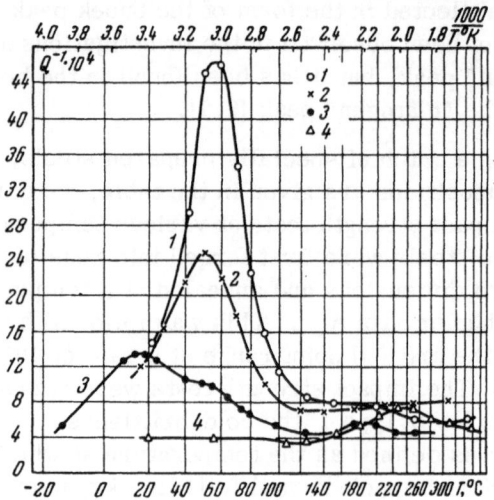

Fig. 2. Temperature dependence of the internal friction of alloys of various compositions: 1) manganese-free alloy; 2) alloy 7; 3) alloy 13; 4) alloy 19.

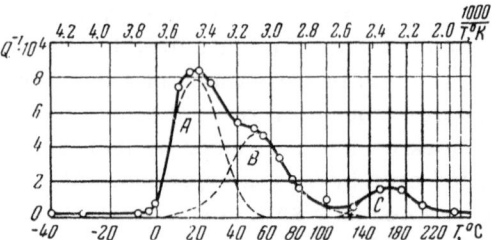

Fig. 3. Internal friction of the alloy 13 in the annealed state.

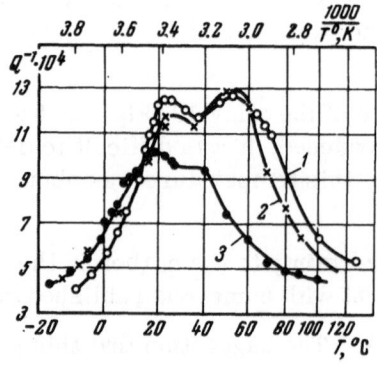

Fig. 4. Internal friction of alloy 13: 1) initial state after cooling in vacuum furnace; 2) after heat treatment at 60°C for 24 h; 3) after heat treatment at 220°C for 24 h.

To determine the nature of peak A, we carried out experiments based on the fact that the rates of precipitation of nitrogen and carbon from a supersaturated α solid solution were different at different temperatures [11]. At low temperatures (below 100°C) the precipitation of nitrogen in the form of the α'' phase ($Fe_{16}N_2$) was much faster than the precipitation of carbon in the form of carbide, while at higher temperatures, for example at 250°C, carbon was precipitated faster than nitrogen. It therefore followed that samples rapidly cooled from high temperatures and held for a long time at two different temperatures (for example, at 60 and 250°C) should, other conditions being equal, exhibit the nitrogen peak in the first case and the carbon peak in the second case, if such peaks were present at all.

To check this, we heated the samples in a vacuum furnace and cooled them at a rate of about 400 deg/h. One sample was kept for 24 h at 60°C, and another was kept for the same period at 220°C. The temperature dependences of the internal friction of samples treated in this way are given in Fig. 4.

It is evident from Fig. 4 that the treatment at 220°C caused considerable changes in the form and amplitude of the peak, but the treatment at 60°C caused no such changes. Hence, we could conclude that both peaks (A and B) were associated with carbon.

The nature of the peak C was not clear. It has been mentioned [9] that quenched steels, alloyed with some elements, have an additional peak at high temperatures (160-200°C), which is ascribed to the migration, in a stress field, of carbon atoms which are in the $Fe-C-Me$ positions. However, the alloys containing 1.27% Mn did not exhibit this peak. If, nevertheless, we were to assume that the peak C was also of carbon origin, then we would find that the solid solution contained several "types" of carbon atoms, differing energetically from one another. Such an effect could not be explained simply by two possible positions of carbon atoms in the iron lattice: $Fe-C-Fe$ or $Fe-C-Me$. Possibly, in such a case, not only the site occupied by carbon atoms in the equilibrium state of the iron lattice, but also the site which carbon might occupy under stress, could become important. Then, we could

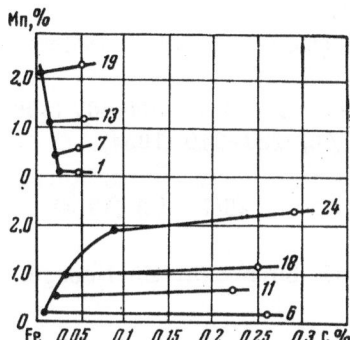

Fig. 5. Variation of the composition of the α-phase (black circles) as a function of the alloy composition (open circles). Numbers by the curves indicate the alloy designations.

distinguish four positions, which would determine the state of carbon. Exchange between these positions could occur in the following ways:

$$Fe/Fe \rightarrow Fe/Fe, \quad Fe/Fe \rightarrow Fe/Me,$$
$$Fe/Me \rightarrow Fe/Fe, \quad Fe/Me \rightarrow Fe/Me.$$

On the basis of these assumptions an increase in the cold brittleness of an alloy with an increase in the manganese content up to 1.2% could be explained by the fact that, in the presence of manganese, the behavior of some interstitial atoms was different. The chemical interaction of these atoms with manganese distributed over the lattice could possibly reduce their blocking ability.

When the manganese content was higher than 2%, the Snoek peak disappeared, and a small peak appeared in the region of 220°C (Fig. 2). The appearance of the latter peak at this temperature is usually ascribed to the presence of dislocations in a solid solution. This peak is found in deformed [12] or quenched [13] metals and is destroyed by annealing. In our case the reason for the appearance of this peak is evidently phase deformation. The alloy with 2% Mn exhibited a characteristic sharp drop in the rate of the $\gamma \rightarrow \alpha$ transformation and, consequently, the rate of cooling during annealing was far too great for the full relief of the stresses after the phase transformation.

Thus, the alloys containing more than 2% manganese had a matrix in the state of phase deformation. However, Fig. 1 shows that even in this state the alloy 19 had a cold brittleness threshold lower than that of the alloys containing less manganese. Consequently, the stresses due to phase deformation, in the degree in which they were present in the matrix of the alloy 19, did not have an unfavorable effect on the cold-brittleness threshold. The internal-friction curve of the alloy 24, containing more than 2% manganese and 0.26% carbon, was similar to the alloy 19. However, the cold-brittleness threshold was shifted toward higher temperatures. Such an effect could be associated with a higher amount of carbon in the matrix of this alloy. As shown in [14], when our method of cooling was used, the solubility limit of carbon in ferrite, found from the carbide analysis, was different for alloys with different carbon contents (Fig. 5).

The fact that the supersaturation of a matrix with carbon did not greatly affect the amplitude of the peak at 220°C could be explained by assuming that in the phase deformed state the dislocation density in the alloys 19 and 24 was the same. In such a case this peak could be stabilized by the amount of carbon present in the α solid solution in the alloy 19.

Literature Cited

1. O. Werner and K. H. Maierhof, Arch. Eisenhüttenw., 34:685-699 (1963).
2. M. A. Krishtal, S. A. Golovin, and V. P. Pudoveeva, Collection: Relaxation Phenomena in Metals and Alloys, Metallurgizdat (1963), p. 120.
3. S. N. Polyakov and K. F. Starodubov, Fiz. Metal. i Metalloved., 6(6):1110-1121 (1958).
4. L. S. Livshits, S. I. Panich, and E. A. Asonova, Fiz. Metal. i Metalloved., 13(4):572 (1962).
5. H. Borchers and W. König, Arch. Eisenhüttenw., 34(6):453-463 (1963).

6. J. Fast, Iron and Coal Trades Rev., 160 : 837 (1950).

7. J. F. Enrietto, Trans. AIME, 224(6) : 1119–1123 (1962).

8. I. A. Azizov, K. V. Popov, and V. F. Vinogradov, Collection: Relaxation Phenomena in Metals and Alloys, Metallurgizdat (1963), p. 72.

9. M. A. Krishtal and V. I. Baranova, Fiz. Metal. i Metalloved., 18(3) : 464–467 (1964).

10. W. Köster and L. Bangert, Arch. Eisenhüttenw., 25(5/6) :231–240 (1954).

11. C. A. Wert, Acta Met., 2(4) : 361–364 (1954).

12. W. Köster, L. Bangert, and R. Hahn, Arch. Eisenhüttenw., 25(11-12) (1954).

13. I. N. Chernikova, Fiz. Metal. i Metalloved., 5(1) : 176.

14. E. S. Nosyreva, L. L. Okhapkina, K. V. Popov, and A. G. Suvorova, Zh. Neorgan. Khim., 9(6) : 1393–96 (1964).

INVESTIGATION OF THE AMPLITUDE DEPENDENCE
OF THE INTERNAL FRICTION OF IRON AT LOW
TEMPERATURES

G. A. Beresnev and V. I. Sarrak

The yield point of metals with the bcc crystal lattice increases strongly on cooling below about 300°K. The resistance to the motion of free dislocations increases to almost the same extent [1, 2]. This phenomenon is important in the practical applications of such metals, since it is responsible for their tendency to brittle fracture (cold brittleness).

In the present investigation the temperature dependence of the resistance to the motion of dislocations in iron was investigated by the method of measuring the amplitude dependence of the internal friction. It was shown in [3, 4] that this method makes it possible to record small displacements of dislocations under relative deformation of 10^{-5}-10^{-4}, and that it is very sensitive to the processes of dislocation pinning by impurity atoms.

We investigated the influence of temperature, plastic deformation, the content of carbon in a solid solution, and of the degree of pinning of dislocations on the amplitude dependence of the internal friction, i.e., on the value of the energy dissipated in the motion of dislocations.

The investigation was carried out using iron of two types: iron A was the usual technical-grade iron containing 0.025% C; iron B contained also 0.35% Ti, which ensured the binding of carbon in titanium carbides (the content of carbon in iron B was 0.025%). Because of the presence of titanium, iron B had practically no carbon in the solid solution state (the temperature dependence of the internal friction did not have a carbon maximum). Earlier investigations have shown that, in titanium-containing iron in the annealed state, the degree of pinning of dislocations by impurity atoms was less than in the usual technical-grade iron and that such iron does not undergo deformation aging [5].

The measurements of the internal friction were carried out using a relaxator of the "inverted torsional pendulum" type [6]. We used wire samples of 0.8-mm diameter. The samples were deformed in the relaxator. Iron, being a ferromagnet, exhibited, in addition to the dislocation damping, also the magnetoelastic damping associated with the motion of domain walls [7, 8]. To suppress the magnetoelastic energy dissipation, the internal friction was measured in a constant longitudinal magnetic field of 325-Oe intensity. The temperature dependence of the energy dissipation due to the motion of dislocations was investigated in the temperature range 120-470°K.

Figures 1 and 2 show the amplitude dependences of the internal friction of iron A and B in the annealed and deformed states at two temperatures. Up to a certain amplitude (ε_{cr}), the

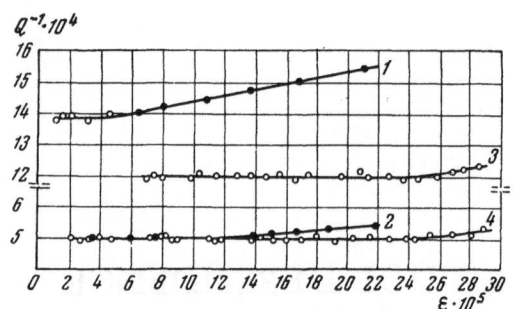

Fig. 1. Amplitude dependence of the internal friction of iron A in the annealed and deformed states at temperatures of +20 and −160°C. 1) 5% deformation, measurements at 20°C; 2) the same, at −160°C; 3) annealing, measurements at +20°C; 4) the same, at −160°C.

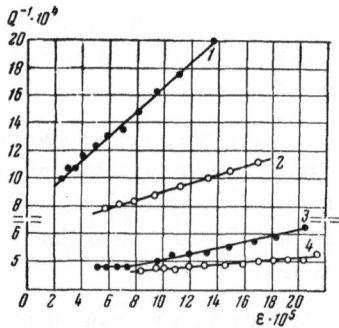

Fig. 2. Amplitude dependence of the internal friction of iron B in the annealed and deformed states at temperatures of +20 and −160°C. 1) 5% deformation, measured at +20°C; 2) the same, at −160°C; 3) annealing, measured at +20°C; 4) the same, at −160°C.

internal friction was independent of the amplitude. When the deformation amplitude was increased above ε_{cr}, the internal friction increased practically linearly with the amplitude. The measurements were carried out in the deformation amplitude range which did not cause any irreversible changes in the internal friction characteristics.

The critical amplitude for iron A in the annealed state reached a very high value, equal to $2.5 \cdot 10^{-4}$. This indicated strong pinning of dislocations in the annealed iron. A 5% deformation reduced ε_{cr} to $0.5 \cdot 10^{-4}$ and considerably increased $\Delta Q^{-1}/\Delta\varepsilon$, due to the generation, by the deformation, of new unpinned dislocations (the measurements on the deformed samples were carried out 30 min after the deformation). Cooling to −160°C reduced the value of $\Delta Q^{-1}/\Delta\varepsilon$.

The value of ε_{cr} for iron B in the annealed state was less ($0.7 \cdot 10^{-4}$) but the value of $\Delta Q^{-1}/\Delta\varepsilon$ was larger, compared with the technical-grade iron A. As already mentioned, this was associated with the decrease in the degree of pinning of dislocations by impurity atoms. Plastic deformation and temperature variation gave rise to the same changes in the amplitude dependence of the internal friction of iron B as in the case of iron A.

The amplitude-dependent component of the internal friction was associated with the dissipation of energy during the motion of dislocations [9]. As is known, the internal friction is equal to the ratio of the energy ΔW dissipated in one cycle to the elastic deformation energy of a cycle W

$$Q^{-1} = \frac{1}{2\pi} \frac{\Delta W}{W}. \tag{1}$$

In view of the relatively slight increase in the internal friction with the amplitude, this expression could be regarded as approximately valid also in the amplitude-dependent region.

This meant that the ratio $\Delta Q^{-1}/\Delta\varepsilon$ was proportional to the value of the energy ΔW dissipated by moving dislocations (when measurements were made in a constant amplitude range):

$$\frac{\Delta Q^{-1}}{\Delta\varepsilon} = \alpha \cdot \Delta W. \tag{2}$$

Consequently, we assumed that the energy ΔW, dissipated during the motion of dislocations, was equal, to within a coefficient proportional to the elastic deformation energy in a cycle $W = G\varepsilon^2/2$ (where G is the shear modulus, ε is the deformation), to the value of the tan-

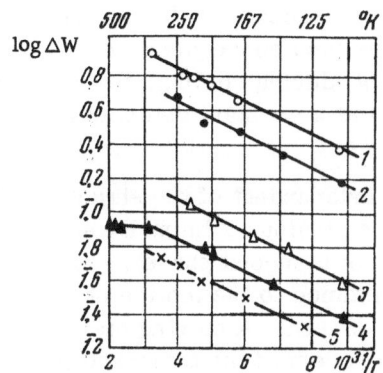

Fig. 3. Temperature dependence of the energy dissipation in the reversible motion of dislocations about their equilibrium positions. 1) Iron B, 5% deformation; 2) iron A ($Q_{max}^{-1} = 15 \cdot 10^{-4}$), 5% deformation at −25°C; 3) iron A ($Q_{max}^{-1} = 48 \cdot 10^{-4}$), 5% deformation at −25°C; 4) iron B, annealing; 5) iron A ($Q_{max}^{-1} = 15 \cdot 10^{-4}$), 5% deformation + aging for 1 h at +60°C.

gent of the angle of the slope of the amplitude dependence of the internal friction $\Delta Q^{-1}/\Delta \varepsilon$.

It should be mentioned that this method of measuring the amplitude dependence of the internal friction makes it possible to determine the energy dissipation in the reversible motion of dislocations about their equilibrium positions at deformations $\stackrel{\sim}{<} 10^{-4}$, i.e., when the dislocation displacements are so small that the dislocations do not intersect in a "forest" and dislocation multiplication does not take place. Figure 3 shows the results of measurements of the amplitude dependence of the internal friction in the form of a graph of the dependence of log ΔW on the reciprocal of temperature. Figure 3 gives the data obtained for iron A and B after annealing, cold plastic deformation, and aging. The results obtained can be described empirically by the equation

$$\Delta W = \Delta W_0 \exp\left(-\frac{H}{kT}\right), \qquad (3)$$

in which the activation energy H, as shown in Fig. 3, is not altered by the plastic deformation and is independent of the pinning of dislocations by impurity atoms or of the concentration of interstitial impurity atoms in the solid solution. In contrast to H, the pre-exponential factor ΔW_0 depends strongly on the structure of iron and is altered by deformation aging. It is evident from Fig. 3 that above about 280°K the energy dissipated in the motion of dislocations in titanium-containing iron in the annealed state became independent of temperature.

Thus, the dissipation of energy by moving dislocations (the value of ΔW) increased with increase of the density of dislocations freshly formed by plastic deformation and decreased due to the pinning of these dislocations by interstitial atoms. However, the activation energy of the process of dislocation motion was independent of such factors as the density and pinning of dislocations and the number of interstitial atoms in the solid solution.

In the experiments on the amplitude dependence of the internal friction we determined the energy dissipated during the motion of dislocations under the action of a constant external stress (represented by the deformation amplitude) during a fixed time interval (represented by the vibration frequency) at various temperatures. Under these experimental conditions, the temperature dependence of the energy dissipated in the motion of dislocations could be explained as follows. At low temperatures, dislocations overcame obstacles by thermal fluctuations. Such a mechanism indicated the existence of a temperature-dependent probability of an elementary act of plastic deformation. Cooling reduced the probability of elementary acts of plastic deformation, i.e., the path traveled by dislocations under the action of constant applied stress (dislocation deformation) became smaller and, consequently, the energy dissipated during the motion of dislocations became less.

In accordance with these representations of the thermally activated motion of dislocations [10], the relationship between the deformation rate $\dot{\varepsilon}$ and the applied stress τ is given by the expression

$$\dot{\varepsilon} = \nu \exp\left(-\frac{H - \upsilon\tau}{kT}\right), \qquad (4)$$

where v is a coefficient known as the activation volume; T is temperature; ν is a factor allowing for the density of dislocations taking part in the deformation process. It follows from Eq. (4) that to maintain a constant velocity of motion of dislocations (and consequently constant rate of dislocation deformation when the applied stress was constant) during cooling, the applied stress should be increased. Thus, in the case of experiments at a constant deformation rate, cooling would increase the resistance to deformation.

The activation energy of the motion of dislocations was independent of the structure state of iron (the density and distribution of dislocations, dislocation pinning by impurity atoms, concentration of interstitial atoms in the solid solution, presence of a second phase). From this we could conclude that the temperature dependence of the resistance to the motion of dislocations was associated with the properties of the crystal lattice of iron. A moving dislocation overcame periodic stress fields in the crystal lattice and this governed the temperature dependence of the resistance to the motion of free dislocations.

Above ~300°K the energy dissipated by dislocations ceased to depend on temperature. This meant that above this temperature the stored thermal energy was sufficient to overcome, by thermal fluctuations, the barriers in the path of moving dislocations, and this governed the temperature dependence of the resistance to deformation. The limiting value of the dislocation deformation was in this case governed by the tensile stresses in dislocation lines and the influence of the impurity atoms pinning the dislocations.

The lack of a dependence of the activation energy of the motion of dislocations on the degree of pinning by interstitial atoms suggested a mechanism for the amplitude-dependent energy dissipation. The existing representations of this mechanism associate the amplitude dependence of the internal friction with the breakaway of dislocation lines from impurity atoms locking them [11]. The breakaway process should be thermally activated [9, 12]. If this mechanism were correct, the processes of dislocation pinning should affect the temperature dependence of the motion of dislocations. The absence of such a dependence probably indicated that, during the motion of dislocations observed in the measurement of the internal friction in the amplitude-dependent region, there was no breakaway of dislocations from impurity atoms pinning them. The mechanism of energy dissipation by dislocations in the measurement of internal friction was not clear; the mechanism was probably associated with the motion of kinks along dislocations in the case of small displacements of dislocations from their equilibrium positions, as suggested by Seeger and Schiller [13]. However, to solve this problem it would be necessary to determine whether it is possible to apply the Seeger−Schiller mechanism to the amplitude-dependent region of the internal friction.

It is known that the effective stress causing the motion of dislocations is $\tau_{eff} = \tau_{app} - \tau_{res}$, where τ_{app} is the applied stress and τ_{res} is the resistance to the motion of dislocations. At stresses smaller than τ_{res}, dislocations do not move. Obviously, the value of τ_{res} governs the value of ε_{cr}. This quantity depends on the dislocation structure of a sample (lengths of dislocation segments, the presence of steps, etc.), the amount of second-phase impurities, the resistance of the crystal lattice, etc. However, this approach predicts no motion of dislocations in the amplitude-independent region. This was evidently the case in the well-annealed state when the critical amplitude was very large (Fig. 1). The conclusion that there was no motion of dislocations in the amplitude-independent region of the internal friction of annealed low-carbon steel was also reached in [14]. However, there is as yet insufficient data on the influence of plastic deformation on the amplitude-independent internal friction.

The large change in the dissipated energy because of the pinning of dislocations by impurity atoms is probably associated with the fall in the number of dislocations moving under the action of applied stresses. The ability of dislocations to move may depend also on the distribution of lengths of dislocation segments.

The results obtained can be used in the investigation of the temperature dependence of the yield point. We may assume that the temperature dependence of the resistance to the motion of free dislocations is governed by the properties of the crystal lattice of iron. However, it is known that the presence of interstitial impurities enhances the temperature dependence of the resistance to deformation. The role of impurities is probably to reduce the density of moving dislocations. At a constant deformation rate the reduction of the density of moving dislocations leads to an increase in their velocity and, consequently, to a stronger temperature dependence of the resistance to the motion of dislocations, compared with high-purity iron.

Literature Cited

1. A.H. Cottrell and B.A. Bilby, Proc.Phys.Soc. (London), A62:49 (1949).
2. H. Conrad and G. Schoeck, Acta Met., 8:791 (1960).
3. S.O. Suvorova, V.I. Sarrak, and R.I. Éntin, Collection: Problems of Metallography and Physics of Metals, Metallurgizdat (1964), Vol.8, p. 125.
4. S.O. Suvorova, V.I. Sarrak, and R.I. Éntin, Fiz.Metal.i Metalloved., 17(1):105 (1964).
5. S.O. Suvorova, V.I. Sarrak, and R.I. Éntin, Izv. Akad. Nauk SSSR, Otd. Tekhn. Nauk Met. i Gorn. Delo, 4:127 (1964).
6. Yu.V. Piguzov, Author's Certificate No. 142452.
7. I.B. Kekalo and B.G. Livshits, Fiz.Metal.i Metalloved., 13(4):599 (1962).
8. G.A. Beresnev and V.I. Sarrak, Fiz.Metal.i Metalloved., 19:5 (1965).
9. D.H. Niblett and J. Wilks, Usp.Fiz.Nauk, 80:1 (1963) [Russian translation of a paper in Advan. Phys., 9:1 (1960)].
10. A. Seeger, Dislocations and Mechanical Properties of Crystals [Russian translation] (1960), p. 179. [English edition: J.R.C. Fisher, ed., Wiley, New York (1957).]
11. K. Lücke and A. Granato, Dislocation and Mechanical Properties of Crystals, J.R.C. Fisher, ed., Wiley, New York (1957), p. 425.
12. J. Friedel, The Relation Between the Structure and Mechanical Properties of Metals, Proceedings of a Conference at Teddington, London (1963), p. 409.
13. A. Seeger and P. Schiller, Acta Met., 10:348 (1962).
14. W.J. Bratina and D.Mills, Acta Met., 10:419 (1962).

INVESTIGATION OF TEMPER BRITTLENESS
BY THE AMPLITUDE-DEPENDENT
INTERNAL FRICTION METHOD

K. F. Starodubov, S. N. Polyakov, and A. S. Kudlai

The temper brittleness of steel can be explained on the basis of a dislocation theory [1-4] and therefore it is interesting to determine experimentally the differences between the dislocation structure of steel in the ductile and embrittled states [10]. The present communication gives the main results of an investigation of the reversible temper brittleness by the amplitude-dependent internal friction method and by analysis of the true elongation diagrams.

The investigation was carried out on samples of Armco iron and steel 30KhGSA. The chemical compositions of the investigated alloys are given in Table 1. For impact tests we used the samples of 11 × 11 × 55 mm dimensions. The heat treatment was as follows: normalization from the temperature of 880-950°C (at which the samples were held for 1 h), cooling in air, quenching in oil (or in water) from 880-950°C (the two temperatures represent, respectively, the treatments of the 30KhGSA steel and of the Armco iron after keeping them at these temperatures for 30 min). After quenching, the samples were subjected to a double tempering cycle [7]:

a) tempering I at 650°C for 2 h, cooling in water; tempering II at 650°C for 2 h, cooling at the rate of 20°C/h to 400°C — brittle state;

b) tempering I at 650°C for 2 h, cooling at the rate of 20°C/h to 400°C; tempering II at 650°C for 2 h, cooling in water — ductile state.

From the samples subjected to this heat treatment, we prepared Menager samples and samples for tensile tests (5-mm diameter and nominal length of 25 mm). The tendency to temper the brittleness was estimated by carrying out batch tests with an average temperature shift [8]. The static tensile tests were carried out on a TsDM4-30 machine at a rate of de-

Table 1

Steel designation	Content of elements, %							
	C	Mn	Si	Cr	S	P	Ni	Cu
30KhGSA	0.32	1	1.04	1	0.014	0.015	0.1	0.16
Armco iron	0.04	0.13	0.18	—	0.027	0.004	—	—

161

Table 2

Steel designation	State	$K_y d^{-\frac{1}{2}}$, kg/mm²	Length of the yield plateau, mm	Impact strength,* kg · m/cm²	Shift of critical embrittlement temp. during embrittlement Δt, °C
30KhGSA	Ductile	7.2	0.30	20	110
	Brittle	8.3	0.24	8	
Armco iron	Ductile	3	0.18	31	64
	Brittle	10	0.38	20	

* At room temperature.

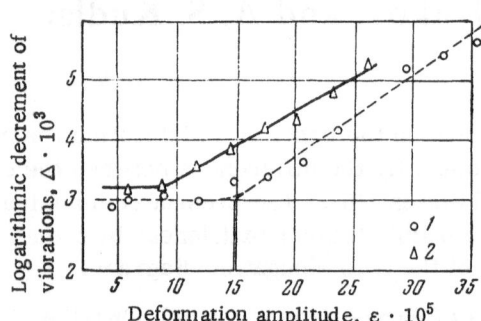

Fig. 1. Amplitude dependence of the internal friction of the steel 30KhGSA in the brittle (1) and ductile (2) states.

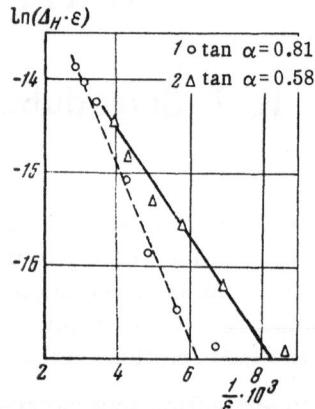

Fig. 2. Value of A_2 for the steel 30KhGSA in the brittle (1) and ductile (2) states.

formation of 0.013 mm/sec. The experimental dependences obtained were used to plot the true elongation diagrams and to determine the parameters in Petch's equation

$$\sigma_y = \sigma_i + K_y d^{-1/2},$$

where σ_y is the yield point (kg/mm²), σ_i is the resistance to the motion of free dislocations (kg/mm²), 2d is the grain diameter (mm),

$$K_y = \sigma_d l^{1/2} \text{ (kg/mm}^{3/2}\text{)},$$

σ_d is the dislocation unpinning stress (kg/mm²), l is the distance from a dislocation cluster to the nearest sources (mm).

The internal friction was determined at room temperature using samples of 0.83-mm diameter and of 127- to 130-mm length in apparatus of the RKF MIS type at a vibration frequency of 2 cps. The samples were subjected to heat treatment in an argon atmosphere in accordance with the sequence described above.

Table 2 lists the results of mechanical tests on ductile and embrittled samples.

It follows from Table 2 that the development of the embrittlement process was accompanied by very great changes in the dislocation structure, which was reflected in the degree of blocking of dislocations (increase of the quantity $K_y d^{-1/2}$ in the brittle state) and in the change of the length of the yield plateau. These results provided the basis for carrying out the investigations by the amplitude-dependent internal-friction method.

The internal friction in the amplitude-dependent region is related to the deformation amplitude by the expression [5]

$$\Delta_H = \frac{A_1}{\varepsilon_0} \exp\left(-\frac{A_2}{\varepsilon_0}\right),$$

where Δ_H is the internal friction in the amplitude-dependent region, A_1 is a constant, ε_0 is the deformation amplitude, A_2 is proportional to the concentration of impurity atoms at dislocation lines.

It was of great interest to determine the so-called critical breakaway amplitude ε_{cr}, which was governed both by the forces binding impurity atoms to dislocations and by the concentrations of these atoms at dislocation lines.

The results of the determination of these quantities for the ductile and brittle states of the steel 30KhGSA is given in Figs. 1 and 2. It can be seen from Fig. 1 that the breakaway amplitude in the brittle state was 72% larger than ε_{cr} in the ductile state. This indicated that the forces binding impurity atoms to dislocations were stronger in the brittle state. The value of A_2 was equal to the tangent of the slope angle of the straight line plotted in the coordinates $\ln(\Delta_H\varepsilon) - (1/\varepsilon)$.

Figure 2 shows that in the brittle state the straight line $\ln(\Delta_H\varepsilon) - (1/\varepsilon)$ had a slope angle 40% larger than in the ductile state, which indicated different concentrations of impurities at the dislocation lines in the brittle and ductile states.

Similar dependences were obtained in an investigation of the Armco iron samples in the ductile and brittle states, from which it followed that the breakaway amplitude in the brittle state was 43% larger than in the ductile state, and the slope angle of the straight line $\ln(\Delta_H\varepsilon) - (1/\varepsilon)$ was 45% larger.

Thus, the development of the embrittlement processes in the steel 30KhGSA and in Armco iron led to a considerable increase in the forces which bind the impurity atoms to dislocations and to an increase of the concentrations of these atoms near dislocations.

The segregation and precipitation during the embrittlement process were intensive near dislocations, i.e., the processes were analogous to those which take place during the final stages of deformation aging [9].

The data obtained in the present investigation by the amplitude-dependent internal-friction method were in good agreement with results reported in [10].

Literature Cited

1. Abdul-Fattah, K. Kaddoh, and P.C. Rosenthal, Trans. Am. Soc. Metals, 52(1): 116 (1960).
2. I.I. Wert and P.C. Rosenthal, Trans. Am. Soc. Metals, 55(1): 439 (1962).
3. A.S. Keh and W.C. Porr, Trans. Am. Soc. Metals, 52(1): 81 (1960).
4. M.A. Krishtal, Yu.V. Piguzov, and S.A. Golovin, Collection: Internal Friction in Metals and Alloys, Metallurgizdat (1964).
5. A. Granato and K. Lücke, Dislocation theory of absorption, in Collection: Ultrasonic Methods of Investigating Dislocations [Russian translation] (1963), p. 27.
6. A.H. Cottrell, Trans. AIME, 212: 192 (1958).
7. S.N. Polyakov, Scientific Proceedings of the Institute for Ferrous Metallurgy, Izd. Akad. Nauk UkrSSR (1957), p. 56.

8.   A.P. Taber, G.F. Thorlin, and G.F. Wallace, Trans.Am.Soc.Metals, 42:1033 (1950).
9.   D.V. Wilson and B. Russell, Acta Met., 8(1) (1960).
10. S.N. Polyakov and A.S. Kudlai, Izv.Akad.Nauk SSSR, Otd. Tekhn.Nauk, Met., Gorn. Delo, No. 6:117-124 (1964).

INVESTIGATION OF THE INFLUENCE OF
A CONSTANT SHEAR STRESS ON THE
AMPLITUDE-DEPENDENT INTERNAL FRICTION

V. V. Khil'chevskii

The present communication considers the amplitude dependence of the internal friction under high stresses. The earlier publications [1-9] show that, in some cases, constant stresses may have a considerable influence on the damping of vibrations. The determination of the degree and nature of this influence, and the possibility of allowing for it in dynamic calculations and explaining its physical aspects, naturally require further accumulation of experimental data. In contrast to all the experiments carried out before, in which free flexural vibrations of cantilevered samples were used, the present author employed free torsional vibrations and the method described in [10].

Figure 1 shows the dependence of the logarithmic decrement δ on the alternating stress amplitude τ_a (for various values of τ_{sh}) for steel, grey cast iron, aluminum alloy D16A, and brass.

The results of the tests were as follows.

1) The application of a constant shear stress τ_{sh} altered the logarithmic decrement in all cases (i.e., for all materials).

2) Depending on the nature of the induced change, the materials (metals) could be divided into two groups: the application of τ_{sh} to ferrous metals reduced the decrement, while the application of τ_{sh} to nonferrous metals increased the decrement. This duality (not reported earlier) in the nature of the influence of τ_{sh} made it impossible to explain the physical basis without considering the magnetic energy dissipation.

3) The degree of the influence of the average shear stress was also different for different materials: it ranges from 15-17% (grey cast iron, globular cast iron, etc.) to 40-50% (brass, aluminum alloy), and even up to 90% (for the ferromagnetic alloy 403 [13]).

This strong influence of a constant stress must, naturally, be allowed for in calculations. In the case of free vibrations we shall proceed as follows. If, using the dependence $\delta = f(\tau_a)$ for various values of τ_{sh}, we plot the dependences $\delta = f_1(\tau_{sh})$ for various values of τ_a, then we find that, in the first approximation, the latter dependences are linear {see, for example, the plots of $\delta = f(\tau_{sh})$ shown in Fig. 2 for the aluminum alloy D16A and the dependence given in [10] for steel}, and, consequently, they can be represented in the form

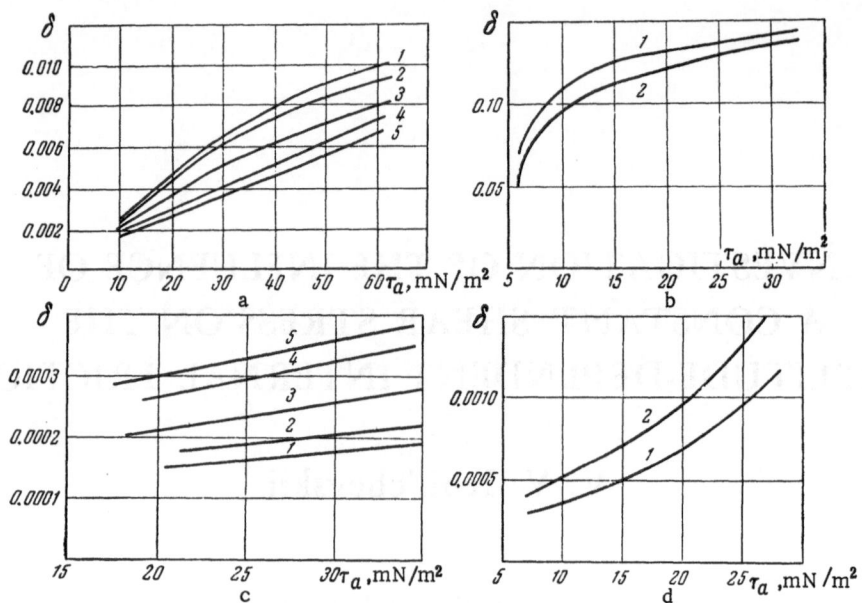

Fig. 1. Dependence of the logarithmic decrement δ on the alternating stress amplitude τ_a. a) Steel: 1) $\tau_{sh} = 0$; 2) 20; 3) 40; 4) 60; 5) 80 mN/m²; b) grey cast iron: 1) $\tau_{sh} = 0$; 2) 75 mN/m²; c) aluminum alloy D16A: 1) $\tau_{sh} = 0$; 2) 17.5; 3) 35; 4) 52.5; 5) 70 mN/m²; d) brass: 1) $\tau_{sh} = 0$; 2) 37.5 mN/m².

$$\delta = \delta_0 + k\tau_{sh}, \qquad (1)$$

where k is an angular coefficient (k > 0 if δ increases with increase of τ_{sh}, but k < 0 if δ decreases with increase of τ_{sh} [10]); δ is the logarithmic decrement for a given value of τ_a in the case of the symmetrical cycle ($\tau_{sh} = 0$), which, in most cases, can be represented in the form

$$\delta_0 = \alpha_0 \tau_a^{n-1},$$

where α_0 and n are constants of the material.

Using the well-known dependence $\delta \approx \Delta A/2A$, and bearing in mind that the absolute value of k increases with increase of τ_a and that $A = C\tau^2/2$, where C is the rigidity of a rod, we obtain the following expression for the energy lost by the rod per cycle:

$$\Delta A = (\alpha_0 + k_1\tau_{sh}) C\tau_a^{n+1}, \qquad (2)$$

or

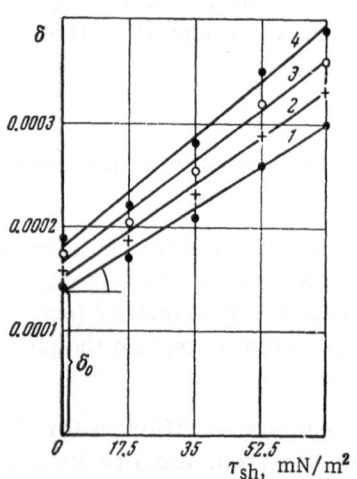

Fig. 2. Dependence of the logarithmic decrement δ on the constant stress τ_{sh} for the alloy D16A. 1) $\tau_{sh} = 20$; 2) 25; 3) 30; 4) 35 mN/m².

$$\Delta A = \alpha C\tau_a^{n+1}, \qquad (3)$$

where α is a characteristic of the material which allows for the application of a constant stress τ_{sh}.

On the other hand, analysis of a vibrogram of the damped vibrations in the case of a sufficiently slow decay shows that, in problems of this type, in the case of metals we have [11]

$$\Delta A \approx C\tau_{a_i}\Delta\tau_{a_i}, \tag{4}$$

and, moreover,

$$\Delta\tau_{a_i} = -T\dot{\tau}(t), \tag{5}$$

where $\tau(t)$ is the equation for the envelope of the vibrogram; T is the vibration period.

Substituting Eqs. (5) and (4), and then comparing the results with Eq. (3), we obtain

$$-T\tau\dot{\tau} = \alpha\tau^{n+1}. \tag{6}$$

Separating the variables and integrating within the limits of the stress amplitude in one cycle, we find

$$t = -T\int_{\tau_0}^{\tau}\frac{d\tau}{\alpha\tau^n}, \tag{7}$$

and hence, after integration, we obtain the equation for the envelope

$$\tau_{a_i} = f(\tau_0, \alpha, T, t, n), \tag{8}$$

where the average stress is allowed for by the coefficient α (τ_0 is the initial stress amplitude at t = 0).

Obviously, in each actual case the expression for ΔA may differ considerably from Eqs. (2) and (3). In particular, mainly because of the differences in the definition $k_1 = f_2(\tau_a)$, the expression for ΔA may be obtained in the form such that the closed-contour integration of Eq. (7) cannot be carried out for an arbitrary value of n. In such cases the envelope should be plotted by means of the recurrence formulas of Nazarov [11, 12].

The treatment given here applies to free vibrations. To determine the resonance stresses in the case of forced vibrations, allowing for the influence of the constant stress τ_{sh}, one should use the hysteresis loop, deforming the loop in such a way that it reflects the influence of the constant stress on the value of the energy dissipated per cycle.

Literature Cited

1. V.V. Khil'chevskii, Calculation of the vibrations of turbine blades of variable cross section in the case of strap joints, allowing for hysteretic losses, Izv. Kiev Polytechnic Institute, Vol. 18 (1955).
2. G.S. Pisarenko, Vibrations of Elastic Systems Allowing for Energy Dissipation in Materials, Izd. AN UkrSSR (1955).
3. G.S. Pisarenko, Energy dissipation in mechanical vibrations, Izd. AN UkrSSR (1962).
4. A.P. Filippov, Forced transverse vibrations of turbine blades with allowance for damping (Proceedings of the Laboratory for Problems of High-Speed Machines and Mechanisms), Izd. AN UkrSSR (1953).
5. E.S. Sorokin, Closed solution of the problem of forced vibrations of rods with hysteresis, in collection: Investigations in the Theory of Buildings, No. 4, Stroiizdat (1949).

6.   Ya.G. Panovko, Allowance for hysteretic losses in problems of applied theory of elastic vibrations, Zh. Tekhn. Fiz., 23(3) (1953).

7.   E.S. Sorokin, Problems of inelastic resistance to deformation of the constructional materials subjected to vibrations, Central Scientific Research Institute of Industrial Structures (1954), Paper No. 15.

8.   Ya.Ya. Ulle, Hysteretic losses in wood subjected to asymmetric cycles, in collection: Problems of Dynamics and Dynamic Strength, No. 2, Izd. AN LatvSSR (1954).

9.   V.I. Shashlov, Relationship between internal energy dissipation and some mechanical properties of carbon steels, Izv. Kiev Polytechnic Institute, Vol. 24 (1957).

10. V.V. Khil'chevskii, Investigation of energy dissipation in a plane stressed state, in collection: Energy Dissipation in Vibrations of Elastic Systems, Kiev (1963).

11. Ya.G. Panovko, Internal Friction in Vibrations of Elastic Systems, Fizmatgiz (1960).

12. A.G. Nazarov, Method for allowing for energy dissipation in elastic vibrations, Dokl. Akad. Nauk ArmSSR, 16(5) (1953).

13. Person and Lazan, The effect of static mean stress on the damping properties of materials, Proc. Am. Soc. Testing Materials, Vol. 50 (1956).

SOME METHODS OF INVESTIGATING THE AMPLITUDE-DEPENDENT ENERGY DISSIPATION IN MATERIALS

A. P. Yakovlev

A description is given of three experimental units representing successive stages in the development of a method for measuring the energy dissipation in materials, caused mainly by microplastic deformation.

Apparatus D-3

This apparatus was intended for the investigation of the damping properties of materials at low temperatures (down to −196°C). The apparatus is shown in Fig. 1.

The schematic diagram of the apparatus is given in Fig. 2. The main components of the apparatus were: a base 1 and a chamber 2, consisting of two cylinders joined by sleeves 12. The chamber was evacuated by joining to it, by a tube 13, a diffusion pump of the TsVL-100 type (14) and a backing pump of the RVN-20 type (15).

Reliable thermal insulation of the inner cylinder of the chamber was ensured by pumping out the space between the walls of the chamber down to 10^{-5} mm Hg. The inner cylinder con-

Fig. 1. External view of the apparatus D-3.

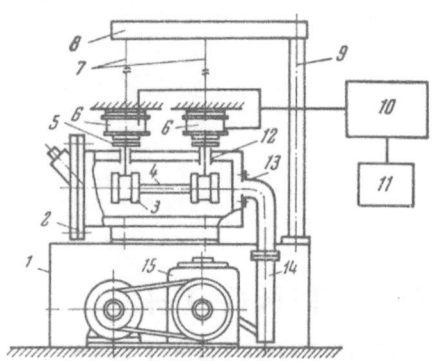

Fig. 2. Schematic diagram of the apparatus D-3.

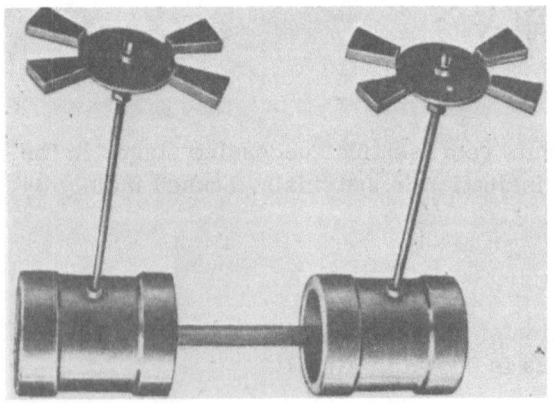

Fig. 3. Vibrating system.

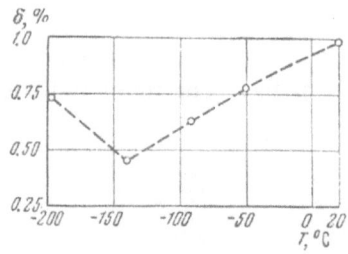

Fig. 4. Diagram showing the temperature dependence of the logarithmic decrement of the alloy AMG-6T.

tained a vibrating system, consisting of a prismatic sample ($2 \times 12 \times 120$ mm) (4), fixed by means of wedge clamps in load-grips 3 supported by spiders 5. The vibrating system (Fig. 3) was suspended by steel wires 7 from a cross-piece 8, fixed to a stand 9. The use of wires made it possible to reduce to a minimum the loss of the vibration energy to the base.

The vibrations were excited by means of electromagnets 6, which were supplied at a controlled frequency from a special infralow-frequency oscillator 11, coupled to a power amplifier of the TU-5-3b type (10).

The infralow-frequency oscillator produced an amplitude-modulated signal of frequencies from 3 to 50 cps.

The procedure in the use of the apparatus was as follows: the vibrating system was placed in the inner cylinder and suspended by the wires. The arms of the spiders and the electromagnet cores were oriented in such a way as to produce twisting moments at the nodes of the vibrating system. Then the chamber was closed and pumped out.

The sample temperature was measured with a Chromel–Alumel thermocouple, which was placed close to the sample or moved away from it by means of a special device. The second junction of the thermocouple was placed in a Dewar flask filled with melting ice. The optical system was adjusted to record the vibrations: the system consisted of a light source, a number of mirrors, and a recorder NS-1. Then the excitation was switched on and the amplitude of the resonant vibrations of the system was raised to the required value. The recording system was switched on and the vibration switched off. Thus, the system executed free damped vibrations, which were recorded on the oscillograph chart. From the vibrograms obtained in this way, the values of the logarithmic decrement of vibrations was determined and this decrement was used as the damping characteristic of the investigated material.

A diagram showing the variation of the logarithmic decrement of vibrations in the alloy AMg-6T during cooling is shown in Fig. 4 as an example. In this figure, the values of the logarithmic decrement correspond to a stress of $\sigma = 1000$ kg/cm^2 in the outermost filament of the sample at all temperatures.

It follows from Fig. 4 that cooling down to $-140°$C reduced the logarithmic decrement considerably; at $-196°$C the decrement increased somewhat, but it still remained smaller than at room temperature.

Fig. 5. External view of the apparatus D-5.

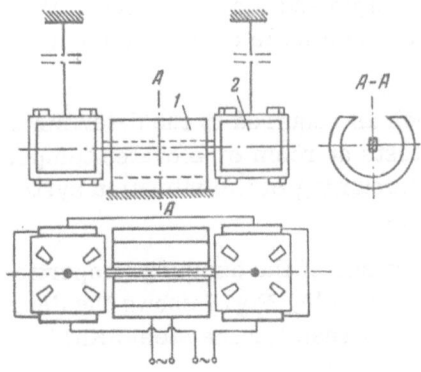

Fig. 6. Schematic representation of the vibrating system together with the heaters.

Apparatus D-5

The apparatus D-5 (Fig. 5) was intended mainly for investigation of the damping properties of materials subjected to flexural vibrations at room temperature and at elevated temperatures up to 200°C. In principle, the apparatus D-5 was similar to the apparatus D-3 described above. Again, the sample had prismatic shape, but was of somewhat different dimensions (4 × 15 × 190 mm), with thicker ends. The ends of the sample were attached by wedge clamps to two weights which had projections above and below. The whole system was suspended by wires. Four electromagnets, whose cores had projections identical in shape with the weight projections, were placed near the loads.

The vibrations were excited by means of these electromagnets in exactly the same way as in the apparatus D-3.

Thus, the apparatus D-5 had the same advantage as the apparatus D-3, namely:

1) the fixing of the sample ensured minimum vibration energy loss to the base;

2) the excitation system ensured that the stress in the sample was close to the fatigue limit;

3) the sample was subjected to pure bending;

4) the optical method of recording the vibrograms was simple, reliable, and did not interfere with the investigated process;

5) the apparatus could be used to test practically any type of structural material (metallic and nonmetallic, magnetic and nonmagnetic, etc.), because the sample did not participate actively in the process of excitation of the vibrations in the system.

The stress in the sample subjected to pure bending was found from the formula

$$\sigma = \frac{Eh}{l}\frac{a}{2R}, \tag{1}$$

where σ is the bending stress in the sample; E is Young's modulus; h is the sample thickness; l is the working length of the sample (the gauge length); a is the deflection (the vibration amplitude) of the light beam on the drum of the recording device; R is the distance from the drum to a mirror fixed to the load.

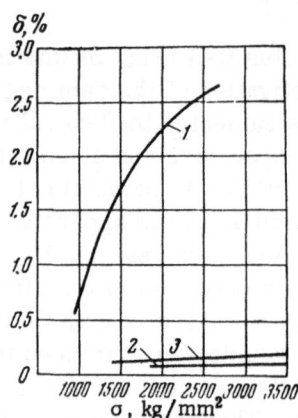

Fig. 7. Dependence of the logarithmic decrement on the stress in the steel 3Kh13. 1) Sample not subjected to heat treatment; 2) sample of HRC = 45-50 hardness; 3) HRC = 50-55.

The same formula was used to calculate the distance R or the amplitude a when the stress σ was known.

To investigate the damping properties of materials at elevated temperatures (up to 200-250°C) the sample was heated by means of special heaters. The vibration system and the heaters are shown schematically in Fig. 6.

The main heater (1) ensured the required high temperature over most of the sample; this heater was in the form of a cylinder of a refractory material with a Nichrome helix wound on its internal surface. The voltage across the helix was controlled by means of an autotransformer. To equalize the temperature distribution along the sample length, additional heaters (2) were used. Each of the four heaters consisted of an asbestos-slate plate, in which a Nichrome helix was fitted in special grooves. The plates were fixed in pairs to each weight of the vibrating system, as shown in Fig. 6. All these plates were connected electrically in series and also supplied from an autotransformer.

In addition to the heaters described, the weights were heated as well and, from the weights, the thicker sample ends. During tests the plate-shaped heaters were removed. Since the heat capacity of the metal loads was very large, the temperature of the sample ends remained practically constant during tests. The temperature distribution over the working length of the sample was checked by means of a Chromel—Alumel couple and a potentiometer PP. A check showed that the difference between the temperatures of the middle of the sample and of its ends did not exceed 1-2°C.

The apparatus D-5 was convenient for mass tests, such as required in the determination of the damping properties of materials or for comparative tests at room or elevated temperatures, since there was no chamber and the mounting and dismounting of the vibration system was quick and allows good access to all parts of the apparatus.

The apparatus D-5 has been used to carry out investigations of the damping properties of more than 70 different constructional materials. By way of example, Fig. 7 shows the dependence of the logarithmic decrement of vibrations on the applied stress for the steel 3Kh13.

The curves in Fig. 7 show that the steel 3Kh13, not subjected to heat treatment, showed a strong dependence of the logarithmic decrement on the stress.

Heat treatment reduced strongly the value of the logarithmic decrement: for a sample with a hardness HRC = 45-50, the logarithmic decrement under the stress $\sigma = 2500$ kg/mm^2 amounted to less than 0.1%, i.e., it was 25 times smaller than for a sample not subjected to heat treatment. The sample with HRC = 50-55 had approximately the same logarithmic decrement at the same stresses. The dependence of the logarithmic decrement on the stress for both samples was weak.

Apparatus D-7

The apparatus D-7 was intended for investigations of the energy dissipation in materials subjected to transverse vibrations in vacuum at room and higher temperatures. A prismatic

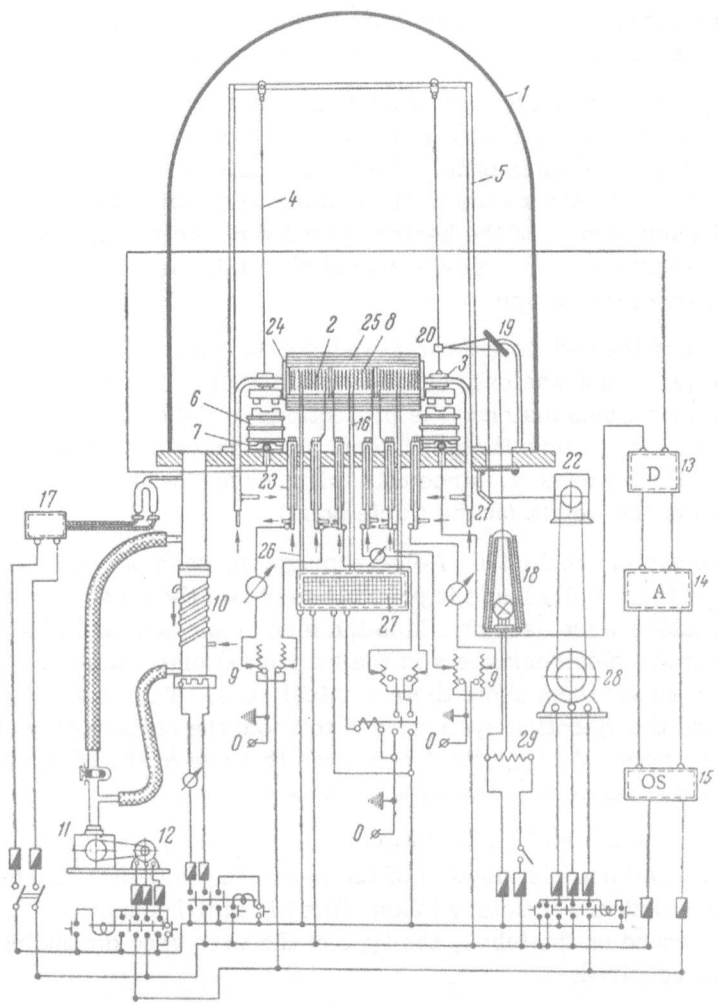

Fig. 8. Schematic representation of the apparatus D-7.

sample was again subjected to pure bending. Test temperatures of 1200-1500°C make it possible to investigate the damping properties of heat-resistant materials.

The main components of the apparatus D-7 were as follows (Fig. 8).

1. A vibrating system, consisting of a sample 2 and of end weights 3. The system was suspended by steel wires 4, fixed to stands 5, placed on a base plate.

2. An excitation system consisting of: two electromagnets 6, placed on special tables 7, by means of which the electromagnets could be moved along two directions; a detector 13, an amplifier 14, and an infralow-frequency oscillator 15, supplying the electromagnets.

3. A heating system, including a three-section heater 8 with a helical winding and three autotransformers 9, by means of which temperature was regulated. Temperature was measured by means of three platino—platinorhodium thermocouples 16.

4. A vacuum system, consisting of a glass (or metal) bell 1, in which the vibrating system, the electromagnets, and the heater were placed; a diffusion vacuum pump 10, a backing pump 11, with an electric drive 12. Vacuum was measured by means of electronic gauges LT-2 and LM-2 and a gauge VIT-1 (16 in Fig. 8).

5. An optical system for recording vibrograms, including light source 18, mirrors (19, 20, 21), and a recorder 22 with a drive 28. The illumination was controlled with a rheostat 29.

The vibrating system was similar to the systems used in apparatus D-3 and D-5. The differences in construction were limited to the shape of the sample and the end weights. In D-7 the sample had elongated thick ends, which were only partly inside the wedge grips of the weights. Large parts of the thick ends of the sample were outside the weight clamps. This was done so that the active zone of the heater extended not only over the working part of the sample but over its thick ends, in order to establish a uniform distribution of temperature along the working part of the sample.

The upper part of the end weight had a wedge grip and enclosed the thick end of the sample. The middle part of the weight was only of 15-mm diameter. The lower part of the weight was thicker with projections which matched the projections of the electromagnet cores and were used to excite resonant vibrations in the system. This shape of weight was used in order to avoid overheating of the lower part, since this might cause deterioration or total loss of the magnetic properties of the weight (at the Curie point).

The excitation system consisted of electromagnets, each of which had four coils, and different connections of these coils made it possible to obtain the most effective exciting force. Moreover, special tables were used on which the electromagnets are placed, and these tables made it possible to move electromagnets in the horizontal plane along two mutually perpendicular directions and to rotate them about their vertical axes. This made it possible to tune the excitation system and the vibrating system, and to avoid the buildup of oscillations of the vibrating system as a whole. An important point was that one of the tables could be moved inside the vacuum bell. A special device was used for this purpose.

A heater used to obtain high temperatures (1200-1500°C) consisted of three sections, necessary to obtain a uniform distribution of temperature along the sample length. Each section represented 13 Alundum refractory tubes, fixed in molybdenum clamps. Molybdenum or Nichrome wire was wound on the tubes, the type of the wire used depending on the maximum sample temperature required.

Each section of the heater was supplied from an autotransformer RNO-250/10 through specially cooled copper current leads 23 (Fig. 8). To reduce heat losses from the working zone, the heater was surrounded by a system of screens 25 made of molybdenum sheet, placed at distances of 3-4 mm from one another, and the ends were protected by special water-cooled copper screens 24. This prevented overheating of the weights and electromagnets by thermal radiation. The temperature of the working part of the sample was found by direct measurement of the temperature of a control sample, which was also placed in the active zone of the heater. Three platino—platinorhodium thermocouples 26 were attached to the control sample and the signals from these thermocouples were applied to the input of a recording potentiometer of the ÉPP-09 type (27). Such a method of measuring temperature required preliminary adjustment and calibration.

The great advantage of the apparatus D-7 was the possibility of carrying out high-temperature investigations in $(2-5) \cdot 10^{-5}$ mm Hg vacuum. This was very important in the case of some special heat-resistant materials, which oxidize rapidly at high temperatures in open atmosphere.

The apparatus D-7 made it possible also to carry out investigations at atmospheric pressure in inert or some other gaseous medium.

Thus, the apparatus D-7 has all the advantages of the apparatus D-3 or D-5 and opened new possibilities of investigating energy dissipation in materials, and of the influence of various factors on this dissipation (stress amplitude, surrounding medium, temperature, structural transformations, etc.). It should be mentioned that the apparatus D-7 was the most universal of the three units built for the investigation of the energy dissipation in materials subjected to transverse vibrations.

INTERNAL FRICTION IN MAGNESIUM CERMET

V. E. Ivanov, V. F. Zelenskii, V. I. Savchenko, S. I. Faifer, and S. M. Zhdanov

An investigation was made of the internal friction of magnesium cermet samples, prepared by a method described earlier in [1].

To obtain different degrees of surface oxidation, magnesium powders were additionally oxidized, which resulted in a range of MgO contents. The internal friction of samples containing 0.3, 2.3, and 5 wt.% MgO was measured using a method described in [2], in which the change in the amplitude of the forced vibrations at resonance was used as a measure of the internal friction. Before the tests, the samples were subjected to annealing for 1 h at 500°C. The presence of even a small amount (0.1–0.3 wt.%) of MgO slowed down the collective recrystallization processes [1], which made it impossible to obtain alloys having identical grain dimensions. The microstructure of the alloys after annealing is shown in Fig. 1. The alloys prepared from oxidized magnesium powders had polycrystalline structure with a uniform distribution of fine oxide particles and layers of various shapes along grain boundaries and in the grains themselves. Such magnesium alloys had high resistance to heat and high thermal stability.

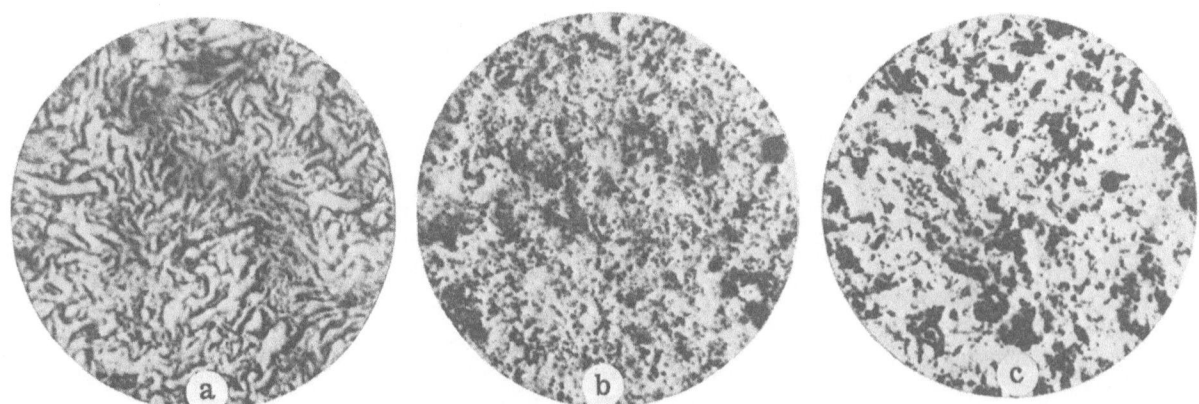

Fig. 1. Microstructure of the investigated magnesium alloys. ×340. a) 0.3 wt.% MgO; b) 2.3 wt.% MgO; c) 5 wt.% MgO.

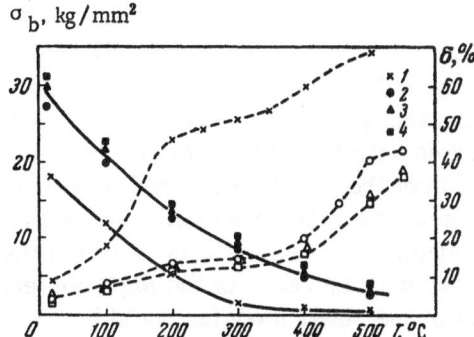

Fig. 2. Temperature dependence of the ultimate tensile strength σ_b and of the relative elongation δ of Mg—MgO alloys. 1) Mg; 2) Mg + 0.3% MgO; 3) Mg + 1% MgO; 4) Mg + 5% MgO.

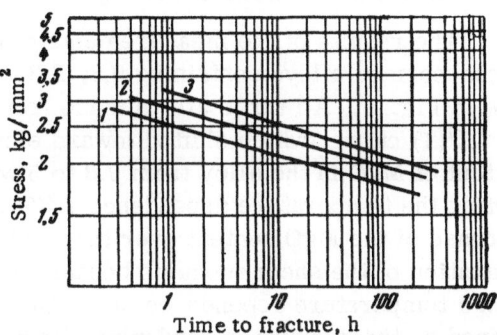

Fig. 3. Long-term strength of Mg—MgO alloys at various temperatures. 1) Mg + 1% MgO at 350°C; 2) Mg + 1% MgO at 300°C; 3) Mg + 5% MgO at 350°C.

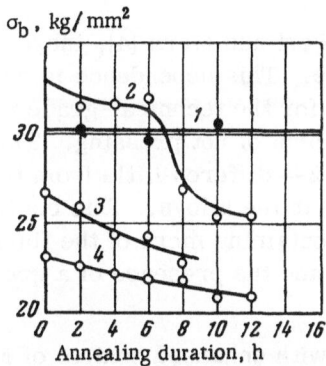

Fig. 4. Dependence of the ultimate tensile stress σ_b on the duration of annealing t at 450°C (1, 2, 4) and at 400°C (3): 1) SMP; 2) VM65-1; 3) VM17; 4) MA8.

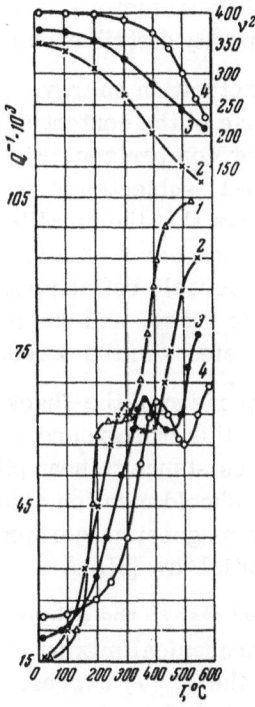

Fig. 5. Temperature dependence of the internal friction $Q^{-1} \cdot 10^3$ (lower curves) and of the square of the vibration frequency ν^2 (upper curves) for Mg—MgO compositions. 1) Technical-grade magnesium Mg-1 deformed by hot pressing; 2) Mg + 0.3 wt.% MgO; 3) Mg + 2.3 wt.% MgO; 4) Mg + 5 wt.% MgO.

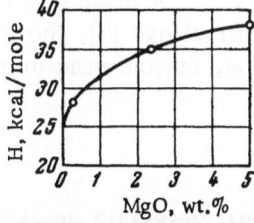

Fig. 6. Dependence of the activation energy H on the percentage content of MgO in an alloy.

Figures 2, 3, and 4 give the strength, as well as the influence of annealing on the strength, for various compositions of the Mg−MgO alloys.

The results of the measurements of the internal friction Q^{-1} and of the shear modulus * of the cermet alloys containing 0.3, 2.3, and 5 wt.% Mg, are given as a function of temperature in Fig. 5. All three alloys made from oxidized magnesium powders exhibited one peak in the $Q^{-1}(T)$ curve, which shifted toward elevated temperatures with increase of the percentage content of MgO in the alloy from 0.3 to 5 wt.%. Comparison of the curves 2, 3, and 4, representing the temperature dependence $\nu^2(T)$, showed some increase of the shear modulus with increase of the MgO content over the whole temperature range. The temperature at which the relaxation of the shear modulus became noticeable increased with increase of the MgO content. Such a temperature dependence of the internal friction and of the shear modulus could be explained by the dispersion hardening effect of the magnesium oxide present in the alloys. Increase of the content of the finely dispersed oxide phase in the alloy prevented the relaxation at a given temperature, and thus suppressed the $Q^{-1}(T)$ peak, caused by the presence of grain boundaries [3]. The shift of the peak toward high temperatures with increase of the MgO content in the alloys could be explained by the hardening effect of the second phase.

In contrast to pure metals, the position of the $Q^{-1}(T)$ peak of the alloys was affected by certain additional factors, such as the state of grain boundaries, the nature of the impurity, and the impurity distribution in the alloy [3].

The activation energy, calculated using the formula of Wert and Marx [4], increased with increase of the content of the finely dispersed oxide phase. This dependence is shown in Fig. 6. For comparison, Fig. 5 gives the dependence $Q^{-1}(T)$ for the technical-grade magnesium MG-1, subjected to a preliminary deformation in the form of hot pressing. This comparison shows that the amplitude of the $Q^{-1}(T)$ peak in curves 2-4 differed little from the peak for MG-1, increasing only slightly for higher contents of MgO in the alloys. This could be due to the considerable refinement of the structure of the alloys containing more of the finely dispersed oxide phase, and due to a greater degree of distortion and the presence of a greater number of defects in the polycrystalline structure.

Comparison of the "background" of $Q^{-1}(T)$ of the alloys with enhanced content of MgO showed that this background decreased with increase in the temperature, indicating an increase in resistance to heat [5] of the alloys containing more MgO. This has been confirmed in [1]. We should mention also the behavior of the "background" of $Q^{-1}(T)$ near the melting point of the magnesium cermets, which (like the MG-1 magnesium), exhibited a deviation from the exponential law [6].

In contrast to the alloys prepared from oxidized powders, the cermet compositions (prepared by mechanical mixing of magnesium and MgO powder followed by hot pressing) had two maxima in the $Q^{-1}(T)$ curves. As in the case of certain other similar alloys [7], the $Q^{-1}(T)$ peaks did not shift in the presence of 2.3 and 5 wt.% MgO in the alloy, but only the amplitude of these peaks varied with the amount of the MgO impurity.

Literature Cited

1.   V. E. Ivanov, V. F. Zelenskii, S. I. Faifer, et al., New Nuclear Materials Including Non-metallic Fuels, Vienna (1963), Vol. 2.
2.   V. E. Ivanov and B. I. Shapoval, Radiation Damage in Reactor Materials, Vienna (1963).

---

* Figure 5 gives the dependence of the square of the frequency of torsional vibrations on temperature.

3. V.P. Elyutin, E.I. Mozzhukhini, et al., Collection: Relaxation Phenomena in Metals and Alloys, Metallurgizdat (1963), p. 243.
4. C. Wert and J. Marx, Acta Met., 1(1) : 113 (1953).
5. V.S. Postnikov, Usp. Fiz. Nauk, 66 : 43 (1958).
6. B.I. Shapoval, Fiz. Metal. i Metalloved., 18(2) : 306 (1964).
7. B.Ya. Pines and Teng Ko-seng, Fiz. Metal. i Metalloved., 8(6) : 867 (1959).

INTERNAL FRICTION IN BERYLLIUM CERMET

V. E. Ivanov, V. F. Zelenskii, S. I. Faifer,
V. I. Savchenko, and V. I. Maksimenko

Investigations of the internal friction of the cermet systems Cu−Fe−Ni, Cu−Mo, Cu−W [1], Ni + Al$_2$O$_3$ [2], SAP [3], and beryllium [4], have shown that the temperature dependence Q^{-1}(T) is affected by the nature of the initial components, the method of preparation of the compacted material, and its structure. The solution of the components and their distribution in the alloy may be quite different in metal powder mixtures, metal−oxide mixtures, or sintered oxidized metal powders. This governs the nature of the behavior of the internal friction and of the Q^{-1}(T) peaks.

We investigated the temperature dependence of the internal friction, Q^{-1}(T), and of the shear and Young's moduli of hot-pressed cermet Be−BeO alloys containing 0.3, 1.5, and 7 wt.% of beryllium.

From brickettes compacted by hot pressing, we prepared cylindrical samples of 20-mm working length and 3-mm diameter which had thick ends for rigid clamping. The tests were carried out using a vacuum relaxator and forced torsional vibrations at resonance. Q^{-1}(T) was determined from the change in the vibration amplitude. We also measured the frequencies in order to plot the temperature dependences of the shear and Young's moduli. The method has been described in detail earlier [5].

Before tests, the samples were annealed in vacuum for 1 h at 1000°C in order to relieve the stresses, remove adsorbed gases, and put the alloy in a state closer to equilibrium.

Such annealing did not yield alloys with grains of equal size because the presence of BeO retarded the process of collective recrystallization. After their annealing, the alloys had grains of average size of 240, 140, and 50 μ, respectively, for the oxide content of 0.3, 1.5, and 7%.

The microstructure is shown in Fig.1. The results of the measurements are given in Fig.2. All the Q^{-1}(T) curves had well-defined maxima: at 500°C for the alloy containing 0.3% BeO, at 600°C for the alloy containing 1.5% BeO, at 870°C for the alloy containing 7% BeO (the maximum degenerated into an inflection).

Comparison of the curves of 1-3 showed that an increase in the percentage content of beryllium oxide in the alloy shifted strongly the peak toward high temperatures and increased slightly the amplitude of the peak, as well as its width. Comparison of the temperature dependences of the shear and Young's moduli showed that at lower temperatures the values of the

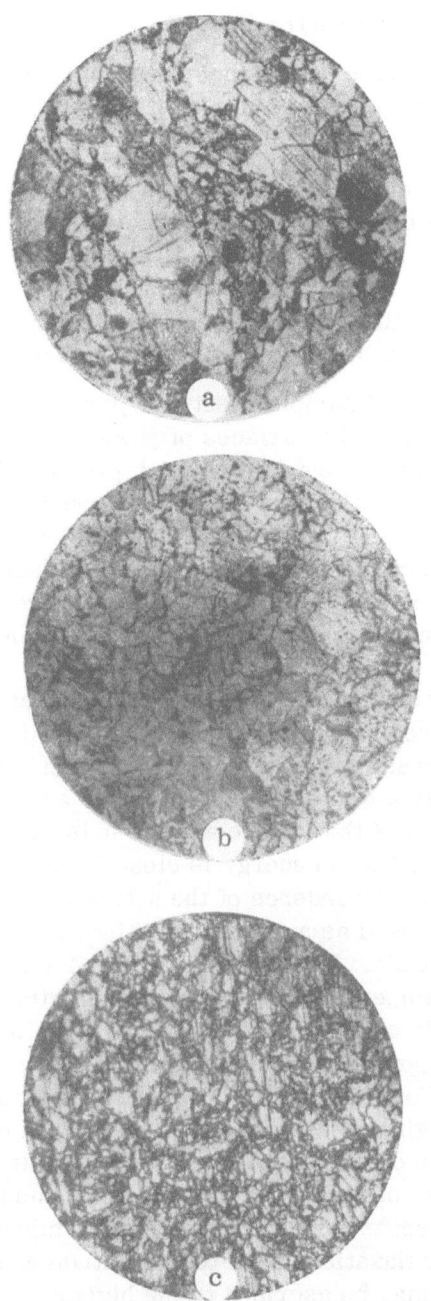

Fig. 1. Microstructure of investigated samples. ×70. a) Be + 0.3% BeO; b) Be + 1.5% BeO; c) Be + 7% BeO.

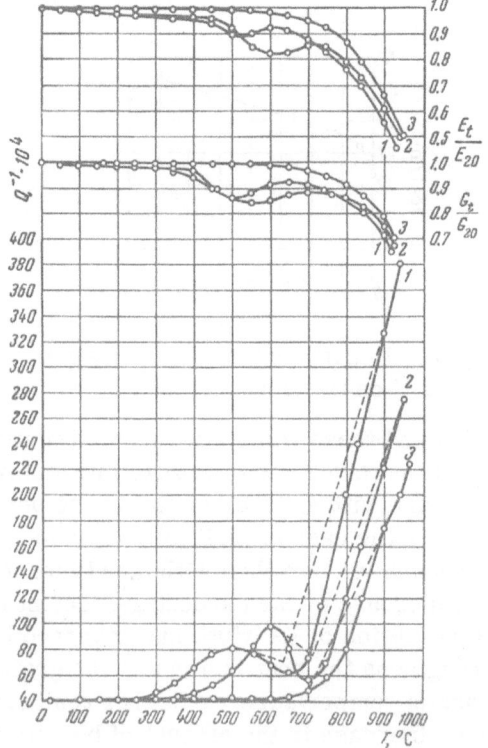

Fig. 2. Temperature dependence of internal friction $Q^{-1} \cdot 10^4$ and of relative values of shear G_t/G_{20} and Young's moduli E_t/E_{20} of Be−BeO cermet compositions. 1) Be + 0.3 wt.% BeO; 2) Be + 1.5 wt.% BeO; 3) Be + 7 wt.% BeO. Dashed curves represent $Q^{-1}(T)$ during cooling.

moduli were little affected by the beryllium oxide content in the alloy, increasing slightly with increase of the amount of BeO, but when temperature was increased, there was a considerable relaxation of the moduli at the same temperatures at which peaks appeared in the $Q^{-1}(T)$ curves. For the alloy containing 7% BeO, the relaxation of the moduli was not observed up to the temperature of the inflection in the curve. The activation energies of the relaxation processes responsible for the appearance of the maxima in the $Q^{-1}(T)$ curves were calculated using the Wert−Marx formula and they were, respectively: 41, 48, and 62 kcal/mole for the alloys containing 0.3, 1.5, and 7% BeO. Figure 3 shows that as the oxide content in the alloy decreased, the activation energy approached the activation energy of self-diffusion along grain boundaries in beryllium [6].

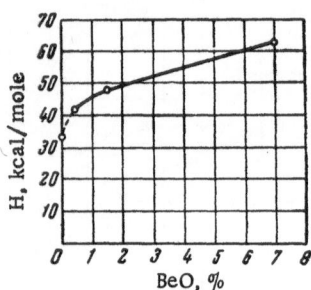

Fig. 3. Dependence of the activation energy H on the BeO content.

Discussion of Results

The temperature dependence of the internal friction in cermets is more complex than in pure metals [1, 2, 7].

The $Q^{-1}(T)$ curve of a mixture of metals (1) exhibits maxima corresponding to relaxation processes at contact surfaces between like and unlike components. The behavior of these maxima depends on the solubility of the components, their concentrations, distributions, and other factors.

Hot pressing of samples destroys partly the oxide film present on the surfaces of powder particles and these films are then distributed mainly along the grain boundaries. Hot-pressing conditions favor rapid seizure and sintering leading to the formation of a single polycrystalline structure with characteristic strong bonds between particles. Probably because of this, there is (as in the case of SAP [3]), only one peak in the $Q^{-1}(T)$ curve, since relaxation takes place at boundaries enriched with beryllium oxide. The shift of the peak toward high temperatures with increase of the percentage content of BeO, the slight increase of the peak amplitude, and the reduction of its width are governed by several factors. Increase of the BeO content strengthens the boundaries and the grains themselves, which suppresses shear processes at a constant temperature. A higher temperature may be required for the relaxation process, because increase in the amount of beryllium oxide favors increase in the strength of the bonds between particles and increase of the strength of the alloy as a whole. This is why the activation energy of the relaxation process increases with increase of the oxide content. It is evident from Fig. 3 that in the case of low oxide content, the activation energy is close to the energy of boundary self-diffusion in Be, reported in [6]. The dependence of the activation energy on the amount of beryllium oxide in the alloy may be used as a criterion of the increase in the strength of alloys containing various amounts of the oxide. It follows from Fig. 2 that when the oxide content is increased from 0.3 to 1.5%, the temperature of the internal friction peak shifts from 500 to 600°C, while an increase of the oxide content to 7% shifts the peak to 870°C. At lower temperatures the relaxation process is suppressed. This is particularly noticeable in the temperature dependence of the shear and Young's moduli. We must allow for the fact that additional reduction of the grain size in the original powder and the retardation of the collective recrystallization in the alloys containing more of the oxide should tend to shift the $Q^{-1}(T)$ peak toward lower temperatures, but the increase of the strength of the alloy due to the presence of beryllium oxide predominates and the peak shifts toward higher temperatures. The decrease of the grain size and the slowing down of the relaxation affect the amplitude and the width of the peak. The reduction of the peak amplitude may be ascribed to the higher structural stability of the alloy containing more oxide. Comparison of the high-temperature "background" of $Q^{-1}(T)$ shows that the temperature at which the curve begins to rise sharply increases with increase of the oxide content, but the slope of the curve decreases. Such a behavior of the "background" may serve as a criterion of the increase in the resistance to heat in the alloy with increase of the oxide content [7].

Literature Cited

1.   B. Ya. Pines and Teng Ko-seng, Fiz. Metal i Metalloved., 8(4) : 599, 867 (1959).
2.   V. P. Elyutin et al., Collection: Relaxation Phenomena in Metals and Alloys, Metallurgizdat (1962), p. 243.

3.   A.I. Dashkovskii et al., Collection: Metallurgy and Metallography of Pure Metals, No. 4 (1963), p. 100.
4.   V.E. Ivanov and B.I. Shapoval, Fiz.Metal.i Metalloved., 11(1) : 52 (1961).
5.   V.E. Ivanov and B.I. Shapoval, Radiation Damage in Reactor Materials, Vienna (1963).
6.   A.A. Kruglykh et al., Fiz.Metal.i Metalloved., 9(1) : 149 (1960).
7.   V.S. Postnikov, Usp.Fiz.Nauk, 66 : 43 (1958).
8.   B.I. Shapoval et al., Fiz.Metal.i Metalloved., 15(5) (1963).

INVESTIGATION OF AERO-ENGINE TURBINE
BLADES BY THE INTERNAL FRICTION METHOD

F. M. Titov and N. P. Devichenskii

The present communication reports the results of an investigation of turbine blades and of refractory alloy samples by the internal friction method [1-5].

In addition to the internal friction method, which was the chief means of investigation, we used also the hardness method and observed changes in the lengths and diameters of the samples.

The initial investigations gave promising results [6-8], indicating that it would be desirable to continue these studies. We investigated aero-engine turbine blades and samples made of nickel alloys.

Blades. We investigated batches of 25-30 blades, whose service life in aero engines ranged from 0 to 4000 h. The logarithmic decrement was measured using apparatus described earlier [9].

The measurements of various properties were carried out on a large number of similar samples. For each batch, we determined the average value and the variance of the logarithmic decrement and the hardness from the formulas

$$\bar{\delta} = \frac{1}{N} \sum_{i}^{N} \delta_i \quad \overline{HB} = \frac{1}{N} \sum_{i}^{N} HB_i, \tag{1}$$

$$S_{\delta}^2 = \frac{1}{N} \sum_{i}^{N} (\delta_i - \bar{\delta})^2 \quad S_{HB}^2 = \frac{1}{N} \sum (HB_i - \overline{HB})^2, \tag{2}$$

where δ_i, HB_i are, respectively, random values of the decrement and hardness of the blades; N is the number of values of the decrement and the hardness obtained in the measurements.

Nickel Samples. Turbine blades of reaction engines work under conditions of rapid and frequent variation of temperature over a wide range. These temperature fluctuations are due to the stopping and starting of the engine and changes in its working conditions.

The repeated changes in the internal conditions intensify the diffusion processes, cause volume changes in the materials, including heat-resistant materials, and this process is accompanied by the appearance of fine and then larger cracks [10-14].

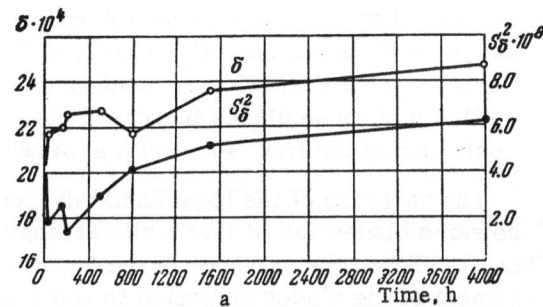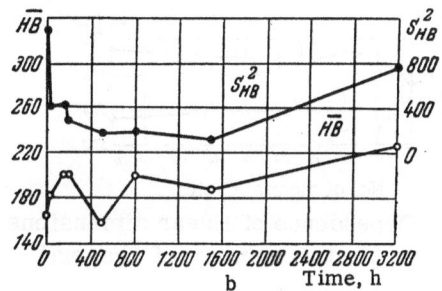

Fig. 1. Dependence of certain properties of a deformable nickel alloy on the service life of blades made from it. a) Logarithmic decrement $\bar{\delta}$ and variance of the decrement S_{δ}^2; b) hardness $\overline{HB}$ and variance of the hardness S_{HB}^2.

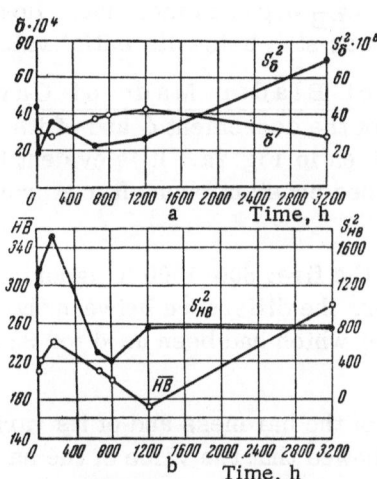

Fig. 2. Dependence of certain properties of a cast nickel alloy on the service life of blades made from it. a) Logarithmic decrement $\bar{\delta}$ and variance of the decrement S_{δ}^2; b) hardness $\overline{HB}$ and variance of the hardness S_{HB}^2.

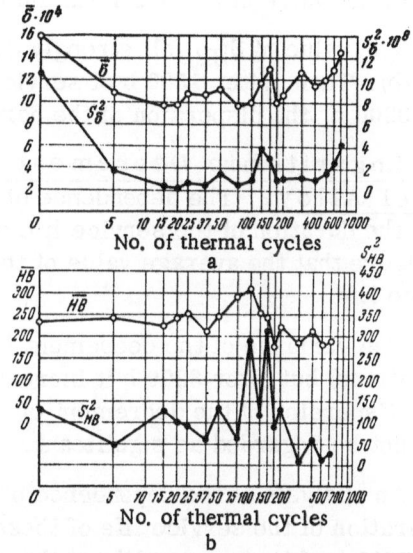

Fig. 3. Dependence of certain properties on the number of thermal cycles. a) Logarithmic decrement $\bar{\delta}$ and variance of the decrement S_{δ}^2; b) hardness $\overline{HB}$ and variance of the hardness S_{HB}^2.

In view of this, we undertook an investigation of the influence of thermal cycling on the internal friction of samples made of a nickel alloy. The thermal cycling was carried out by means of apparatus which made it possible to test up to 23 samples under identical conditions. The following procedure was followed during thermal cycling: 500, 800, 500°C, the whole cycle lasting about 6.5 min. After a certain number of cycles, we measured the decrement, hardness, and the linear dimensions of the samples, and then, as in the case of blades, we calculated the statistical parameters of these quantities.

Results of Experiments

Logarithmic Decrement and Hardness of Blades Made of a Deformable Nickel Alloy. Graphic representation of the average value of the logarithmic decre-

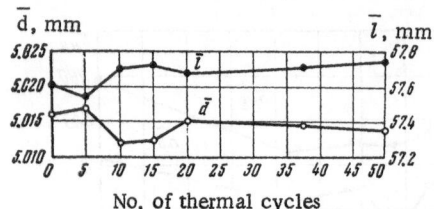

Fig. 4. Dependence of linear dimensions of samples on the number of thermal cycles.

ment $\overline{\delta}$ and of its variance S_{δ}^2 as a function of the duration of the service life of nickel-alloy blades is given in Fig. 1a. It is evident from this figure that $\overline{\delta}$ of these blades increased considerably only during the first 400-500 h in service.

The variance of the logarithmic decrement of the same blades fell at first quite strongly (Fig. 1a). Thus, the variance of the logarithmic decrement of the blades subjected to 200 h service was less than half the variance for new blades. Next, the variance increased somewhat, reaching a maximum in the region 1500-1600 h, and beyond this region it remained practically constant up to 4000 h.

The dependence of the average value of the hardness $\overline{HB}$ and of the variance of the hardness S_{HB}^2 of the blades on the duration of service life is given in Fig. 1b, which indicates that the average value of the hardness did not vary much during the service life.

The value of S_{HB}^2 fell strongly during the first 400 h of the service life of the blades (Fig. 1b). Then, after 1500 h of service life, the variance S_{HB}^2 began to increase. However, after 3200 h, the dispersion of the hardness of the vanes was still below its initial value.

Logarithmic Decrement and Hardness of Blades Made of Cast Nickel Alloy. The dependence of the average value of the decrement $\overline{\delta}$ and of its variance S_{δ}^2 on the duration of the service life of these blades is given in Fig. 2a. It is evident from this figure that the average value of the decrement remained constant even after a prolonged service life.

The variance of the decrement fell initially (during the first 800-1000 h, ignoring initial fluctuations) but after 3200 h it increased again. However, the difference between the values of the dispersion of the decrement of new blades and those which had been used for 3200 h could not be regarded as significant.

An analysis of the dependence of the average value of the hardness and of its variance on the duration of the service life of these blades (Fig. 2b) showed that the value of the hardness after 3200 h of the service life of the blades changed little, without any definite tendency as far as the direction of this change was concerned.

The variance of the hardness (Fig. 2b) fell considerably during the first 800-1000 h of the service life of the blades. The blades which had 800 h of operation were characterized by a variance of the hardness almost three times smaller than the value for new blades. Further increase of the service life left the variance of the hardness unchanged up to 3200 h.

Decrement, Hardness, and Linear Dimensions of Nickel Samples. The results of an investigation of the dependence of the decrement $\overline{\delta}$ and of the variance S_{δ}^2 of the nickel samples on the number of thermal cycles (in accordance with the 500, 800, 500°C procedure) are given in Figs. 3a and 3b. It is evident from Fig. 3a that after the first 15-20 cycles the average value of the decrement fell strongly, and after 150 cycles a maximum appeared. After 200 cycles the average value of the decrement became greater. However, after 700 cycles the average value of the decrement was still below its initial value.

The changes in the average value of the hardness $\overline{HB}$ and in the variance of the hardness S_{HB}^2 during the thermal cycling are given in Fig. 3b, from which it follows that up to 50 cycles the average value of the hardness remained practically unchanged. A hardness maximum was observed in the region of 100 cycles, followed by a drop.

Figure 4 shows the dependence of the linear dimensions of the samples (the length $\overline{l}$ and the diameter $\overline{d}$) on the number of thermal cycles: a considerable change in both these quantities took place during the first 15-20 thermal cycles. Further increase of the number of cycles altered these parameters slightly.

Discussion of Results

The data on the internal friction and the hardness of the blades are given in Fig. 1.

One of the causes of the increase in the average value of the decrement during the first 400-500 h of service life could be the hardening of the material of the blades by the forces acting on it, which has been observed for other materials [15]. The working temperature of the investigated blades was below the recrystallization temperature and the initial stress amplitude in the measurement of the decrement was relatively small.

The decrement could have increased also as a result of additional aging of the alloy, because the alloy was used in its metastable state. The working temperatures of these blades was below the temperature at which aging occurred, but the stresses acting during the service life of the blades would favor further precipitation of the solid solution.

Some increase in the average value of the hardness of the blades compared with $\overline{HB}$ of new blades (Fig. 1b) supported both the first and the second hypotheses.

A similar result has been reported for the internal friction of turbine blades of a reaction engine [8]. The average value of the decrement was again found to be practically independent of the service life of the blades.

The variance of the decrement (Fig. 1b) decreased strongly after the first 200 h in service and after 4000 h it reached its initial value again. This showed that the ability of the material to dissipate energy under working conditions was rapidly recovered and during further service life remained at its initial value.

We must mention that the dependence of S^2_{HB} of the blades on the duration of the service life was almost of the same nature as the dependence of the variance of the decrement (Fig. 1b). This allowed us to assume that the blades were working under favorable stress conditions because the properties of the material of the blades rapidly recovered and remained stable during subsequent service life. The initial period of the service life of the blade was evidently a kind of running-in process, with favorable effects on the ability of the blades to withstand working conditions subsequently.

This nature of the dependence of the average values of the decrement and of the hardness on the duration of the service life of the blades made of deformable nickel alloy was, in general, true also of the analogous dependences obtained for the blades made of the cast-nickel alloy (Fig. 2).

The less well-defined running-in period (Fig. 2a) indicated by a slight fall in the variance of the decrement and in the variance of the hardness (Fig. 2b) in the first stage of the service life of the blades, and the increase in the values of these quantities after 3200 h service life were due to the differences between the materials of the blades and the more difficult working conditions of the blades in the latter case.

We may assume that the constancy of the variance of the decrement during a prolonged service life indicates a sufficiently large margin of the ability of these blades to withstand the working conditions. They show no defects during the service life, in contrast to the results obtained in an earlier investigation [8], when an increase in the decrement was accompanied by the appearance of cracks.

We shall now consider the results obtained in the investigation of the internal friction, hardness, and linear dimensions of the samples subjected to thermal cycling. The average value of the logarithmic decrement of the samples (Fig. 3a) decreased considerably during the first 15-20 thermal cycles. In the 25-50 cycle range, the decrement remained practically constant.

The change in the average value of the length $\bar{l}$ and the diameter $\bar{d}$ of the working part of the samples (Fig. 4) also took place during the first 15-20 thermal cycles. These dimensions changed little during the subsequent cycles, in agreement with the results reported by Stepanov [14].

Stepanov's suggestion [14] of the hardening of the investigated alloys by microplastic deformation and the formation of an intragrain mosaic in the first stage of thermal cycling may be used to explain the nature of the dependence of the average value of the decrement on the number of thermal cycles.

The hardening reduces the internal friction, if such friction is measured at relatively high stress amplitudes, which was indeed the case in our tests. The initial stress amplitude in the samples during the measurements of the logarithmic decrement was 17 kg/mm². There were probably also other causes of the fall in the logarithmic decrement under the action of thermal cycling.

The dependence of the variance of the decrement (Fig. 3a) on the number of thermal cycles indicates that, after a small number of cycles, the alloy recovers its initial value of the internal friction and probably the values of the associated mechanical characteristics of various samples in a batch. Further increase of the number of thermal cycles does not produce any scatter in the properties of various samples.

The dependence of the variance of the hardness on the number of thermal cycles confirms, to some extent, the suggestions put forward here. The appearance of the peak in the hardness variance curve (Fig. 3b) requires further consideration.

The average value of the hardness of the nickel samples is not affected by thermal cycling. Some increase of the average value of the hardness may be explained by the hardening of the material under the action of thermal stresses and possibly final aging of the alloy, because the thermal-cycling temperature was higher than the initial aging temperature. The results obtained in the investigation of the blades should be compared with the results obtained for the nickel samples.

The service life of the blades differed from the conditions under which the nickel samples were tested. However, they had much in common. During their service life the blades were acted upon by stresses caused by mechanical forces and by thermal stresses due to the starting, stopping, and changes in the working conditions of the engine. The nickel samples were acted upon solely by thermal stresses but of higher intensity.

For this reason the curves representing the dependence of the average value of the decrement on the service life of the blades and the dependence of the average value of the decrement on the number of thermal cycles were basically similar (Figs. 1, 2, and 3). The fall of the decrement with increase of the duration of the service life in the former case and the increase with increase in the number of thermal cycles in the latter case could be due to different conditions during the internal-friction measurements.

The same can be said about the variance of the decrement. The action of mechanical stresses and of a certain number of thermal cycles during the initial relatively short period equalized the internal friction of the blades belonging to different batches, i.e., it reduced the variance of the decrement in the same way as the action of a small number of large changes in thermal conditions equalized the internal friction of different samples tested under the same conditions. We may assume that further continuation of thermal cycling would lead to the accumulation of damage in the samples, due to thermal cycling, and the dispersion of the decrement would start to increase.

Literature Cited

1. R.S. Lebedev and V.S. Postnikov, Fiz.Metal.i Metalloved., 7(3) (1959).
2. V.S. Postnikov, Internal friction and strength, in collection: Fatigue Strength of Metals, Izd. Akad. Nauk SSSR (1962).
3. I.L. Mirkin, V.Z. Tseitlin, and G.G. Morozova, Investigation of internal friction and shear modulus of binary nickel alloys, in collection: Investigation of New Heat-Resistant Alloys for Power Engineering, Mashgiz (1961).
4. I.L. Mirkin, V.Z. Tseitlin, and G.G. Morozova, Investigation of the process of aging of nickel alloys, using changes in temperature dependence of the internal friction, in collection: Investigations of New Heat-Resistant Alloys for Power Engineering, Mashgiz (1961).
5. V.S. Postnikov, I.V. Zolotukhin, and G.A. Gorshkov, Investigations of mechanical and thermal fatigue of metals by the internal friction method, in collection: Fatigue Strength of Metals, Izd. Akad. Nauk SSSR (1962).
6. D.N. Vidman and É.S. Ginzburg, Dependence of δ of stainless steel on the structure state and mechanical properties, in collection: Service Reliability of Metals in Steam Power Plant, Gosénergoizdat (1959).
7. L.A. Glikman and M.I. Grinberg, Zh.Tekhn.Fiz., No.9 (1946).
8. F.M. Titov and N.P. Devichenskii, Internal friction determination of the technical state of turbine blade materials in the maintenance of aero engines, Tr. Rizhskogo Inzhenernogo Inst.GVF im.Leninskogo Komsomola (1962).
9. N.P. Devichenskii, F.M. Titov, and V.S. Fastritskii, this volume, p. 211.
10. L.B. Getsov, Behavior of heat-resistant materials under cyclic temperatures and stresses, in collection: Heat-Resistant Alloys under Varying Temperatures and Stresses, Gosénergoizdat (1960).
11. A.A. Bochvar and P.K. Novikov, Dokl.Akad.Nauk SSSR, 112(6) (1957).
12. A.A. Bochvar and E.B. Brovchenko, Dokl.Akad.Nauk SSSR, Vol.127 (1957).
13. S.T. Kishkin and A.A. Klypin, Metalloved. i Term. Obrabotka Metal., No. 5 (1959).
14. V.M. Stepanov, Changing shape of heat-resistant alloys subjected to thermal cycling, Tr. LKVVIA im. Mozhaiskogo, No. 219 (1958).
15. L.A. Glikman and E.A. Khein, Zh.Tekhn.Fiz., No.3 (1954).

INVESTIGATION OF THE ENERGY DISSIPATION
BY VIBRATIONS IN HEAT-RESISTANT ALLOYS
AT HIGH TEMPERATURES

S. O. Tsobkallo, V. A. Chelnokov,
V. I. Timofeeva, I. G. Val'ter,
V. D. Nikitin, and L. A. Privalov

The solution of the various problems, associated with the investigation of the logarithmic decrement of constructional materials, is a pressing matter in physical metallography. The solutions to these problems should be sought in the following areas.

1. Development of an effective method for the measurement of the decrement of materials for machine construction, which makes it possible to carry out measurements both at relatively high stresses, corresponding to the stresses in machine parts, as well as in a wide range of temperatures, corresponding to the requirements of modern technology (in particular, power plant construction).

2. Systematic investigation of the logarithmic decrement of the main machine-construction materials over a wide range of stresses and temperatures for the purpose of comparative determination of their ability to dissipate energy.

3. Investigation of the influence of various physical and technological factors (deformation, temperature, chemical composition, casting conditions, etc.) on the decrement of machine-construction materials. Search for optimum conditions of preparation and treatment of these materials from the point of view of energy by vibrations.

4. Analysis of the nature of changes in the logarithmic decrement under the action of these factors from the point of view of the structural state of materials.

The results of an investigation representing the first of these directions were presented in earlier papers of the present authors [1, 2]. The data given below are concerned with the work in the last three categories.

Three groups of alloys were investigated:

1) the nickel alloy ÉI437B;

2) the austenitic steels ÉI696M and ÉI787;

3) the pearlite steel ÉI415.

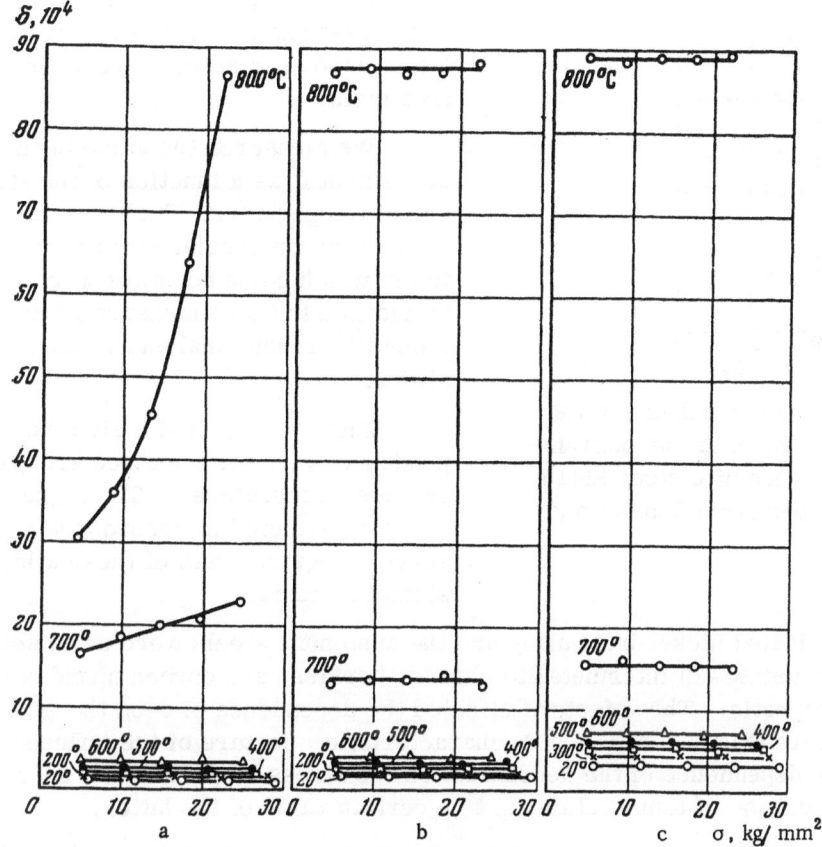

Fig. 1. Amplitude dependence of the logarithmic decrement at various temperatures for the nickel alloy ÉI437B, subjected to various treatments: a) hot-deformed, aged at 700°C for 8 h; b) cast, aged at 750°C for 16 h; c) cast, aged at 700°C for 16 h. In all cases the aging was preceded by quenching in air after heating for 8 h at 1080°C.

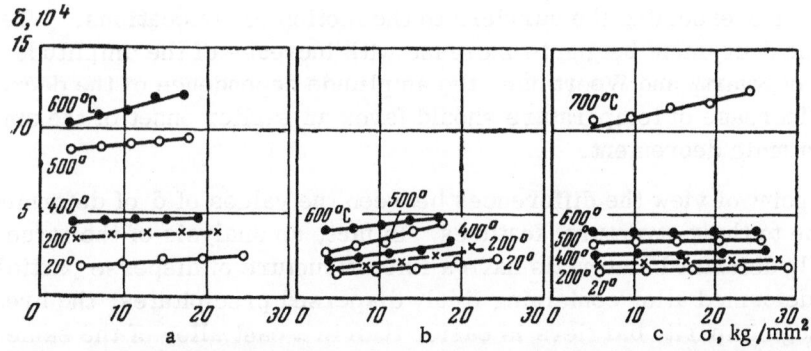

Fig. 2. Amplitude dependence of logarithmic decrement at various temperatures for austenitic steels, subjected to various treatments: a) steel ÉI696M, hot-deformed, quenched in air after heating for 2 h at 1080°C; aged at 750°C for 16 h; b) steel ÉI696M, cast, heat-treated as before; c) steel ÉI787, hot-deformed, quenched in air after heating for 2, 5, and 4 h at 1180 and 1050°C, respectively, aged at 780°C for 16 h.

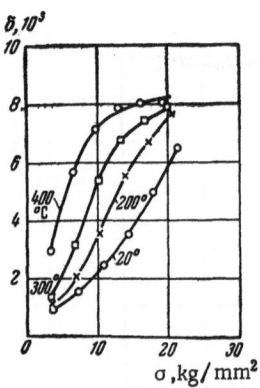

Fig. 3. Amplitude dependence of the logarithmic decrement at various temperatures for the pearlite steel ÉI415, hot-deformed, aged from 1058°C, and tempered at 680°C.

Some of these materials were supplied both in the cast and in the hot-deformed states. In the latter case all materials were subjected to a heat treatment.

We measured the value of the logarithmic decrement δ as a function of the stress amplitude and of temperature. Such measurements were carried out on special samples and using apparatus in which pulse counters were employed [1, 2]. In addition to these measurements, we carried out a metallographic analysis of the structure of the alloys.

Figures 1, 2, and 3 give the amplitude dependences of δ for the three groups of alloys at various temperatures. The highest test temperature was selected in accordance with the possible service requirements of these alloys as heat-resistant materials.

The investigated nickel-base alloy and the austenitic steels were solid solutions hardened by disperse precipitates of intermetallic phases, borides, and carbon nitrides [3]. These alloys were nonmagnetic. These factors affected the dependence of δ on the stress amplitude and of temperature (Figs. 1 and 2). A characteristic feature of the influence of these factors was a weak dependence of the decrement on the stress amplitude and its relatively small increase with increase of temperature up to a certain value of the latter.

These observations can be analyzed qualitatively from the point of view of dislocation theories of the amplitude-dependent internal friction. The theory of Swartz and Weertman [4], which represented further development of the work of Granato and Lücke [5], has attracted most attention. These theories are based on a model according to which a dislocation loop is pinned not only at branching points but also by impurity atoms. The application of a tangential stress in the plane of motion of a dislocation breaks it away from such pinning points. However, further motion of dislocations is impeded by neighboring impurity atoms. The amplitude dependence of the logarithmic decrement does not appear until the stress reaches the value necessary for overcoming the barriers to the motion of dislocations. When such barriers are overcome, the decrement begins to increase with increase of the amplitude. Thus, according to the theory of Swartz and Weertman, the amplitude dependence of the decrement should have a plateau. Increase of temperature should favor an earlier onset of the amplitude dependence of the logarithmic decrement.

From this point of view the differences between the values of δ of deformed and cast alloys should be due to their structural features. In fact, an analysis of the structure shows that the plastically deformed materials have a finer structure of disperse particles than the cast ones. In a deformed alloy containing finely dispersed precipitates, the breakthrough of dislocations through impurity barriers is easier than in a cast alloy of the same composition, but having coarse precipitates in the form of colonies. Therefore, the decrement of a cast alloy and its amplitude dependence (particularly at elevated temperatures) should be smaller than for a deformed alloy (Figs. 2 and 3).

Lowering of the aging temperature reduces the intermetallic phase content and makes these phases more disperse. Therefore, in the alloy ÉI473B, aged at 700°C, the breakthrough of dislocations through barriers should be easier than in the same material aged at 750°C. Consequently, the logarithmic decrement for this alloy is less than for the alloy aged at 700°C (Fig.1).

Comparison of deformed austenitic steels ÉI696M and ÉI787 (Fig. 2) shows that the value of δ is smaller for the steel ÉI787. This can be understood by bearing in mind the higher degree of alloying of the steel ÉI787 compared with the steel ÉI696M and, correspondingly, the presence of a larger number of obstacles to the motion of dislocations. Moreover, the microstructure of the ÉI787 steel contains coarser precipitates of disperse phases, compared with such in the case of the ÉI696M steel.

All the observations discussed so far refer to alloys which were initially homogeneous solid solutions in which dispersion hardening took place later. The logarithmic decrement of the pearlite steel ÉI415 (Fig. 3) was an order of magnitude   larger than for the other alloys discussed above. This could be associated with the heterogeneous structure and magnetic properties of the steel ÉI415.

Literature Cited

1.   S. O. Tsobkallo and V. A. Chelnokov, Zh. Tekhn. Fiz., 24 : 499 (1954).
2.   S. O. Tsobkallo and V. A. Chelnokov, this volume, p. 204.
3.   F. F. Khimushin, Heat-Resistant Steels and Alloys, Izd. "Metallurgiya" (1964).
4.   J. C. Swartz and J. Weertman, J. Appl. Phys., 32 : 1860 (1961).
5.   A. Granato and K. Lücke, J. Appl. Phys., 27 : 583 (1956).

REMOTE-CONTROLLED APPARATUS FOR MEASURING THE INTERNAL FRICTION AND YOUNG'S MODULUS OF RADIOACTIVE MATERIALS

Yu. I. Pokrovskii, V. N. Perevezentsev, and V. I. Vikhrov

The results of investigations of the influence of neutron irradiation on the internal friction and Young's modulus of some metals have been reported in [1-4].

The present communication describes a remote-controlled apparatus for the measurement of the internal friction and Young's modulus, * by means of which the cited investigations have been carried out, as well as the results of investigations of the internal friction due to the thermoelastic effect. The results of the thermoelastic effect investigations are of methodological importance.

The apparatus described here can be used to measure the internal friction ranging from values of $5 \cdot 10^{-2}$ to $5 \cdot 10^{-5}$ in the frequency band 60-600 cps using both flexural and torsional vibrations at room temperature, at elevated temperatures (up to 700°C), and at low temperatures (−196°C). The apparatus can also be used to measure Young's modulus and the shear modulus, as well as to carry out fatigue tests. All the measurements can be carried out over a wide range of vibration amplitudes.

Two methods of measuring the internal friction (the method of damped vibrations and the resonance method) are unified in the apparatus described. The measurements by the damped vibrations method are carried out automatically over a wide range of vibration amplitudes, the results being recorded on a potentiometer chart in the form of an amplitude dependence of the internal friction.

Measurement Method

The influence of radiation on the internal friction and Young's modulus is determined by a comparative method: the same samples are investigated before and after irradiation. To ensure the repeatability of the measurements, which is important in any comparative method, and also to be able to carry out remote-controlled measurements of the irradiated materials, the samples of special shape [1] with thick ends are used.

---

*Author's Certificate No. 823088/26-25 dated March 5, 1963.

Either flexural or torsional vibrations are excited in a sample. As the measure of the internal friction, we use Q^{-1}, which is the logarithmic decrement of the vibrations (δ) divided by π. Higher values of δ (from $5 \cdot 10^{-2}$ to $5 \cdot 10^{-4}$) are determined by the resonance method. In this case Q^{-1} is found from the expression

$$Q^{-1} = \frac{\delta}{\pi} = \frac{2\Delta f}{f_0},\qquad\qquad(1)$$

where f_0 is the natural (resonance) frequency of the vibrations of a sample; $2\Delta f$ is the width of the resonance curve at a level at which the amplitude is 0.7 of the maximum value.

Small values of δ (from $5 \cdot 10^{-3}$ to $5 \cdot 10^{-5}$) are determined by the method of damped vibrations

$$\delta = \frac{\ln A_0/A_n}{n},\qquad\qquad(2)$$

where A_n is the amplitude of an n-th vibration; A_0 is the initial amplitude of the vibrations.

If $A_n = A_0/2$, then δ can be calculated from the number of vibrations n:

$$\delta = \frac{\ln 2}{n} = \pi \; \frac{0.22}{n}.\qquad\qquad(3)$$

If δ depends on the vibration amplitude, which decreases from A_0 to A_n, the average value of the decrement, given by Eq. (2), is used.

The true value of the decrement (n = 1) can be found by plotting the dependence of $\ln A$ on n, and then the true decrement for any value of the amplitude is found from the slope of the tangent to this dependence [5].

Using the apparatus described here, the value of δ can be measured by the method of damped vibrations over a variable range of amplitudes in which the average decrement is determined. For this purpose a discriminator switches on an electronic pulse counter, which measures the number of vibrations n and the time interval during which the initial value of the vibration amplitude (A_0) decreases to $0.5 A_0$, $0.7 A_0$, or $0.9 A_0$. When the vibration amplitude decreases from A_0 to $0.9 A_0$, the value of δ approaches the true value.

Young's modulus (E) is found from the resonance frequency of flexural vibrations, the geometrical dimensions of the sample, and the sample density [3].

General Description

The remote-controlled apparatus consists of a unit employed to excite vibrations in a sample and two sets of electronic instruments. One of these sets is used in the measurement of the internal friction by the resonance method, and the other is used in the method of damped vibrations. The electronic measurement (indicating) instruments are common to both groups. These instruments form a control panel, which is placed in a laboratory. The unit employed to excite vibrations is placed in a protective box and is used to clamp a radioactive sample and to excite vibrations of required amplitude in it. A block diagram of the unit used to excite vibrations and of the protective box with remote controls is given in Fig. 1. The box is fitted with a viewing system 4, manipulators 3, and rubber gloves 5. To measure the vibration amplitude of a sample 1, a remote-controlled microscope 2, constructed from a microscope of the MBS-2 type, is employed. To attach ferromagnetic pieces 10 to a nonmagnetic sample 1, screw fixings 6 are employed. To heat a sample to 700°C, a dc furnace 9 is used. At low temperatures (down to −196°C), a different variant of the apparatus 7 is used with a liquid-nitrogen cryostat 8. The placing of this unit in a lightened protective box with gloves makes it

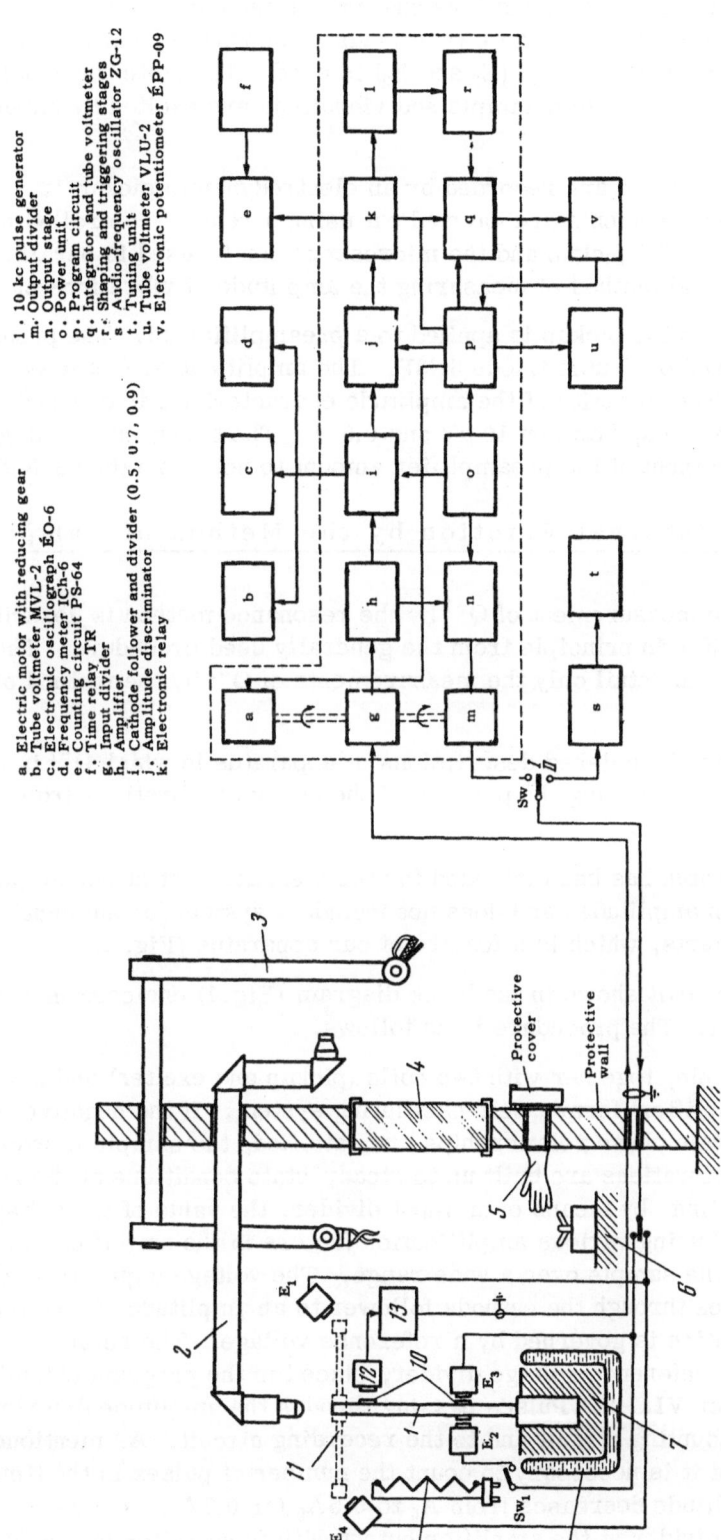

a. Electric motor with reducing gear
b. Tube voltmeter MVL-2
c. Electronic oscillograph ÉO-6
d. Frequency meter ICh-6
e. Counting circuit PS-64
f. Time relay TR
g. Input divider
h. Amplifier
i. Cathode follower and divider (0.5, 0.7, 0.9)
j. Amplitude discriminator
k. Electronic relay

l. 10 kc pulse generator
m. Output divider
n. Output stage
o. Power unit
p. Program circuit
q. Integrator and tube voltmeter
r. Shaping and triggering stages
s. Audio-frequency oscillator ZG-12
t. Tuning unit
u. Tube voltmeter VLU-2
v. Electronic potentiometer ÉPP-09

Fig. 1. Block diagram showing the protective box with an exciting system and the apparatus used to measure the internal friction and Young's modulus. I) Position of switch Sw for the measurement of the internal friction by the method of damped vibrations; II) position of the switch Sw for the measurement by the resonance method. Part of the circuit enclosed by a dashed line represents the unit employed in the automatic measurement of the internal friction by the method of damped vibrations.

possible to carry out measurements on samples of activities up to 100 microcuries at elevated and low temperatures. Both transverse and torsional vibrations can be excited. The transverse vibrations are excited by means of an electromagnet exciting system E_3. To carry out fatigue tests a differential exciting system (E_3 and E_2) is used. To excite torsional vibrations, a special attachment (11) is fitted onto a sample and vibrations are excited by means of systems E_1 and E_4.

The vibrations of the system are recorded by an electrodynamic pickup 12. At high vibration amplitudes the measurements are carried out using a remote-controlled microscope with an ocular micrometer. The pickup and the microscope can be used in parallel, and thus we can calibrate the electrical method of measuring the amplitude of vibrations of a sample.

The voltage signal from the pickup is applied to a preamplifier 13. The preamplifier consists of two stages, based on double triode 6N3P. The amplification factor is 20, and the pass band is 60-600 cps. The linearity of the amplitude characteristic is checked by varying the voltage signal from the pickup between 10 μV and 0.5 V. The background voltage and the stray voltages reaching the input of the preamplifier amount to not more than 2-3 μV.

Measurement of the Internal Friction by the Method of Damped Vibrations

The apparatus for the measurement of Q^{-1} by the resonance method is described in [3] and the method does not differ in principle from the generally used procedure. Therefore, we shall consider in this paper in detail only the measurements of Q^{-1} by the method of damped vibrations.

There have been several published descriptions of apparatus in which Q^{-1} is measured by electronic counting of pulses during the process of the decay of vibrations from A_0 to $A_0/2$ [6-8].

However, such apparatus has been intended for the measurement at one amplitude or in a narrow range of vibration amplitudes and does not include a system for automatic recording of the results of measurements, which is a feature of our apparatus (Fig. 1).

Part of the general circuit shown in the block diagram (Fig. 1) enclosed in a dashed box represented a separate unit. The procedure is as follows.

The investigated sample, together with two coils (pickup and exciter) and amplifiers (preamplifier and main amplifier) form a self-oscillating system. If the negative feedback is sufficiently strong and certain phase relationships are satisfied, the sample is excited at its natural frequency and the vibrations are built up to steady-state conditions at the required vibration amplitude. Regulating, by means of an input divider, the value of the voltage applied to the exciting system, and using various amplification factors in the amplifier, we can vary the vibration amplitude of the sample over a wide range. The voltage signal from the output of the main amplifier passes through the cathode follower to an amplitude discriminator and the threshold of discrimination is governed by a reference voltage. The reference voltage is applied from a logarithmic reference voltage divider, placed in the program circuit and controlled with a tube voltmeter VLU-2. Pulses passing through the amplitude discriminator are shaped and applied to the counting circuit and to the recording circuit. As mentioned earlier, to determine the decrement it is necessary to count the number of pulses in the time interval in which the vibration amplitude decreases from A_0 to $0.5 A_0$ (or $0.7 A_0$, or $0.9 A_0$). For this purpose there is a voltage divider at the amplifier output with fixed ratios of 0.5, 0.7, 0.9, and 1.0 of the value of the voltage applied to it; the divider is switched over to the required position by means of a fast relay.

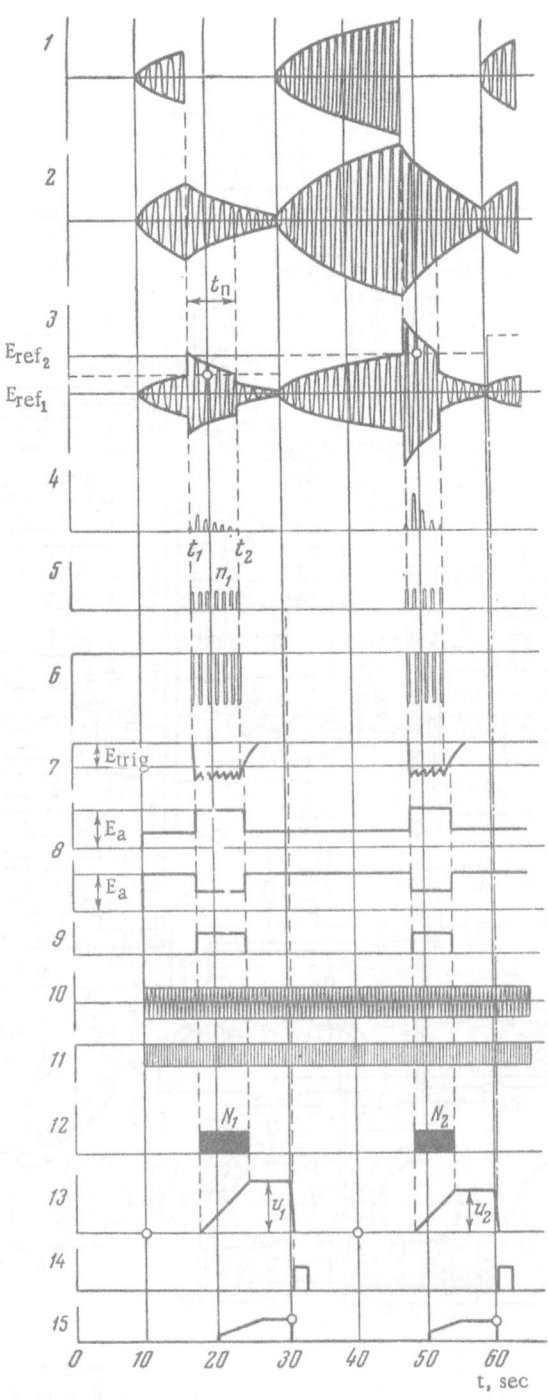

Fig. 2. Simplified dependences of voltage on time at various points in the circuit: t_n) time interval in which vibrations decay from initial amplitude A_0 (at time t_1) to $A_0/2$ (at time t_2); n_1) number of pulses passing through discriminator; N_1) number of pulses (F = 10 kc) reaching integrator; U_1) output voltage of integrator, directly proportional to n_1, i.e., to logarithmic decrement. 1) Exciting system input; 2) amplifier output; 3) discriminator input; 4) discriminator output; 5) PS-64 input; 6) detector input; 7) detector output; 8) anode voltage in electronic relay tubes; 9) electromagnetic relay; 10) 10-kc pulse generator; 11) pulse shaper; 12) integrator input; 13) integrator output; 14) step-by-step switch winding; 15) ÉPP-09 input.

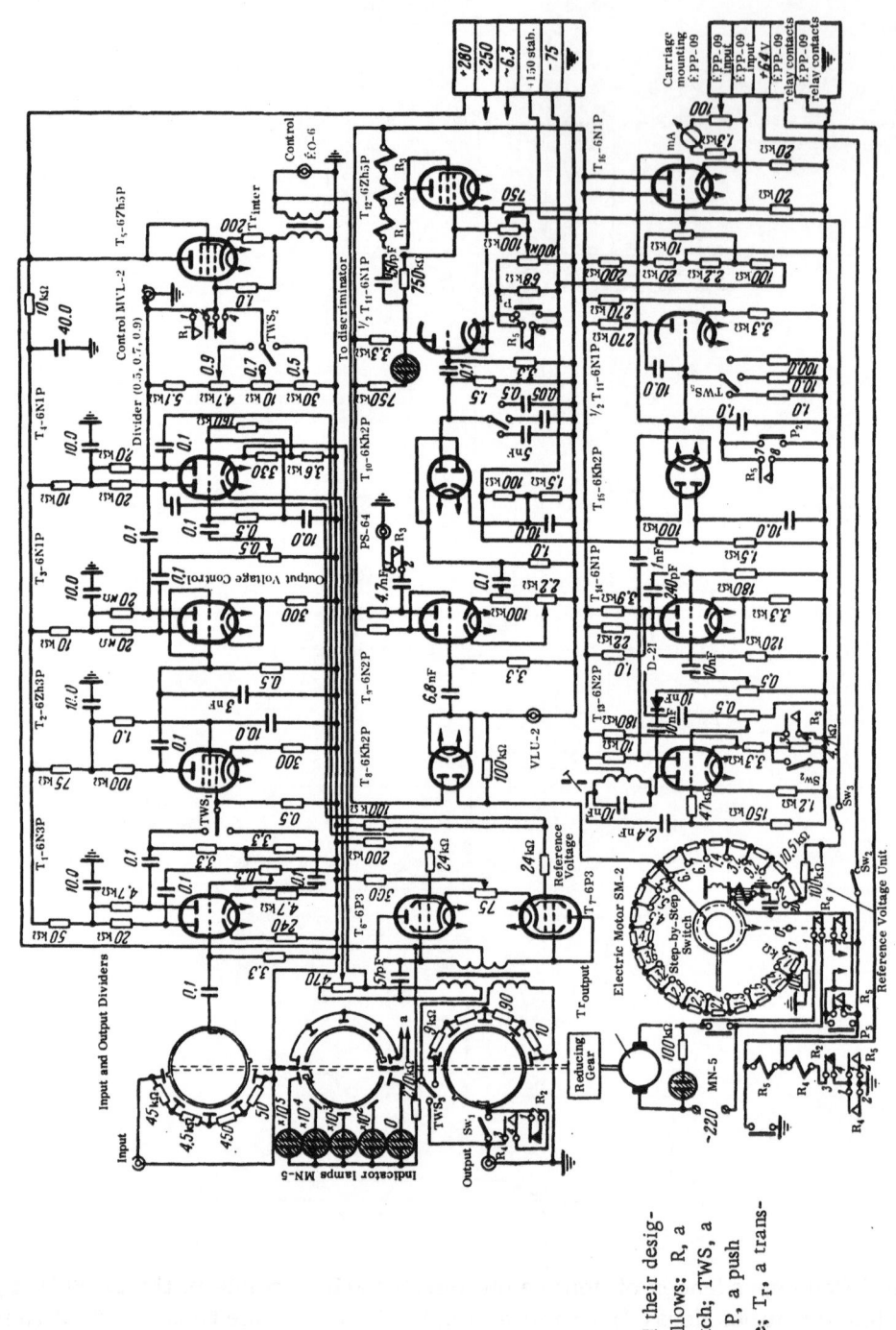

Fig. 3.   Basic electronic circuit for the automatic measurement of the internal friction by the method of damped vibrations.

The symbols and their designations are as follows: R, a relay; Sw, a switch; TWS, a two-way switch; P, a push switch; T, a tube; T_r, a transformer.

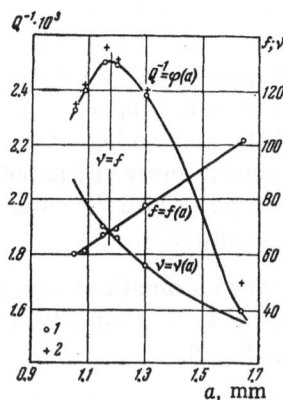

Fig. 4. Dependence of the thermoelastic internal friction, the frequency of vibrations of a sample (f), and of the relaxation "frequency" (ν) on the sample thickness (a). 1) Calculated values of the internal friction; 2) experimental values of the internal friction.

Two operations are possible: semiautomatic with the results being recorded by means of the counting system, and automatic, with the amplitude dependence of the internal friction being recorded on the chart of a potentiometer.

We shall consider the operation of the apparatus when the vibration amplitude decays to $0.5 A_0$, because the operation in the case of decay to $0.7 A_0$ (or $0.9 A_0$) is in principle completely identical. The system operates as follows. The excitation is switched on after applying the necessary value of the reference voltage, corresponding to the required vibration amplitude. The sample begins to vibrate at an increasing amplitude and the voltage at the amplifier output increases correspondingly. The relay controlling the voltage divider is in a position such that half of the output voltage of the amplifier is applied to the discriminator. When the sample vibrations reach the amplitude equal to A_0, the voltage at the discriminator input becomes equal to the reference voltage, the discriminator operates, and the output pulse from the discriminator trips the electronic relay. This relay switches off the exciting system and the voltage divider is switched over to a position such that the total voltage is applied to the discriminator. The sample vibrations begin to decay and when their amplitude decreases to half, i.e., to the reference voltage, no more pulses are produced at the discriminator output. The electronic relay operates again, and again half the voltage from the divider is applied to the discriminator. The exciting system is switched on again, and the whole cycle (described above) is repeated.

Thus, the counting system counts the number of pulses which correspond to the decay of vibrations from A_0 to $A_0/2$. Setting a new value of the reference voltage, we can measure the number of pulses for a different vibration amplitude. Knowing the number of pulses, we can calculate the value of the internal friction from Eq. (3).

Simplified curves representing voltages at various points in the circuit explain the operating principle of the apparatus (Fig. 2).

When the automatic mode of operation is used, the reference voltage at the discriminator is obtained from a program circuit, connected to an automatic recording potentiometer ÉPP-09, which controls the duration of the recording cycle. After one cycle, the reference voltage increases in a discrete step from 10 to 100 V. The reference voltage is varied logarithmically. This is because the amplitude dependence $Q^{-1} = \varphi(A)$ is usually plotted in semilogarithmic coordinates ($\ln A$ is plotted along the abscissa). When the reference voltage reaches its maximum value (100 V) the next amplification level is switched on. The input voltage divider is operated by means of a small electric motor (in a later variant of the apparatus a step-by-step switch is used).

Four amplification levels are used (10^2, 10^3, 10^4, 10^5). At each amplification level the reference voltage is applied in 20 discrete steps. Thus, the total number of measured and recorded points in the amplitude characteristic is 80.

The results are recorded as follows. The time interval for the vibrations to decay to half the amplitude, i.e., the interval from A_0 to $A_0/2$, is filled with a "packet" of pulses supplied by an internal pulse generator at 10 kc to the shaping and triggering stages. The pulse "packet" is applied to a frequency integrator, which transforms the packet into a constant voltage. This voltage is memorized and applied through a tube voltmeter to the input of a potentiometer ÉPP-09 capable of recording three curves. At the end of the measuring cycle, this voltage is recorded in the form of a point on the paper chart. The second input of the potentiometer is used for control record of the value of the reference voltage, while the third input may be used to check the integrator zero position or to record temperature.

Figure 3 shows the basic circuit of the apparatus as used for the automatic measurement of the internal friction by the method of damped vibrations. The numbers 1-15 in Fig.3 correspond to the simplified curves representing the voltages shown in Fig.2.

The voltage signal from the preamplifier is applied to the input of the step-by-step voltage divider and then to a voltage amplifier, which includes tubes T_1, T_2, T_3; a tube T_4 is the basic element of a phase inverter, and tubes T_6 and T_7 represent a power amplifier with two stages. The amplifier is connected to several feedback circuits. The amplified voltage from the secondary winding of the output transformer is applied through an output divider to the exciting system. The power amplifier supplies a power of not less than 5 W in the frequency band 60-600 cps to the exciting system. A tube T_5 forms part of a second output stage. The grid circuit includes the divider $0.5-0.7-0.9-1.0$, which is operated by means of contacts in a relay R_1. The use of a step-up transformer in the cathode circuit makes it possible to obtain an output voltage up to 200 V with a good linearity of the amplitude characteristic of the amplifier.

The diode discriminator includes a double diode (tube T_8). The reference voltage is applied to the discriminator from a logarithmic divider, based on a step-by-step switch R_6.

The pulses from the discriminator are applied to a multivibrator with a cathode coupling in a tube T_9. From the anode of the tube T_9, pulses of positive polarity are applied to a counting circuit and negative pulses are applied from the cathode to a diode detector based on a tube T_{10} with an RC storage circuit. The trigger includes tubes T_{11} and T_{12}. The trigger is operated by the rise and fall fronts of a negative pulse reaching a storage capacitor. The anode of the tube T_{12} is connected to relays R_1, R_2, R_3, which ensure the necessary switching in the circuit.

A tube T_{13} forms a part of the 10-kc frequency pulse generator and of the triggering stage. A tube T_{14} forms part of the shaping stage.

The integrator includes a tube T_{15} and half of T_{11}. The stage including the tube T_{11} is used to increase the effective value of the circuit time constant RC by a factor of $k + 1$. The voltage from the integrator is applied to a tube voltmeter, which includes a tube T_{16}. The voltage from a 100-Ω resistor in the cathode circuit is applied to the input of the potentiometer ÉPP-09.

The necessary commutation in the circuit is achieved by means of relays R_4 and R_5, which act after the closing of a contact relay included in ÉPP-09.

The accuracy of the measurement of the internal friction by the resonance method for Q^{-1} from $5 \cdot 10^{-2}$ to $5 \cdot 10^{-4}$ is not less than 4% and in the method of damped vibrations the accuracy is not less than 4% for Q^{-1} from $5 \cdot 10^{-3}$ to $5 \cdot 10^{-4}$, but for values from $5 \cdot 10^{-4}$ to $5 \cdot 10^{-5}$ it is not less than 2%. (The accuracy was determined from the scatter of individual points in the curve representing the amplitude dependence of the internal friction in the region $\sigma \leq \sigma_{cr}$.)

The accuracy of the measurement of the natural frequency of vibrations of a sample in the comparative method of measuring Young's modulus is $\pm 0.2\%$.

Control measurements of the internal friction due to the thermoelastic effect were used to check the operation of the apparatus and to estimate the internal-friction background.

The thermoelastic internal friction, investigated by us, was due to heat flow across a sample. The relaxation effect could be associated with the establishment of thermal equilibrium in a time interval τ

$$\tau = \frac{1}{v} = \frac{a^2}{1.57D}, \quad \text{where } D = \frac{\varkappa}{\rho c},$$

$\varkappa$ is the thermal conductivity; c is the specific heat; ρ is the density; a is the sample thickness.

The value of the internal friction depends on the relationship between τ and the vibration period $(1/\omega)$

$$Q^{-1} = \Delta \frac{v}{\omega^2 + v^2}, \quad \text{where } \Delta = \frac{E_\infty - E_0}{E_0}.$$

When $\omega = v$, an internal-friction maximum Q^{-1}_{max} is observed.

When the vibration frequency is varied, Q^{-1} of the sample lies on an intersection of two curves

$$\omega = \omega(a), \quad v = v(a).$$

Figure 4 shows, for magnesium, the calculated curve $Q^{-1}(\omega)$, as well as the curve plotted from the experimental results; the position of Q^{-1}_{max} corresponds to the intersection of two curves: $f = f(a)$ and $v = v(a)$.

These results for magnesium are in good agreement with the results of Barducci [9], i.e., the internal friction due to the thermoelastic effect appears mainly in the frequency range 20-100 cps.

Literature Cited

1. S.T. Konobeevskii, N.F. Pravdyuk, and V.N. Kutaitsev, Investigations in Geology, Chemistry, and Metallurgy (Papers presented by the Soviet Delegation at the First International Conference on Peaceful Uses of Atomic Energy, Geneva, 1955), Izd. AN SSSR (1956), p. 165.
2. N.F. Pravdyuk, S.T. Konobeevskii, A.D. Amaev, Yu.I. Pokrovskii, Papers Presented by Soviet Scientists, Vol. 3, Proceedings of the Second International Conference on Peaceful Uses of Atomic Energy, Geneva, 1958, Izd. AN SSSR (1959), p. 610.
3. N.F. Pravdyuk, Yu.I. Pokrovskii, and V.I. Vikhrov, Collection of Papers Presented at All-Union Conference on the Effects of Nuclear Radiations on Materials (1962).
4. N.F. Pravdyuk, Yu.I. Pokrovskii, and V.I. Vikhrov, Soviet Journal of Atomic Energy, 10(4): 334 (1962).
5. Elasticity and Anelasticity of Metals (collection of papers, S.V. Vonsovskii, ed.), IL (1954), p. 201.
6. A.E. Fedorovskii, Zavodsk. Lab., No. 12: 1517 (1958).
7. A. Smith, Sci. Inst., 28(4): 106 (1951).
8. S.O. Tsobkallo and V.A. Chelnokov, Zh. Tekhn. Fiz., 24(3): 499 (1954).
9. I. Barducci, Aluminium, Vol. 4 (1950).

APPARATUS FOR THE MEASUREMENT OF THE ENERGY DISSIPATION BY VIBRATIONS IN MATERIALS AT HIGH TEMPERATURES AND STRESSES

S. O. Tsobkallo and V. A. Chelnokov

The present communication describes a method developed earlier [1], modified for the measurement of the logarithmic decrement at high temperatures.

The measurements of the decrement were carried out employing flexural vibrations of samples 300 mm long (Fig. 1). A sample consisted of three parts: the middle part made of the material being tested, and two outer parts which were ferromagnetic soft steel pieces, attached by cold-friction welding. It should be mentioned that this method makes it possible to weld the steel ends to most metals and alloys (heat-resistant steels, aluminum and copper alloys, etc.).

The middle part of the tested material was recessed to form a cylindrical piece of length less than one tenth of the total composite sample. This ensured that, during flexural vibrations of the composite rod, the energy dissipation was concentrated mainly in the material of which this cylindrical part was made. Moreover, practically constant stresses were acting along the length of the cylindrical piece.

During measurements, the composite sample 2 (Fig. 2) was fixed at two nodal points by a suspension 3 and could undergo transverse vibrations. The vibrations were excited by means of a pickup and an exciter placed at the free ends of the rod. These units were coupled through an amplifier, forming a self-oscillatory system. The amplitude of the vibrations of a sample a was measured by means of a microscope. From these measurements we calculated the highest stresses σ_{max} in the external filament of the working part of the sample

$$\sigma_{max} = Ead/2ll_c, \tag{1}$$

where E is Young's modulus; d and l are, respectively, the diameter and length of the sample; l_c is the distance from the end of the sample to a vibration node.

The logarithmic decrement was calculated from the formula

$$\delta = \frac{1}{N} \ln \frac{a_n}{a_{n+N}}, \tag{2}$$

Fig. 1. Construction of a sample used in the measurement of the logarithmic decrement of materials at high temperatures. 1) Part made of the investigated materials; 2) welded joint; 3) lengthening pieces made of ferromagnetic material.

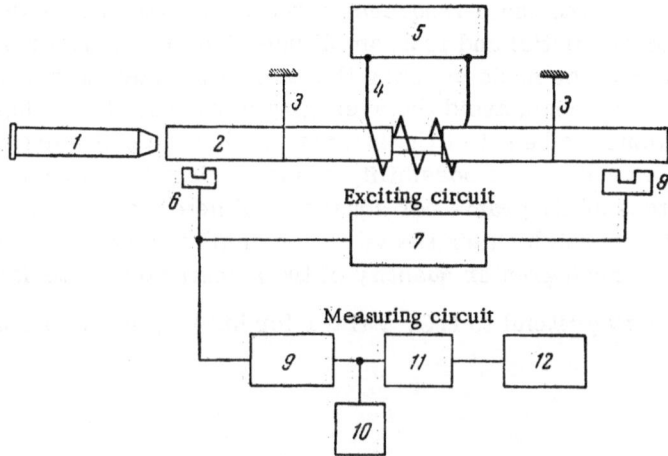

Fig. 2. Schematic diagram of the apparatus used to measure the energy dissipation by vibrations in materials at high temperatures. 1) Microscope; 2) sample; 3) suspension; 4) inductance; 5) high-frequency heating units; 6, 8) pickups; 7) exciting system amplifier; 9) measuring amplifier; 10) tube voltmeter; 11) discriminator; 12) counting circuit.

where N is the number of vibrations in the time interval in which the amplitude decreases from the value a_n to a_{n+N}. The value of N was found by means of a special device in the measuring circuit of the apparatus (Fig. 2). For this purpose the electrical oscillations excited in the pickup were applied, after amplification, through a discriminator to a counting circuit, which recorded these oscillations if the amplitude was greater than the discrimination threshold a_{n+N}.

An important part of the apparatus was a heating unit 4-5 (Fig. 2), consisting of a high-frequency source LGZ-10A and an inductance. The power consumption of this unit was up to 8 kW at a frequency of 300-450 kc.

During the determination of the logarithmic decrement, we measured and controlled the sample temperature with a thermocouple made of wire not more than 0.1 mm thick. The hot junction of the thermocouple was soldered to the sample. To minimize the energy losses through the thermocouple, the point at which the thermocouple was soldered and its leads were located in the neutral plane of bending of the sample.

The temperature of the sample was controlled by means of a special circuit in which the pickup was the thermocouple referred to above. The emf of the thermocouple acted on an electronic potentiometer. A potentiometer relay governed the value of the negative potential across a grid of an oscillator tube in the high-frequency unit, which altered the power of the generated electrical oscillations and the sample temperature.

The apparatus could be used to carry out measurements at any high temperature of practical interest; for modern heat-resistant alloys this could be done up to 800–900°C [2].

The apparatus could be employed to determine the true value of the logarithmic decrement; the suggested shape of the sample made it possible to ensure a constant stress along its working length. Moreover, the radio-frequency circuit ensured that the decrement was measured for a small ratio of the initial and final amplitudes (10/9). Further approach to the true value of the decrement could be made by using the calculation method developed by the present authors [1], using which one could avoid the averaging of the amplitude-dependent decrement by varying the stress along the height of the sample which was undergoing flexural vibrations. These circumstances, as well as the possibility of measuring the decrement at high stresses (amounting to several tens of kilograms per square millimeter) distinguish the proposed apparatus from others [3, 4], in which either the vibration amplitude is small or the working part of the sample requires a much greater quantity of the investigated material.

The authors are very grateful to I.G. Val'ter for his help in the preparation of the special-shaped samples.

Literature Cited

1. S.O. Tsobkallo and V.A. Chelnokov, Zh. Tekhn. Fiz., 24:499 (1954).
2. S.O. Tsobkallo, V.A. Chelnokov, et al., this volume, p. 190.
3. M.G. Lozinskii and A.E. Fedorovskii, Izv. Akad. Nauk SSSR, Otd. Tekhn. Nauk, No. 6:19 (1958).
4. G.S. Pisarenko, Energy Dissipation in Mechanical Vibrations, Izd. Akad. Nauk UkrSSR (1962), p. 72.

APPARATUS FOR THE MEASUREMENT OF
THE INTERNAL FRICTION IN THE KILOCYCLE
RANGE OF FREQUENCIES

V. I. Razumov and V. S. Postnikov

The apparatus is intended for the measurement of the internal friction, the elasticity moduli, and the velocity of propagation of sound in solids. The working frequency range is from 6 to 60 kc. The low-frequency limit of the range is governed by the capability of a differential discriminator used in the apparatus; the upper limit is set by the maximum frequency of an oscillator used to excite a sample. The apparatus can be adjusted easily to vary the vibration amplitude and to measure the amplitude dependence of the internal friction; the maximum vibration amplitude can be made to rise from the minimum value by not less than two orders of magnitude. The temperature range from liquid nitrogen temperature to 200°C can be used.

The internal friction is measured by the method of damped free vibrations. It is also possible to determine the internal friction from the width of the resonance curve (for $Q^{-1} < 5 \cdot 10^{-3}$). The advantage of the apparatus described here, compared with other apparatus of similar type, is the simplicity of the radio-frequency units because all the measurements are carried out at the vibration frequency of the sample without any frequency conversion. This increases the reliability of the apparatus and of the results. The disadvantages of the apparatus are the effects of fine wires used to excite and record the vibrations of a sample and the presence of natural resonance frequencies of these wires, which demands great care in the selection of the frequencies at which the measurements are carried out.

Figure 1 shows the basic circuit. The apparatus works as follows.

A voltage from an oscillator 1 is applied to a changeover switch to a piezoelectric transducer 2. Mechanical vibrations of the piezoelectric crystal are applied through a thin elastic wire 4 to the end of a sample, exciting longitudinal acoustic waves in it. When the frequency of the exciting stress is identical with the natural resonance frequency of the rod, the amplitude of vibrations of the latter increases sharply. For longitudinal vibrations the fundamental frequency of a rod clamped in the middle is

$$f_{\mathrm{r}} = \frac{v}{2l} , \qquad (1)$$

where v is the velocity of propagation of acoustic vibrations in the rod, and l is the length of the rod.

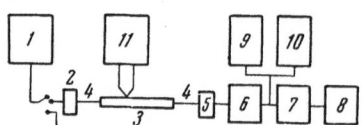

Fig. 1. Schematic diagram of the apparatus. 1) Oscillator; 2) piezoelectric exciter of vibrations; 3) sample; 4) thin wires; 5) piezoelectric receiver of vibrations; 6) electronic amplifier; 7) discriminator; 8) counting unit; 9) tube voltmeter; 10) cathode-ray oscillograph; 11) potentiometer and thermocouple.

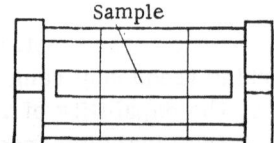

Fig. 2. Suspension of a sample.

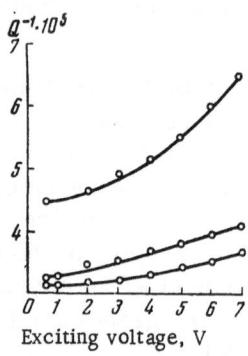

Fig. 3. Dependence of the internal friction of pure aluminum on the vibration amplitude for three samples. Frequency $\nu = 50$ kc. The vibration amplitude is given in relative units.

For the first harmonic, this frequency is twice as high.

Vibrations of the rod are recorded at the other end by means of a piezoelectric receiver, similar to the exciting unit.

The sample is in the form of a cylindrical rod with planar ends perpendicular to the rod axis. The sample length is 4-12 cm and its diameter is 0.4-1.0 cm. Samples of different cross sections can also be used. For longitudinal resonance vibrations the sample is clamped at the nodal points (this clamping should be carried out carefully). Depending on the method of suspension of the sample, the measured internal-friction values may differ by more than one order of magnitude. Prolonged search resulted in the selection of a suspension method so that the sample was supported at two nodal points (Fig. 2). This method ensured a sufficiently reproducible fixed position of the sample in the working chamber and had small damping influence on the vibrations in the rod. The first resonance frequency is then

$$f_{\mathrm{r}} = \frac{v}{l}.$$

Mechanical vibrations of the rod are transformed by the piezoelectric receiver into electrical oscillations which are amplified by an electronic amplifier to a value of 30 V, necessary for the normal operation of the differential discriminator. The amplifier has to satisfy the following requirements: a sufficiently high amplification factor (not less than 10^4); the lowest possible level of intrinsic noise, which (at maximum amplification) should not exceed $\frac{1}{20}$ of the value of the output signal; a good stability of the amplification factor during many-hour operation; a continuous or step-wise adjustment of the amplification factor over a range not smaller than 60 dB; an exact calibration of the amplification conditions, which is necessary for recording the amplitude dependence of the internal friction; a sufficiently wide pass band of the amplifier within the range of frequencies employed. The amplifier constructed by us satisfied all these requirements. The operation of the amplifier was checked by means of a cathode-ray oscillograph and a tube voltmeter.

The amplified voltage is applied from the amplifier output to a differential discriminator. The differential discriminator passes only the oscillations whose amplitude a is within the limits between a_1 and a_2. The fixed amplitudes a_1 and a_2 are selected in accordance with the requirements of the investigator, within the limits from 3 to 50 V.

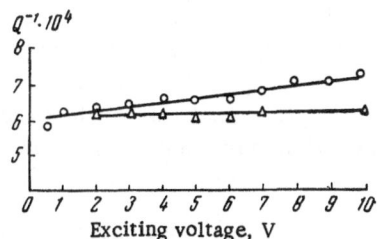

Fig. 4. Dependence of the internal friction of pure copper on the vibration amplitude; $\nu = 38.7$ kc.

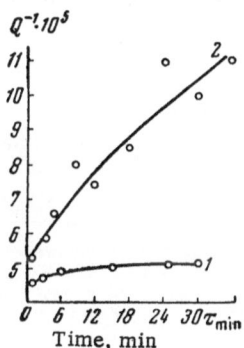

Fig. 5. Dependence of the internal friction of aluminum, measured at a constant vibration amplitude, on the duration of excitation of a sample. Frequency $\nu = 50$ kc. 1) Exciting voltage 7 V; 2) exciting voltage 10 V.

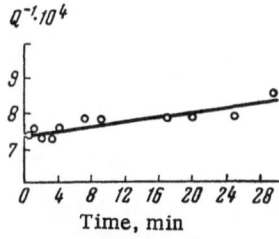

Fig. 6. Dependence of the internal friction on the duration of excitation of a copper sample (at a constant vibration amplitude). Frequency $\nu = 38.7$ kc.

Each period of the alternating voltage of amplitude satisfying the conditions stated above produces, at the discriminator output, a single short pulse. The requirements that the discriminator has to satisfy are:* sharp discrimination thresholds (fixed to within not less than 5 mV); the stability of the discrimination thresholds during prolonged tests; a flat response over the whole investigated range of frequencies for sinusoidal input signals; the shape of output pulses independent of the amplitude and frequency of the input signal.

The pulses from the discriminator output are applied to a counting unit and are counted.

Thus, if the discrimination thresholds are set, for example, at 10 and 20 V, the vibrations of the sample are increased to an amplitude giving a voltage of 30 V at the amplifier output, and the excitation of vibrations is suddenly switched off, the vibrations of the sample become free and decay with time. The counter counts the number of vibrations during a time interval such that the amplitude decreases from 20 to 10 V, i.e., to half its original value. The internal friction is found from the well-known expression for the case of damped free vibrations

$$Q^{-1} = \frac{\ln a_{\text{init}}/a_{\text{fin}}}{n}, \qquad (2)$$

where n is the number of counted vibrations.

The internal friction is measured with an accuracy of 1%. In the case of strong damping the internal friction can be determined from the width of the resonance curve.

In this case we record the resonance curve with an electronic potentiometer and the tuning dial of the main oscillator is rotated by means of a synchronous motor through a reducing gear which slows down the motor considerably. The internal friction is then

$$Q^{-1} = \frac{\Delta f}{f_{\text{r}} \sqrt{3}}, \qquad (3)$$

where Δf is the width of the resonance curve at half the amplitude; f_{r} is the resonance frequency. The internal friction can be measured by this method with an accuracy of up to 5%.

The elasticity moduli are determined from the resonance frequencies of a cylindrical rod subjected to longitudinal or flexural vibrations; the determination is accurate to within 0.1%.

---

*The discriminator constructed in our laboratory has been working well for more than a year.

The frequencies are measured exactly by means of an electronic counter.

The sample should be cooled in vacuum, because if it is cooled in air, good repeatability is not obtained. If the experiments are carried out sufficiently carefully, the results are repeatable. Figures 3 and 4 show the results of an investigation of the amplitude dependence of the internal friction at room temperature. It is clear from these figures that the method can be used to investigate the internal friction in the amplitude-independent region of some metals only.

For example, in the case of aluminum (Fig. 3) the measurement of Q^{-1} was carried out in the amplitude-dependent region. To go over to the amplitude-independent region of Q^{-1} of aluminum it was necessary to reduce the vibration amplitude and to increase the sensitivity of the amplifier, which was difficult because of the internal noise, which limited the amplification factor.

In carrying out the experiments it was necessary to allow also for the considerable sensitivity of the internal friction to the total duration of the excitation of the sample (Figs. 5 and 6), which was due to fatigue effects.

APPARATUS FOR SEMI-AUTOMATIC MEASUREMENT OF THE LOGARITHMIC DECREMENT OF FREE VIBRATIONS OF GAS-TURBINE BLADES

N. P. Devichenskii, F. M. Titov, and V. S. Fastritskii

The apparatus described here is intended for the measurement of the logarithmic decrement of vibrations of turbine blades and it has a special unit for the transformation of mechanical vibrations into electrical oscillations, so that the process of measurement is made automatic and the value of the decrement can be found in several minutes. Descriptions [1, 2] have been published of apparatus for the automatic measurement of the logarithmic decrement of free vibrations, but this apparatus was intended for measurements on samples of simple shape, made of thermomagnetic materials.

The basic layout of the apparatus is shown in Fig. 1. The apparatus consists of the following units: a hydraulic press for holding the blade under investigation, a hydraulic system for supplying the hydraulic press, a mechanism for the excitation of vibrations in the blade, a unit for semiautomatic determination of the logarithmic decrement of vibrations. The mechanical part of the apparatus is mounted on a rigid foundation.

The hydraulic press is in the form of a massive cast body, in which a steel cylindrical sleeve is fitted. The cylinder contains a piston with rubber sealing rings made of oil-resistant rubber. Oil is used as the working substance in the press. To clamp and to free a blade, oil is applied under pressure into the cylinder through connecting pipes at the top cover of the cylinder and at the back wall of the cast body.

The pressure in the press cylinder can be varied over a wide range up to a maximum value of 400 atm, which corresponds to a clamping force of 45 tons. The force, applied by the piston through a rod and a spherical thrust bearing, is exerted on a locking device which clamps the blade.

The hydraulic system supplies the press. This system is on a separate stand and it consists of the following main elements: a pump, driven by an electric motor; a hydraulic accumulator; a filter; a two-position valve; a manometer; flexible hoses. The clamping and freeing of a blade is carried out (with the hydraulic system connected up) by rotating the valve into the required position, which sends the oil into the space in the cylinder either below or above the piston.

The mechanism for excitation of vibrations of a blade consists of a two-armed lever, fixed to a movable bracket. The axis of rotation of the lever is located eccentrically, so that

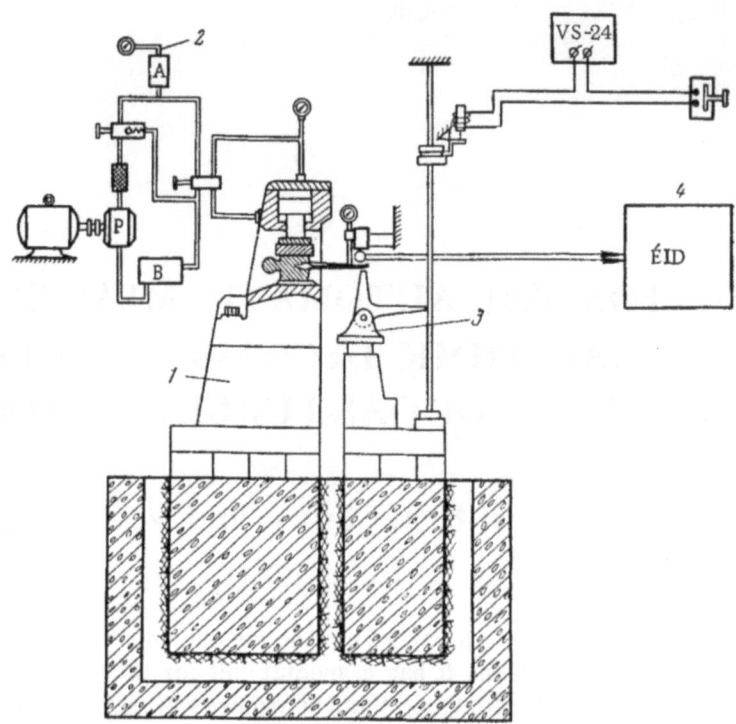

Fig. 1. Schematic diagram of the apparatus used to measure the logarithmic decrement of vibrations in turbine blades. 1) Hydraulic press; 2) hydraulic supply system; 3) device for exciting vibrations; 4) electronic apparatus for measuring the decrement (ÉID). P) Pump; A) hydraulic accumulator; B) booster.

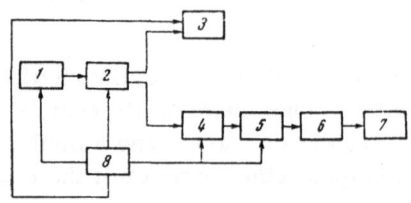

Fig. 2. Block diagram of ÉID-1.

by rotating it we can move the lever up or down and thus apply an initial bending force to the end of a blade.

To make it possible to investigate blades from various turbines of different dimensions, the movable bracket to which the lever is attached is placed on a horizontal table which is fixed to the upper runners of a co-ordinator. The latter can be moved in three directions: transversely and longitudinally by means of screws, and vertically by means of a worm gear and a coupled screw in the cast body. The lever can be rotated about an axis parallel to the axis of rigidity of the clamped turbine blade.

To excite vibrations in a blade, the latter is deflected by means of this lever from its neutral position. At the required moment the blade is released by dropping a freely falling disk on the lever.

To transform mechanical vibrations into electrical oscillations, and to count the number of vibrations, two variants of electronic measuring units (ÉID-1 and ÉID-2) were developed to measure rapidly the logarithmic decrement of free vibrations of reaction turbines. The ÉID-1 and ÉID-2 units have a measuring oscillator with a pickup based on the eddy current effect. The use of an eddy-current pickup, which is a characteristic feature of both units, makes it possible to carry out measurements without contact with a blade (the pickup is not in mechanical

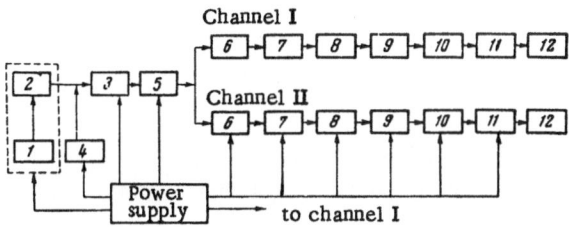

Fig. 3. Block diagram of ÉID-2.

contact with the surface of a vibrating blade) and the measurements can be carried out on parts made of magnetic and nonmagnetic materials; the linearity and the sensitivity of the pickup are satisfactory when the gap between the pickup and the investigated part is of the order of 1.0-2.0 mm. The pickup was designed in accordance with the formulas of Levi-Katetskaya [3] and subsequent experimental tests showed a satisfactory agreement with the theoretical formulas. One variant (ÉID-1) consists of the following elements (Fig. 2): an oscillator 1 with a measuring detector pickup 2; a two-stage low-frequency amplifier 4; an adjustable tube voltmeter 3; a power unit 6; a thyratron amplitude discriminator 7; a counting unit PS-10000 (8).

The procedure is as follows. The pickup is placed at a distance of 1.5 mm from a blade. The pickup is supplied from a high-frequency oscillator (f_{oper} = 1 Mc).

When the distance between the pickup and the blade varies periodically, the parameters of the pickup vary accordingly and, therefore, the carrier frequency voltage becomes amplitude-modulated. The modulation frequency is governed by the natural frequency of the investigated sample, in our case a blade. The modulated low-frequency vibrations are detected and applied through a divider, to the input of a two-stage low-frequency amplifier. The output voltage from the low-frequency amplifier is applied to the amplitude discriminator, which includes thyratrons TG-1-0.1/ 1.3.

The amplitude discriminator circuit is that proposed by Fedorovskii [2]; pulses emerge from the output only when the voltage at the discriminator input is within the range U_{max} to $(\frac{1}{2}) U_{max}$.

Thus, the number of output pulses represents a period during which the amplitude of the vibrations of a blade falls to half its initial value.

The pulses so obtained are applied to the input of the electronic counter PS-10000. The whole measurement process, involving tenfold repetition of the vibration cycle, lasts about 3-5 min.

Using the apparatus described above, the decrement can be measured about three hundred times as fast as using a strain gauge or the induction method, in which the vibration pattern is recorded by means of a loop oscillograph.

The disadvantage of the apparatus described here is that the number of pulses obtained can only correspond to a fixed ratio of the amplitudes U_{max}/ U_{min} = 2. This, in fact, presupposes the existence of an exponential law of the decay of the vibration amplitude or the constancy of the logarithmic decrement, which, generally speaking, is not true. Moreover, the thyratron circuit cannot in fact ensure a constant high accuracy (because of unequal firing times of the thyratrons and because of aging of these tubes).

To eliminate these disadvantages, we developed a new circuit to measure the decrement; the new apparatus was called ÉID-2. The main difference between the new apparatus and ÉID-1 was the ability of the former to allow the ratio of the amplitudes to be varied, between which the number of pulses was measured. The ratio of the amplitudes could be taken up to 10/ 9 without a marked decrease of the accuracy. This made it possible to measure the true and not the average value of the logarithmic decrement.

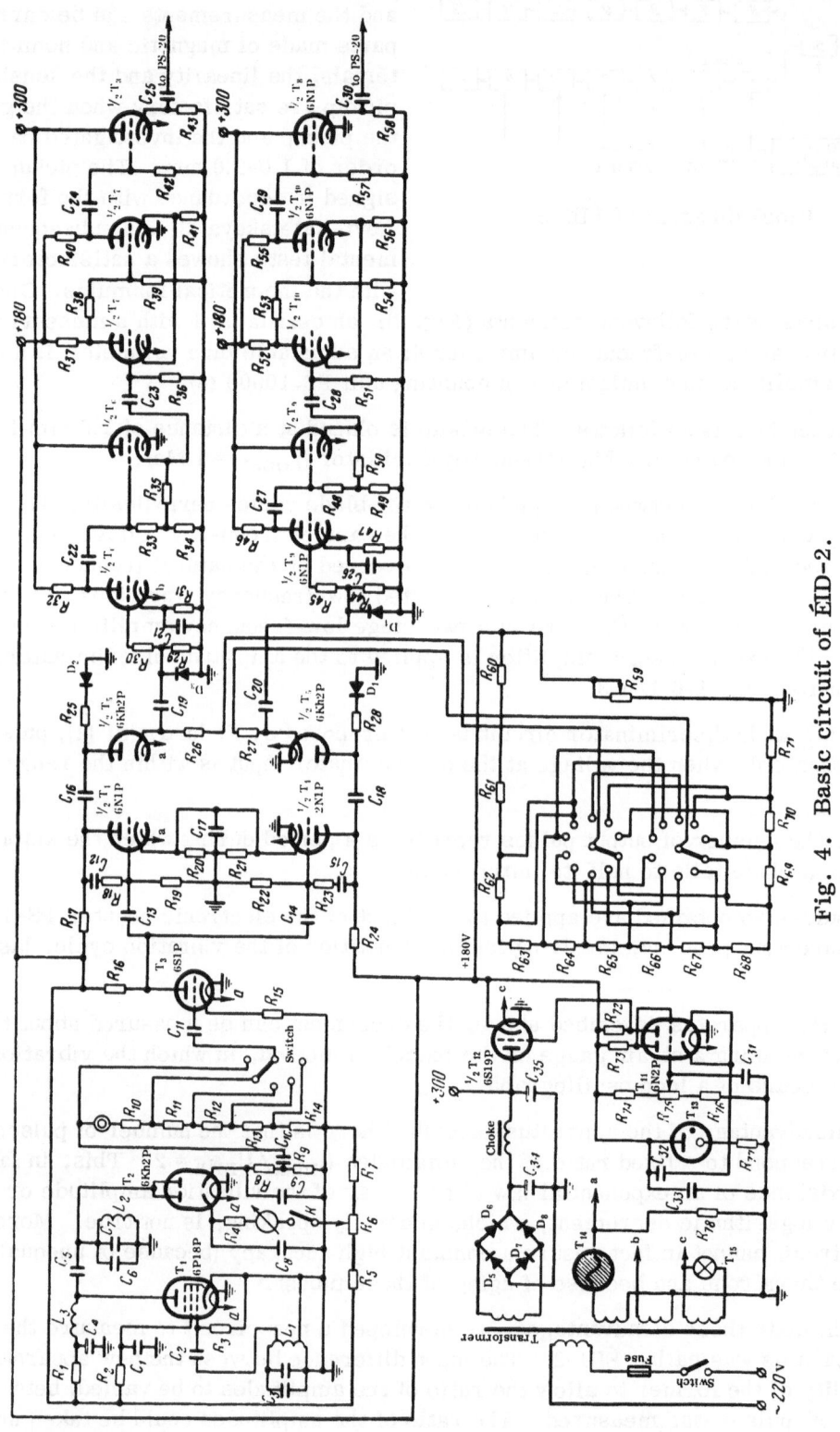

Fig. 4. Basic circuit of ÉID-2.

Moreover, in the new apparatus the amplitude discriminator was constructed using diode limiters, which gave a higher accuracy compared with the circuit in which thyratrons were used.

The block diagram of ÉID-2 is given in Fig. 3. This apparatus consists of the following elements: a high-frequency oscillator 1 with a pickup 2; a detector 3; an adjustable tube voltmeter 4; a preliminary amplification stage 5 (tube 6S1P); two channels in an amplitude discriminator. These channels consist of the following elements: a final amplification stage 6; a diode limiter, an amplifying and peaking circuit with a limiter 7, a cathode follower 8, a Schmitt trigger circuit 10, another cathode follower 11. Moreover, the apparatus includes an electronically stabilized power supply and a counting unit PS-20 (12).

The basic circuit of the electronic measuring unit (without PS-20) is shown in Fig. 4.

The high-frequency oscillator (tube 6P18P) is of the twin-circuit Shembel' type.

In contrast to the measuring oscillator in ÉID-1, a pickup L_2 is not connected to the control grid circuit, but to the anode circuit. This makes it possible to increase the stability of the measuring oscillator quite considerably.

The pickup L_2, together with a capacitor C_6, forms an anode circuit, detuned with respect to the operating frequency of the oscillator, which is governed by the internal circuit parameters.

The detuning of the anode circuit is such as to ensure the linearity of the variation of the voltage in the circuit in a selected range of vibration amplitudes to be measured. This is ensured by setting the working point in the sloping part of the resonance curve. To be able to tune the circuit in this way, the apparatus includes an adjustable tube voltmeter [tube $(\frac{1}{2})T_2$], including a dc bridge consisting of resistors and an indicator (M24 for 100 μA). Using this tube voltmeter, the linearity of the readings of the instrument can be checked easily, the sensitivity can be found fairly easily, the necessary gap can be set, and the tuning can be carried out when the measurements involve samples of different materials.

The operating frequency of the oscillator is f_{oper} = 1 Mc. The frequency range of the measured vibrations is from 35 cps to 20 kc. The amplitude of the measured vibrations can be varied from 0.02 to 0.2 mm without departing from the linearity. The vibrations of a turbine blade alter the parameters of the pickup and amplitude-modulate the carrier voltage of the oscillator.

The modulated low-frequency vibrations pass through the detector [$(\frac{1}{2})T_2$] to the preamplifier (T_3), which has the required pass band and dynamic linearity of the output voltage U characteristic in the investigated range of amplitudes.

Beyond the preamplifier, the whole circuit is divided into two identical channels. Each channel consists of a power amplifier [the second amplification stage of $(\frac{1}{2})T_4$], a diode limiter [$(\frac{1}{2})T_5$], an amplifying and peaking circuit [$(\frac{1}{2})T_6$ and T_9] followed by a limiter, one cathode follower [$(\frac{1}{2})T_6$ and T_9], a Schmitt trigger circuit (T_7 and T_{10}), and an output cathode follower (T_8).

The diode limiter is supplied with a voltage whose value varies in accordance with the measurement conditions, by means of a divider ($R_{62}\ldots R_{71}$). A stabilized voltage U = 180 V is supplied to the divider.

Thus, having set a certain threshold in the limiters of channels I and II, we obtain a different number of output pulses N_1 and N_2 and thus we find the quantity we need $\Delta N = N_1 - N_2$. The presence of the Schmitt trigger circuit (tubes T_7 and T_{10}) makes it possible to obtain pulses of constant amplitude ($U_{pulse} \approx 50$ V) with a short rise time ($\tau \approx 2$-3 μsec).

The output cathode follower lowers somewhat the amplitude of the voltage (to about 30 V) and this ensures better matching with the input stage of the counting unit PS-20. The pulses reaching PS-20 are of positive polarity. The shape and the amplitude of the output pulses are such that any counter can be used in the apparatus.

The power supply unit of the ÉID-2 unit has two output anode voltages of +300 V and +180 V. The +180 V is stabilized by means of an electronic stabilizer, consisting of tubes T_{11}, T_{12}, and T_{13}. The stabilized voltage is applied to the oscillator, to both stages of the low-frequency amplifier, to the detector, to the Schmitt trigger, and to the $R_{62} \ldots R_{71}$ divider (which supplies the diode limiters).

A current regulator tube 0.85B5.5-12 is used to stabilize the heating voltages of the oscillator, the first detector, the two amplifying stages, and the diode limiters.

The rate at which measurements can be carried out using the ÉID-2 unit is somewhat lower than in the case of the ÉID-1 unit, because it is necessary to find the difference between readings of two counting units, while in ÉID-1 the difference is found directly. However, this disadvantage is fully compensated by the higher accuracy of the measurements and by the possibility of using various values of A_0/A_1, where A_0 is the initial vibration amplitude and A_1 is the amplitude of turbine-blade vibrations after a certain time interval from the beginning of excitation.

Moreover the advantages of the ÉID-2 unit include the possibility of using relatively simple and small counting units (instead of the bulky and expensive counter PS-10000).

A check of the operation of the apparatus described here has shown that it can be used with success in nondestructive testing of samples (blades) made of various materials, both under laboratory and industrial conditions.

Literature Cited

1.   S. O. Tsobkallo and V. A. Chelnokov, Zh. Tekhn. Fiz., 24(3) (1954).
2.   A. E. Fedorovskii, Zavodsk. Lab., No. 12 (1958).
3.   A. P. Katetskaya, Plated Transducers Based on the Use of Eddy Currents and Some Methods of Design of Such Transducers, Proc. Conf. RII GVF (1962).

APPARATUS FOR THE INVESTIGATION
OF THE INTERNAL FRICTION
AND SHEAR MODULUS

A. I. Efimov

The apparatus described is in the form of the standard torsional pendulum (a relaxator). It can be used to investigate the logarithmic decrement and the shear modulus (the square of the frequency) in the temperature range between 20°C and the melting point of a given material. A sample is heated to 1200°C by means of a molybdenum furnace. To obtain higher temperatures, a sample is heated directly by passing a current through it. The investigation is carried out on samples of 150-mm maximum length and 0.5- to 2-mm diameter in the frequency range from 1 to 50 cps. The logarithmic decrement can be determined visually or by photographic recording. The use of the photographic recording is desirable in investigations of the internal friction in the frequency range ≥ 2 cps, as well as in investigations of the damping at high temperatures, of the amplitude-dependent magnetoelastic damping, etc. The maximum value of the shear deformation of the surface of the sample varies, with the selected vibration amplitude and a sample dimension, from 10^{-3} to 10^{-6}.

To investigate the influence of plastic deformation and high-temperature creep on the internal friction, the apparatus is provided with a stretching device.

To eliminate the acoustical losses and to prevent the oxidation of samples at high temperatures, the heating is carried out in 10^{-4}-10^{-5} mm Hg vacuum.

The apparatus consists basically of a torsional pendulum, a recording unit, and a control panel.

Figure 1 shows schematically the torsional pendulum. Two II-shaped brackets 14 are attached to a massive iron base 13, which is placed on two channel bars built into a solid wall. Stands 15 with a cross piece 1 are attached to the brackets. A molybdenum rod carrying the upper clamp 2 is screwed into the cross piece. The cross piece and the molybdenum rod carry an attachment for placing a sample 4 in a vertical plane. The sample 4 is fixed in the upper and lower clamps and a rod attached to the lower clamp extends into a damping medium 9. To reduce the weight of the suspension system of the pendulum, all parts of it are made of titanium. Since a sliding contact is required for direct heating by passing a current through the sample, gallium is used as the damping medium. Investigations have shown that the damping introduced by gallium is not greater than the damping of an oil damper. The magnitude of the damping in refractory alloys at room temperature does not exceed $(7-8) \cdot 10^{-5}$. We may assume that the intrinsic background introduced by the apparatus is even smaller than this value.

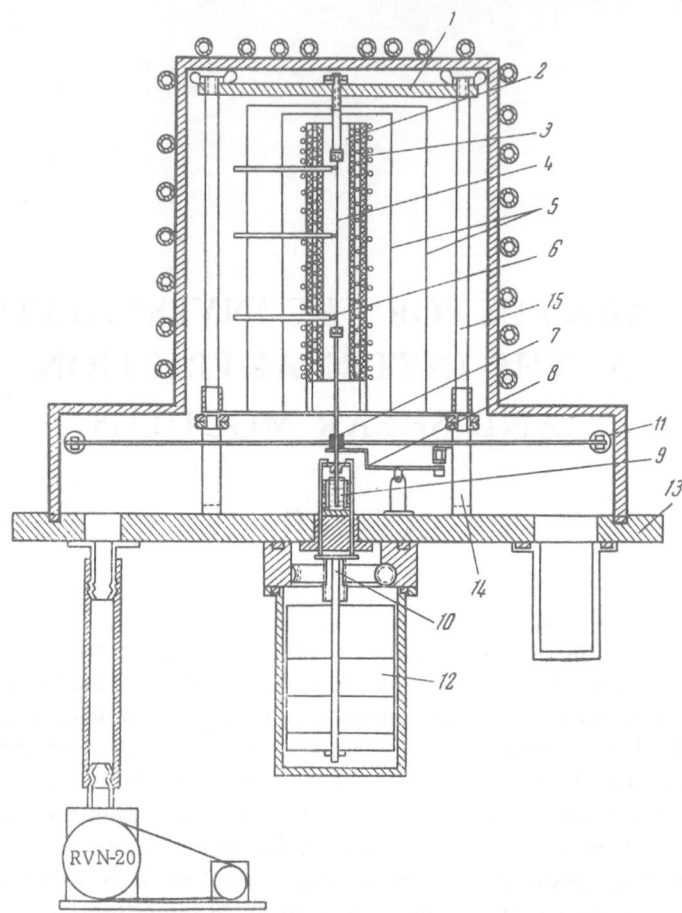

Fig. 1. Schematic diagram of the apparatus.

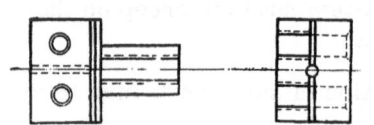

Fig. 2. Construction of a clamp.

Since the melting point of gallium is 29.6°C, a special small heater is used to keep gallium in the liquid state. The crucible in which gallium is contained is made of molybdenum. No noticeable interaction between gallium and molybdenum is found at temperatures of 75-100°C. The vapor pressure of gallium is low. In contrast to mercury (which is sometimes used as sliding contact [1]), gallium is not a toxic material.

A mirror 7 is attached to the lower rod and this mirror has an adjustment for moving the light spot in a vertical plane; the twisting system is also attached to the lower rod. In spite of the fact that the suspension of the pendulum is made as light as possible, axial stresses amount to 30-35 g/mm². During heat treatment in the relaxator or during an investigation of the internal friction at high temperatures in pure metals, such stresses may lead to structural changes in grains, and this may affect the internal friction [2]. To relieve the axial load during heat treatment, a special device 8 is used. A horizontal beam with a fork at one end is used to balance, by means of a weight, the suspension of the torsional pendulum. During measurements of the internal friction, this beam can be lowered by means of an electromagnet.

To ensure a uniform temperature along the sample length, the molybdenum furnace 3 has three windings, on Alundum tubes. The inner winding is uniform. The windings on the outer

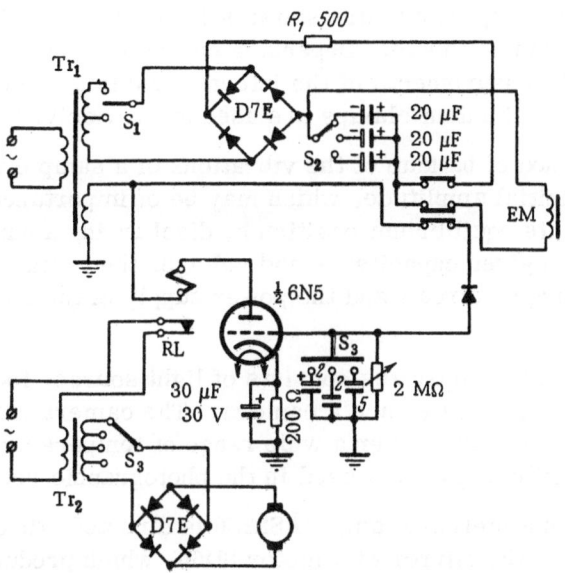

Fig. 3. Basic circuit of power supply of the electromagnets and time relay controlling the camera motor. Tr_1, Tr_2) Transformers; S_1, S_2, S_3) switches; EM) electromagnet; RL) relay.

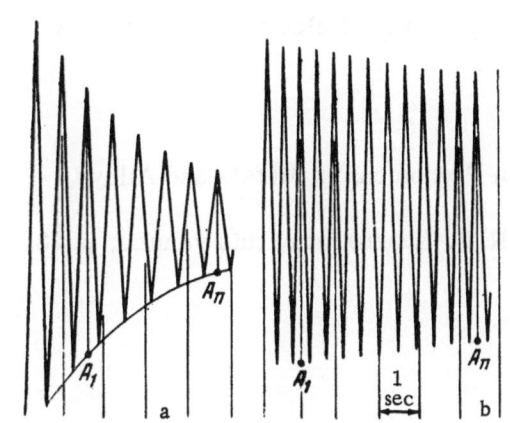

Fig. 4. Oscillograms of vibration damping in electron-beam melted nickel at temperatures of: a) 150°C, $Q^{-1} = \frac{\ln(A\pi/A_1)}{\lambda n} = \frac{\ln(65/24.5)}{\pi 5}$

$= 620 \cdot 10^{-4}$; b) 360°C, $Q^{-1} = \frac{\ln(76.5/65.5)}{\lambda 10} = 50 \cdot 10^{-4}$.

tube are bunched at the ends of the furnace. The windings are supplied separately from autotransformers through step-down transformers. Further leveling of temperature is achieved by means of an iron cylinder screen placed inside the furnace.

The furnace is located within metal screens 5. The nonuniformity of the temperature along the sample length does not exceed ±1°C. The sample temperature is measured by means of three thermocouples 6, placed in the middle of the sample and at its ends. When the sample is heated by passing current directly through it, the temperature is measured using the method described in [1].

The stretching device (10, 12) makes it possible to deform a sample at high temperatures by applying a constant load or by increasing a load at a fixed rate. The operating principle of the device is as follows. A disk is attached to the lower end of the rod which is part of the suspension of the pendulum; by adding weights, a load from 100 g to 20 kg can be applied to this rod. In investigations of the high-temperature plastic deformations the load is lifted during internal-friction measurements by means of a worm gear. Plastic deformation can be applied at a wide range of deformation rates, also by means of a worm gear set in motion by a dc motor.

The clamps used in the apparatus are shown schematically in Fig. 2. Such clamps are simple to construct and reliable in use. Depending on the investigated metal or alloy and on the temperature range of measurements of the internal friction, the clamps are made of nickel or molybdenum. However, titanium clamps are used most widely.

When the usual method of buildup of the vibrations of a sample is used, it is difficult always to obtain the same initial amplitude, which may be of importance. In our apparatus the sample is deflected from its equilibrium position by discharging a capacitor bank through an electromagnet coil. For a given capacitance and voltage, the initial amplitude is always the same. The capacitor charging circuit and the power supply of the electromagnets are shown in Fig. 3.

The recording unit of the apparatus consists of light source, focusing mirrors, a cylindrical lens, a semitransparent scale, and a camera. The camera makes it possible to record photographically the damping curves over a wide range of temperatures without reloading. The maximum width of the oscillograph chart used in the photographic recording is 18 cm.

A beam of light from a mercury lamp DRSh250 is directed, through a splitting device, to the pendulum mirror and to the mirror of a motor SD60, which produces time marks. The frequency is calculated from the known number of vibrations and the known time interval. The drive of the drum carrying the oscillograph chart is started synchronously with the switching on of the electromagnets which are used to start the excitation. The scanning time is governed by a relay (Fig. 3). A dc motor is used to alter the scanning rate when the frequency of vibrations is changed.

Calculations show that the error in the measurement of the internal friction at low temperatures is 1% and at high temperatures it is 3-5%.

Figure 4 shows typical oscillograms of the damping curves of electron-beam-melted nickel at temperatures of 150°C and 360°C.

Literature Cited

1.   L. N. Aleksandrov and V. S. Mordyuk, Relaxation Phenomena in Metals and Alloys, Metallurgizdat (1963).
2.   O. A. Belous, V. N. Gridnev, A. I. Efimov, and N. P. Kushnareva, this volume, p. 23.

RELATIONSHIP BETWEEN THE INTERNAL FRICTION, THE ELASTIC AFTEREFFECT, AND THE CREEP OF A STANDARD LINEAR BODY

T. D. Shermergor, G. N. Pachevskaya, and S. I. Meshkov

The principal relaxation characteristics of a continuous medium are the average relaxation time τ and the modulus defect Δ_μ. The internal friction is used widely to determine these characteristics, and in some cases also the broadening parameter of the relaxation spectrum [1, 2]. The low- and high-frequency methods of measuring the internal friction Q^{-1} cover the frequency range from 1 cps to 10^3 kc [3]. Since the determination of τ is based on the measurement of the relaxation peak for which $\omega\tau = 1$, the internal friction method is inapplicable at low relaxation frequencies $s = 1/\tau$. In practice, the internal friction method is not used for $\tau > 5$-10 sec.

However, in some cases it is of interest to measure τ over a wide range of temperatures [4], which may include large values of τ. To determine large values of τ we can measure the elastic aftereffect or the nonstationary creep. Under strong stresses the nonstationary creep is described by a nonlinear hereditary theory of creep [5, 6]. We shall consider below the elastic aftereffect and the creep only for weak stresses, when the system can be regarded as linear.

We shall consider only the simplest hereditary medium, for which the shear stresses and deformations are described by the standard linear body model. The deviator s_{ik} of the stress tensor σ_{ik} for such a medium has the form

$$s_{ik} = 2\mu_0 e_{ik} + 2\Delta\mu \int_{-\infty}^{t} \dot{e}_{ik}(t') \exp\frac{t'-t}{\tau}\, dt', \tag{1}$$

where e_{ik} is the deviator of the deformation tensor, μ_0 and $\mu_\infty = \mu_0 + \Delta\mu$ are the relaxed and unrelaxed shear moduli. For an infinite medium, Eq. (1) yields the following expressions for the internal friction, the elastic aftereffect, the creep, and the relaxation:

internal friction

$$\tan\delta = \Delta_\mu \frac{\omega\tau_0}{1 + \omega^2\tau_0^2}, \tag{2}$$

221

elastic aftereffect

$$e_{ik} = \frac{1}{2} s_{ik} [j_\infty + \Delta j \exp(-t/\tau_\sigma)], \tag{3}$$

creep

$$e_{ik} = \frac{1}{2} s_{ik} [j_0 - \Delta j \exp(-t/\tau_\varepsilon)], \tag{3'}$$

relaxation

$$s_{ik} = 2e_{lk} [\mu_0 + \Delta\mu \exp(-t/\tau_\varepsilon)], \tag{4}$$

where

$$\tau_\varepsilon = \tau, \ \ \tau_\sigma/\tau_\varepsilon = \mu_\infty/\mu_0, \ \ \tau_0 = (\tau_\sigma \tau_\varepsilon)^{1/2},$$

$$\Delta\mu = \mu_\infty - \mu_0, \ \ \Delta j = j_0 - j_\infty, \ \ j_0 = \frac{1}{\mu_0}, \ \ j_\infty = \frac{1}{\mu_\infty}, \Delta_\mu = \frac{\mu_\infty - \mu_0}{\sqrt{\mu_\infty \mu_0}}. \tag{5}$$

The expressions (3) and (4) define the modulus defect and the relaxation (τ_ε) and retardation (τ_σ) times. However, Eqs. (3) and (3') cannot be used directly to determine the relaxation characteristics of finite bodies. Thus, if the elastic aftereffect is found by means of a torsional pendulum [7] (which is convenient because the same apparatus can be used to determine the internal friction, elastic aftereffect, and creep), then at the moment of application of an external couple and at the moment when this couple ceases to act, damped vibrations are excited in this pendulum (in addition to the elastic aftereffect in the former case and the creep in the latter case) [8]. We shall deal below with the influence of these damped vibrations on the elastic aftereffect and nonstationary creep curves for a medium described by the standard linear body model.

We shall consider a cylindrical torsional pendulum whose upper end is fixed and the lower free end has an attached inertial load to which, beginning from the moment t = 0, a couple M(t) is applied. Then, the solution of the problem can be represented in the form of a convolution

$$\varphi(z, t) = \int_0^t G(z, t - t') M(t') dt', \tag{6}$$

where $\varphi(z, t)$ is the angle of twist, G is the Green's function of the corresponding problem, i.e., the twisting angle if the excitation of the lower end of the pendulum is described by a δ-function.

For low-amplitude vibrations, when the hypothesis of plane sections can be used, the Green's function is [9]

$$G(z, t) = \frac{1}{M_0} \Phi(z, t), \tag{7}$$

$$\Phi(z, t) = \sum_{n=1}^k \Phi_n^a(z, t) + \sum_{n=k+1}^\infty \Phi_n^v(z, t), \tag{8}$$

where $\Phi(z, t)$ is the twisting angle; Φ_n^a and Φ_n^v describe the aperiodic and harmonic damping, respectively; M_0 is the amplitude of the applied couple $M = M_0 \delta(t)$ in the case of pulse excitation.

The twisting angles Φ_n^a and Φ_n^V are given by the expressions

$$\Phi_n^a(z, t) = \frac{2M_0}{I} c_n(z) \sum_{i=1}^{3} H_{in} \exp(-\gamma_{in} t),$$ (9)

$$\Phi_n^v(z, t) = \frac{2M_0}{I} c_n(z) [A_{1n} \exp(-\alpha_{1n} t) + A_{2n} \exp(-\alpha_{2n} t) \sin(\omega_n t - \psi_n)].$$ (10)

The notation used here is:

$$c_n(z) = \sin(b_n z/l) [b_n \cos b_n + (1+a) \sin b_n]^{-1}, \quad b_1 \approx \sqrt{a}; \quad b_{n+1} \approx n\pi,$$
$$a \equiv \pi R^4 \rho l / 2I \ll 1, \quad H_{in} = (s - \gamma_{in}) [(\gamma_{in} - \gamma_{jn})(\gamma_{in} - \gamma_{kn})]^{-1}, \quad i \neq j \neq k,$$
$$\gamma_{1n} = \frac{s}{3} + 2r_n \cos\frac{\theta_n}{3}, \quad \gamma_{2n, 3n} = \frac{s}{3} - 2r_n \cos\frac{\pi \mp \theta_n}{3},$$
$$A_{1n} = 2\alpha_{2n} [\omega_n^2 + (\alpha_{1n} - \alpha_{2n})^2]^{-1}, \quad A_{2n} = A_{1n} \operatorname{cosec}\Psi_n, \quad \alpha_{1n} = s - 2\alpha_{2n},$$
$$\alpha_{2n} = \varkappa_n + \frac{s}{3}, \quad \tan\psi_n = 2\alpha_{2n}\omega_n (\alpha_{1n}^2 - \alpha_{2n}^2 + \omega_n^2)^{-1}.$$ (11)

R denotes the radius of the cylinder; l is the length; ρ is the density of the cylinder; I is the moment of inertia of the torsional suspension; $s = 1/\tau$ is the relaxation frequency. For a given value of n, the damping is aperiodic or harmonic, depending on the sign of the discriminant D_n

$$D_n = q_n^2 + m_n^3,$$ (12)

where

$$q_n = \frac{s^3}{27} + \frac{s}{2}\omega_{n\infty}^2 \left(\frac{\mu_0}{\mu_\infty} - \frac{1}{3}\right),$$ (13)

$$m_n = \frac{\omega_{n\infty}^2}{3} - \frac{s^2}{9}.$$ (13')

Here $\omega_{n\infty}$ denotes the natural frequency of vibrations for $\tau \to \infty$, which is given by $\omega_{n\infty}^2 = \mu_\infty b_n^2 / \rho l^2$.

If $D_n > 0$, then damped harmonic vibrations of Eq. (10) are obtained, and if $m_n > 0$, the quantities $\varkappa_n$ and ω_n are found from the formulas

$$\varkappa_n = -r_n \sinh\theta_n, \quad \omega_n = \sqrt{3} r_n \cosh\theta_n,$$

$$r_n = \pm\sqrt{m_n}, \quad \sinh 3\theta_n = q_n r_n^{-3},$$ (14)

and if $m_n < 0$, but $q_n^2 > |m_n|^3$, then we find $\varkappa_n$ and ω_n from the formulas

$$\varkappa_n = -r_n \cosh\theta_n, \quad \omega_n = \sqrt{3} r_n \sinh\theta_n,$$

$$r_n = \pm\sqrt{|m_n|}, \quad \cosh 3\theta_n = q_n r_n^{-3}.$$ (14')

The sign of r_n is always selected to be identical with the sign of q_n. If $D_n < 0$, we obtain the aperiodic damping of Eq. (9). In this case the parameter θ_n is calculated from the formula $\cos\theta_n = q_n r_n^{-3}$.

Let us now assume that the applied external couple acts for a finite time interval t, i.e., the action of the couple on the torsional pendulum has the form of a rectangular pulse

$$M = M_0 [1 (t) - 1 (\theta)], \tag{15}$$

where $\theta = t - T$ and 1(t) is a unit function, which is equal to zero when t < 0 and equal to unity t ≥ 0, where $(d/dt) 1(t) = \delta (t)$. Then, in the weak damping case, the solution of the problem reduces, according to Eq. (6), to integration of Eq. (10).

After integration and appropriate transformations, we obtain

$$\varphi_n (z, t) = \frac{B_n}{\alpha_{1n}} \{[1 - \exp(-\alpha_{1n}t)] 1 (t) - [1 - \exp(-\alpha_{1n}\theta)] 1 (\theta)\} +$$

$$+ \frac{B_n \sin \eta_n}{\sin \psi_n \sqrt{\alpha_{2n}^2 + \omega_n^2}} \left[1 (t) - 1 (\theta) \right] - \frac{B_n}{\sin \psi_n \sqrt{\alpha_{2n}^2 + \omega_n^2}} \{\exp(-\alpha_{2n}t) \sin (\beta_n t +$$

$$+ \eta_n) 1 (t) - \exp(-\alpha_{2n}\theta) \sin (\beta_n \theta + \eta_n) 1 (\theta)\}. \tag{16}$$

The parameters in Eq. (16) are, as before, found from Eqs. (11)-(14). The quantities B_n and η_n are given by

$$B_n = \frac{2M_0}{I} \left[\frac{1}{9} - \left(1 - \frac{\alpha_{1n}}{\alpha_{2n}}\right)^{-2} \right]^{-1},$$

$$\tan \eta_n = \frac{\omega_n - \alpha_{2n} \tan \psi_n}{\alpha_{2n} + \omega_n \tan \psi_n}. \tag{17}$$

If $a \ll 1$, the contribution of higher harmonics to the total motion is negligibly small and we need consider only the first harmonic. The condition $a \ll 1$ is always true in the case of a torsional pendulum with an inertial load, as used in the measurement of the internal friction [10]. To measure the elastic aftereffect or the nonstationary creep, one uses a current-carrying frame placed in a magnetic field. We can easily see that the condition $a \ll 1$ is again satisfied. For example, for R = 0.5 mm, l = 100 mm, ρ = 8 g/cm³, we have the requirement I ≫ 10⁻³ g/cm², which is always satisfied.

In the approximation considered we have $c_1(z) = z/2l$ for the first harmonic. Bearing this in mind, we find the nonstationary creep φ_c for 0 < t < T and D < 0 (weak damping)

$$\varphi_c = \frac{2Mz}{Il} [C_0 - C_1 \exp(-\alpha_1 t) - C_2 \exp(-\alpha_2 t) \sin (\omega t + \eta)], \tag{18}$$

where the index n = 1 is omitted and the following notation is used:

$$C_0 = [\tau \alpha_1 (\alpha_2^2 + \omega^2)]^{-1}, \quad C_1 = \frac{2\alpha_2}{\alpha_1} [\omega^2 + (\alpha_2 - \alpha_1)^2]^{-1},$$

$$C_2 = \frac{\alpha_1}{\omega} \left\{ \frac{C_0 C_1}{2\alpha_2 s} [\omega^2 + (\alpha_2 - s)^2] \right\}^{1/2}. \tag{19}$$

The expression for the elastic aftereffect is obtained by recording a single pulse when the nonstationary creep dies away. This leads to the requirement $\alpha_1 T \gg 1$. Thus, for sufficiently high values of T and D < 0, we obtain the angle of twist which governs the elastic aftereffect φ_a

$$\varphi_a = \frac{2Mz}{Il} [C_1 \exp(-\alpha_1 \theta) + C_2 \exp(-\alpha_2 \theta) \sin (\omega \theta + \eta)], \tag{20}$$

where the amplitudes C_1 and C_2 are again defined by Eq. (19).

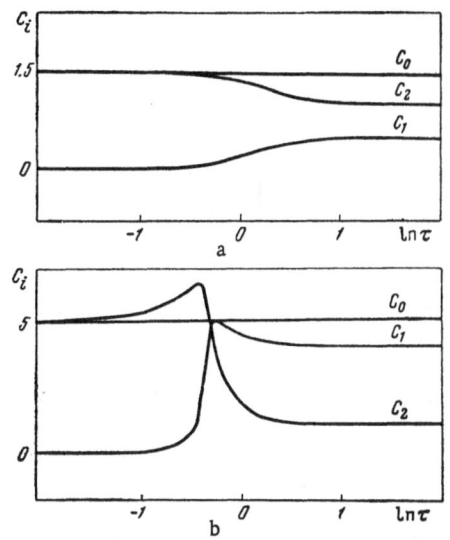

Fig. 1. Temperature dependence of the amplitudes governing the contribution of the aftereffect or creep and of the damped harmonics. a) $m = \frac{2}{3}$; b) $m = 0.2$.

The damping coefficient α_1 in Eqs. (18) and (19) governs the monotonic parts of the curves φ_c and φ_a, while α_2 is related to the damped harmonics. We should mention that the damping coefficients α_i and the corresponding amplitudes c_i are identical for the elastic aftereffect and for the creep. This makes it possible to carry out a unified analysis of the curves $\varphi(z, t)$.

We shall find the conditions under which we can determine, from the known curve $\varphi(z, t)$, the relaxation time τ and the modulus defect Δ_μ or the equivalent quantity $\mu_0/\mu_\infty = m$, which is related to Δ_μ by Eq. (5).

To avoid marked distortion of the aftereffect and creep curves by the vibrations excited on the application of a couple and on removal of this couple, the contribution of the terms corresponding to these vibrations in Eqs. (18) and (20) should be negligibly small. If the amplitudes c_i are of the same order of magnitude, then the coefficient α_2, which governs the damping of the vibrations, should be considerably larger than α_1.

We shall consider first the determination of the modulus defect. To find this defect from the experimental curves $\varphi(z, t)$, we can use the internal friction method and determine Q^{-1} from the damped harmonics. Since the medium is characterized by only one relaxation time, then $\Delta_\mu = 2Q_{max}^{-1}$, where Q_{max}^{-1} is a maximum of the internal friction. If this internal friction maximum occurs at very low frequencies ($\omega \ll 1$ rad/sec), then, as mentioned earlier, the internal friction method cannot be used.

A different method is based on a comparison of an instantaneous deflection φ_∞ and of the relaxed value $\varphi_0 \equiv \varphi(t \to \infty)$ of the angle of twist. We can easily see that in the case of creep, neglecting the inertia forces, we should have the relationship

$$\frac{\varphi_\infty}{\varphi_0} = \frac{\mu_0}{\mu_\infty} = m. \qquad (21)$$

For the elastic aftereffect the same relationship also is valid, but the angle φ_∞ corresponds to the beginning of the measurements. Since the end of the measurements gives the equilibrium value of the angle of twist [in the absence of the stationary creep $\varphi(t \to \infty) = 0$], additional experiments are needed to find φ_∞ in the same way as in the investigations of the creep.

While φ_0 can be measured directly in both cases considered, the measurement of φ_∞ is difficult because of the inertia of the system. To allow for the influence of the inertial forces, we shall draw attention to the following point: if the inertia of the rod ($\rho \to 0$) is neglected, the inertial term in the equation of motion disappears. However, since the inertial term $I\ddot{\varphi}$ remains in the boundary conditions for the lower end of the rod, the solution will differ from that given above only by the absence of higher harmonics. The approach to zero of the moment of inertia of the suspension $I \to 0$ implies increase of the natural frequency $\omega \to \infty$, i.e., transition to the region $\omega\tau \gg 1$. [In the special case $I = 0$ and $\rho = 0$, the system will be described by Eqs. (2)-(4), as in the case of an infinite medium.] Therefore, in the region where $\omega\tau \gg 1$, Eq. (21) should be applicable. Figure 1 shows the curves of C_i as a function of $\ln \tau$ for

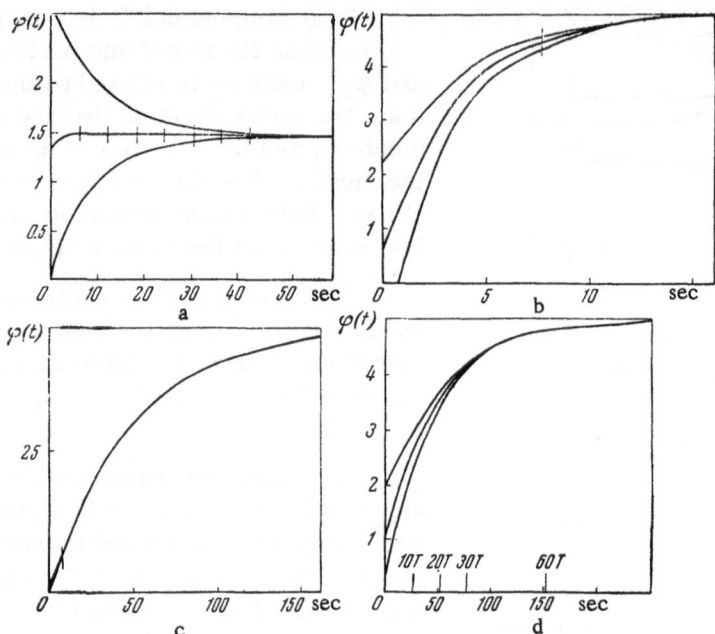

Fig. 2. Reaction of a torsional pendulum, having hereditary properties of a standard linear body, to the application of a constant couple. Vertical segments indicate vibration periods. a, b, c) $\tau = 1$ and m = $^2/_3$, 0.2, and 0.02, respectively; d) $\tau = 10$ and m = 0.2.

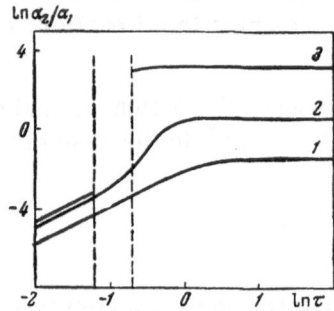

Fig. 3. Temperature dependence of the ratio of the damping coefficients. 1) m = $^2/_3$; 2) 0.2; 3) 0.02. Dashed vertical lines denote the aperiodic region for the third curve.

two values of the modulus defect. In the plotting of these curves, we have assumed that $\omega_\infty = 1$. It follows from Fig. 1 that in the $\omega\tau \gg 1$ region the following relationship is valid: $(C_0 - C_1)/C_0 = \mu_0/\mu_\infty$.

If the moment of inertia is finite, there is no discontinuity of φ at t = 0 because of the term $C_2 \sin \eta$ in Eqs. (18) and (20). If the average line about which the damped oscillation vibrations take place is extrapolated to t = 0, then C_1 can be determined. A typical curve of this type in the case of creep is given in Fig. 2d.

In the region $\omega\tau \approx 1$, the relaxation is strong during a time interval of the order of one vibration period and, therefore, it is difficult to determine the modulus defect.

The creep curves $\varphi_C(t)$ for various values of m when $\omega_\infty = \tau = 1$ for three values of the modulus defect are given in Figs. 2a, 2b, and 2c. It is evident from these figures that if m $\ll$ 1, the nonstationary creep is obtained, while if m $\lesssim$ 1, the $\varphi_C(t)$ curve exhibits damped vibrations.

A more complete representation of the relationship between the nonstationary creep and the damped harmonics can be obtained by comparing the contributions of the corresponding terms with the values of the damping coefficients α_1 and α_2. The dependence of $\ln(\alpha_2/\alpha_1)$ on

$\omega\tau$	m	C_1/C_2	α_1/α_2	Effect
$\gg 1$	2/3	0.5	1	Mixed
$\gg 1$	0.2	4	1	Creep predominates
$\gg 1$	0.02	50	0.02	Creep
$\ll 1$	2/3	0	1	Damped vibrations
$\ll 1$	0.2	0	1	» »
$\ll 1$	0.02	0	1	» »

$\ln\tau$ for $m = 2/3$, 0.2, and 0.02 is given in Fig. 3. The dashed vertical lines define the region of aperiodicity for $m = 0.02$. It is evident from Fig.3 that when $\omega\tau \ll 1$, the condition $\alpha_2 \ll \alpha_1$ is satisfied, i.e., the nonstationary creep should die away rapidly. However, such creep does not appear at all because in this region $C_1 \approx 0$. In the $\omega\tau \gg 1$ region the relationships between the coefficients α_1 and α_2 may be different, but all the coefficients C_i have the same order of magnitude.

The relationship between the creep and the damped harmonics is illustrated in the table. It follows from the table that when $\omega\tau \gg 1$, the nonstationary creep is superimposed on the damped vibrations or the nonstationary creep is active alone. The vibrations in this region can be easily suppressed by means of an electromagnetic or a mechanical damper, which is always possible because of the condition $\omega\tau \gg 1$. (We should note that the use of an external damper may make it difficult to measure the retardation time.) On the other hand, when $\omega\tau \ll 1$, only the damped vibrations are observed.

We have considered the possibility of determining the modulus defect from the ratio φ_0/φ_∞. Another relaxation characteristic which should be determined is the relaxation time τ_ε or the retardation time τ_σ. It follows from Eqs. (18) and (20) that the exponential curves of the nonstationary creep and of the elastic aftereffect are governed by the damping coefficient α_1. Therefore, τ must be determined from the known value of α_1. In the region considered, $\omega\tau \gg 1$, and it follows from Eqs. (13) and (14) that $\alpha_1\tau = \mu_0/\mu_\infty$.

Comparing the expression obtained with Eq.(5), we find that $1/\alpha_1 = \tau_\sigma$ is the retardation time. As expected, this quantity is identically equal to the characteristics of the creep and of the elastic aftereffect of an infinite medium, in accordance with Eq.(3).

Comparing the internal friction, the creep, and the elastic aftereffect methods of measuring the relaxation parameters, we should note one further point. If a system is described by a spectrum of relaxation times, the internal friction peak is found to be lower compared with the peak for a standard linear body. Consequently, the ratio of the moduli m is not measured directly, so that it is necessary to use approximate methods, based on the measurement of the relaxation peak width [11]. The methods based on the measurement of the creep and of elastic aftereffect make it possible to determine m directly if the experimental conditions are suitably selected.

The creep method requires half the time needed in the elastic aftereffect method, since, in the latter case, it is necessary to apply a constant couple and then remove it when the relaxation processes die out. Therefore, the recording of the elastic aftereffect curve cannot begin earlier than the recording of the creep curve. Consequently, if a system does not exhibit the stationary creep, the creep method is to be preferred. However, if the stationary creep is possible in our system, then the elastic aftereffect method is to be preferred, because it gives a "pure" curve from which the retardation time can be determined directly, while in the case of the creep it is necessary to subtract first the stationary contribution.

Literature Cited

1.   T.D. Shermergor, Izv. Vuzov., 1:77 (1961).
2.   A.S. Nowick and B.S. Berry, Acta Met., 10:134 (1962).
3.   C.M. Zener, Collection: Elasticity and Anelasticity of Metals [Russian translation], IL (1954). [English edition: University of Chicago Press, Chicago, Illinois (1948).]
4.   K.M. Shtrakhman and Yu.V. Piguzov, Fiz. Tverd. Tela, 6(8):2274 (1964).
5.   Yu.N. Rabotnov, Vestn. Mosk. Gos. Univ., No. 10 (1948).
6.   Yu.N. Rabotnov, Izv. Akad. Nauk SSSR, Otd. Tekhn. Nauk, No. 6 (1948).
7.   T.S. Ke, Collection: Elasticity and Anelasticity of Metals [Russian translation], IL (1954). [English edition: University of Chicago Press, Chicago, Illinois (1948).]
8.   N.S. Fastov, Collection: Problems of Metallography and Physics of Metals, No. 5 (1958), p. 583.
9.   S.I. Meshkov, V.S. Postnikov, and T.D. Shermergor, Izv. Akad. Nauk SSSR, Mekhan. i Mashinostr., 3:90 (1964).
10. N.S. Fastov, Dokl. Akad. Nauk SSSR, 138(5):1069 (1961).
11. T.D. Shermergor, Collection: Relaxation Phenomena in Metals and Alloys, Metallurgizdat (1963), p. 33.

INFLUENCE OF THE SECOND VISCOSITY ON THE RELAXATION INTERNAL FRICTION

S. I. Meshkov, T. D. Shermergor, and V. S. Postnikov

Investigations of the relaxation phenomena in gases and liquids have shown the necessity of introducing, in addition to the shear viscosity, the so-called second or dilatational viscosity [1-3]. In solids the deviator part of the stress tensor is usually described by a Maxwell or a standard linear body rheological model [4-8] and the shear viscosity is replaced with the shear modulus, the relaxation time τ, and (in the case of a standard linear body) the modulus defect Δ_μ. The relaxation parameters τ and Δ_μ can be measured, for example, using damped torsional vibrations of solids. However, the investigation of the relaxation characteristics in the case of a volume deformation is difficult because of the complexity of the experimental conditions under which only the damped volume vibrations would be observed. In the case of a shear deformation, it is usually possible to ascribe a rheological model to each relaxation process, but this is difficult to do in the case of a volume deformation because of the absence of reliable experimental data. Obviously the most general model characterized by a single relaxation time is, within the framework of the linear theory, the standard linear body model. At low frequencies this model can be replaced by a simpler Voigt model, which can be characterized by the introduction of the bulk modulus and of the second (dilatational) viscosity. The possibility of using the second viscosity in the description of inelastic solids has been pointed out by Frenkel' [9], Gubanov [10], and Slezkin [11]. The introduction of a dissipative function of the type [3]

$$\Psi = \eta \left(v_{ik} - \tfrac{1}{3} \delta_{ik} v_{ll} \right)^2 + \tfrac{1}{2} \zeta v_{ll}^2,$$

(where v_{ik} is the deformation rate tensor, while η and ζ are, respectively, the first and second viscosities) to describe the energy dissipation in solids is also equivalent to the assumption of the rheological model of Voigt for the volume part of the stress tensor.

In this connection, it is of interest to investigate the hereditary behavior of isotropic solids, in which the volume stresses obey the Voigt model. We shall consider a medium in which the volume stresses are described by the Voigt model and the shear stresses are described by the Maxwell model. In the latter case the shear stresses are completely relieved, which does take place at high temperatures because of the vacancy diffusion [12-14].

The stress tensor of the medium has the form

$$\sigma_{ik} = K\varepsilon_{ll}\delta_{ik} + \zeta\dot{\varepsilon}_{ll}\delta_{ik} + 2\mu \int_{-\infty}^{t} \dot{e}_{ik}(t') \exp\frac{t'-t}{\tau_\varepsilon} \, dt', \tag{1}$$

where μ is the unrelaxed shear modulus; K is the bulk modulus; $e_{ik} = \varepsilon_{ik} - \frac{1}{3}\delta_{ik}\varepsilon_{ll}$ is the deviator of the deformation tensor ε_{ik}; τ_ε is the relaxation time of the shear stresses; a dot above a quantity represents the first derivative with respect to time.

The application of Eq. (1) to small-amplitude torsional vibrations makes it possible to describe the temperature and frequency dependence of the internal friction background, i.e., its monotonic rise with increasing temperature or with decreasing frequency [4].

The influence of the second viscosity on the internal friction background will be considered, by way of example, for damped longitudinal vibrations of a rod made of a material which is described by Eq. (1). To solve the problem, we shall express the deformation tensor in terms of the stress tensor, assuming zero initial conditions. The Laplace transform of the stress tensor is, according to Eq. (1),

$$\bar{\sigma}_{ik} = (K + p\zeta)\,\delta_{ik}\bar{\varepsilon}_{ll} + 2\mu\left(1 + \frac{1}{p\tau_\varepsilon}\right)^{-1}\bar{e}_{ik}. \tag{2}$$

Hence

$$\bar{\varepsilon}_{ik} = \frac{\delta_{ik}\bar{\sigma}_{ll}}{9\,(K + p\zeta)} + \frac{1}{2\mu}\left(1 + \frac{1}{p\tau_\varepsilon}\right)\left(\bar{\sigma}_{ik} - \frac{1}{3}\delta_{ik}\bar{\sigma}_{ll}\right). \tag{3}$$

We shall assume that the axis of the rod is the z axis. No forces act on the lateral surface of the rod, i.e., $\sigma_{ik}n_k = 0$, where $\vec{n}$ is a unit vector normal to the rod surface. Since $n_z = 0$, all the components of σ_{ik}, with the exception of σ_{zz}, vanish in the uniform deformation case.

We can then see from Eq. (3) that all the components of ε_{ik} with $i \neq k$ vanish and the diagonal components have the form

$$\bar{\varepsilon}_{xx} = \bar{\varepsilon}_{yy} = -\frac{1}{3}\left[\frac{1}{2\mu}\left(1 + \frac{1}{p\tau_\varepsilon}\right) - \frac{1}{3K}\left(1 + p\tau_\sigma\right)^{-1}\right]\bar{\sigma}_{zz},$$

$$\bar{\varepsilon}_{zz} = \frac{1}{3}\left[\frac{1}{3K}\left(1 + p\tau_\sigma\right)^{-1} + \frac{1}{\mu}\left(1 + \frac{1}{p\tau_\varepsilon}\right)\right]\bar{\sigma}_{zz}, \tag{4}$$

where, for convenience, the second viscosity ζ is replaced by the volume retardation time $\tau_\sigma = \zeta/K$.

The quantity

$$\bar{E}(p) = \frac{9K\mu\,(1 + p\tau_\sigma)}{\mu + 3K\,(1 + p\tau_\sigma)\left(1 + \frac{1}{p\tau_\varepsilon}\right)} \tag{5}$$

is Young's modulus in the Laplace transform space.

Equation (4) can also be used to find easily Poisson's ratio $\bar{\nu}(p)$ in the Laplace transform space

$$\bar{v}(p) = -\frac{\bar{e}_{xx}}{\bar{e}_{zz}} = \frac{1}{2}\frac{(1+p\tau_\sigma)(1+p\tau_\varepsilon)-2\chi p\tau_\varepsilon}{(1+p\tau_\sigma)(1+p\tau_\varepsilon)+\chi p\tau_\varepsilon}.$$

(6)

The substitution $p \to i\omega$ in Eqs. (5) and (6) gives the complex Young's modulus $E^*(\omega) = E'(\omega) + iE''(\omega)$ and the complex Poisson's ratio $\nu^*(\omega) = \nu'(\omega) + i\nu''(\omega)$

$$E^* = 3\mu\frac{(1+\chi+\omega^2\tau_\sigma^2)+i\left(\frac{1}{\omega\tau_\varepsilon}+\frac{\omega\tau_\sigma^2}{\tau_\varepsilon}+\chi\omega\tau_\sigma\right)}{\left(\frac{1}{\omega\tau_\varepsilon}-\omega\tau_\sigma\right)^2+\left(1+\chi+\frac{\tau_\sigma}{\tau_\varepsilon}\right)^2},$$

(7)

$$\nu^* = \frac{1}{2}\frac{\left(\frac{1}{\omega\tau_\varepsilon}-\omega\tau_\sigma\right)^2+\left(1-2\chi+\frac{\tau_\sigma}{\tau_\varepsilon}\right)\left(1+\chi+\frac{\tau_\sigma}{\tau_\varepsilon}\right)-i\left(\frac{1}{\omega\tau_\varepsilon}-\omega\tau_\sigma\right)}{\left(\frac{1}{\omega\tau_\varepsilon}-\omega\tau_\sigma\right)^2+\left(1+\chi+\frac{\tau_\sigma}{\tau_\varepsilon}\right)^2},$$

(8)

where

$$\chi = \frac{\mu}{3K}.$$

Knowing $E^*(\omega)$, we can easily find the loss-angle tangent for the medium considered

$$\tan\delta = \frac{1}{\omega\tau_\varepsilon}\frac{\chi\omega^2\tau_\varepsilon\tau_\sigma+(1+\omega^2\tau_\sigma^2)}{\chi+(1+\omega^2\tau_\sigma^2)}.$$

(9)

We shall now consider the solution of the boundary problem. We shall discuss longitudinal vibrations of the rod along the z axis. Substituting into the equation of motion in the Laplace transform space

$$\rho p^2\bar{u}_z = \frac{\partial\bar{\sigma}_{zk}}{\partial z}$$

(10)

the value of the only nonzero component of the stress tensor

$$\bar{\sigma}_{zz} = \bar{E}(p)\frac{\partial\bar{u}_z}{\partial z}$$

(11)

we obtain

$$\frac{d^2\bar{u}_z}{dz^2} = \lambda^2\bar{u}_z, \qquad \lambda \equiv \frac{\rho p^2}{\bar{E}(p)}.$$

(12)

The boundary conditions for this problem are as follows. One end of the rod ($z = 0$) is fixed and the other end ($z = l$), to which an additional load of mass M is attached, is acted upon by an impulse. This gives

$$u_z|_{z=0} = 0, \quad [\sigma_{zz}S + M\ddot{u}_z]|_{z=l} = F\delta(t),$$

(13)

where F is the value of the impulse and $\delta(t)$ is the Dirac delta function, which describes the instantaneous action of the applied force, and S is the cross-sectional area of the rod.

In the Laplace transform space the conditions of Eq. (13) assume the form

$$\bar{u}_z\big|_{z=0} = 0, \left[\bar{E}(p)\frac{d\bar{u}_z}{dz} + \frac{M}{S}p^2\bar{u}_z\right]\Big|_{z=l} = \frac{F}{S} \equiv f. \tag{14}$$

Solving Eq. (12) with the conditions given by Eq. (14), we obtain

$$\bar{u}_z(p) = \frac{F}{M}\frac{\lambda l}{p^2}\frac{\cosh \lambda z}{\lambda l \sinh \lambda l + a \cosh \lambda l} \tag{15}$$

or, after expansion into a series,

$$\bar{u}_z(p) = \frac{2F}{M}\sum_{n=1}^{\infty} c_n(z)\left[p^2 + \frac{b_n^2\bar{E}(p)}{\rho l^2}\right]^{-1}, \tag{16}$$

where

$$c_n(z) = \frac{\sin(b_n z/l)}{b_n \cos b_n + (1+a)\sin b_n}, \quad b_1 \approx \sqrt{a}, \quad b_{n+1} \approx n\pi. \tag{17}$$

The expression (16) can be rewritten in the form

$$\bar{u}_z(p) = \frac{2F}{M}\sum_{n=1}^{\infty} c_n(z)\frac{Q_{1n}(p)}{Q_{2n}(p)}, \tag{18}$$

where

$$Q_{1n}(p) = p^2 + s_{1n}p + s_\varepsilon s_\sigma, \quad Q_{2n}(p) = p^4 + s_{1n}p^3 + s_{2n}^2 p^2 + s_{3n}^3 p,$$

$$s_{1n} = s_\varepsilon + s_{3n}^3 \beta_{n\infty}^{-2}, \quad s_{2n}^2 = s_\varepsilon s_\sigma + s_{3n}^3 \tau_\sigma, \quad s_{3n}^3 = \frac{3}{2}\beta_{n\infty}^2 s_\sigma(1+\nu_\infty)^{-1},$$

$$\beta_{n\infty}^2 = \frac{E_\infty b_n^2}{\rho l^2}. \tag{19}$$

Here, ν_∞ and E_∞ are the unrelaxed values of Poisson's ratio and Young's modulus, respectively.

Going over, in Eq. (18), from the transform to the original function, we obtain the required solution to the problem

$$u_z(z, t) = \frac{2F}{M}\sum_{n=1}^{\infty} c_n(z)\sum_{k=1}^{4}\frac{Q_{1n}(p_k)}{\prod_{k\neq j}(p_k - p_j)}\exp(p_k t), \tag{20}$$

where p_k, p_j are the roots of the polynomial $Q_{2n}(p)$.

Thus, the final solution of the problem depends on the explicit form of the roots of the polynomial $Q_{2n}(p)$ which are governed by the sign of the discriminant

$$D_n = q_n^2 + m_n^3. \tag{21}$$

Here,

$$q_n = \frac{s_{1n}^3}{27} - \frac{s_{1n} s_{2n}^2}{6} + s_{3n}^3, \quad m_n = \frac{s_{2n}^2}{3} - \frac{s_{1n}^2}{9}. \tag{22}$$

If $D_n > 0$, then the two roots of the polynomial $Q_{2n}(p)$ are complex and the solution represents damped vibrations. If $D_n < 0$, then all the roots are real and the solution describes an aperiodic process. Bearing in mind that the solution $u_z(z, t)$ may describe both vibrations and an aperiodic process, it is convenient to represent this solution in the form

$$u_z(z, t) = \sum_{n=1}^{k} u_{nz}^a(z, t) + \sum_{n=k+1}^{\infty} u_{nz}^v, (z, t), \tag{23}$$

where the aperiodic terms are given by the expressions

$$u_{nz}^a = \frac{F s_\varepsilon}{3\mu} z - \frac{2F}{M} c_n(z) \sum_{i=1}^{3} H_{in} \exp(-\gamma_{in} t), \tag{24}$$

$$H_{in} = \frac{\gamma_{in}^2 - s_{1n} \gamma_{in} + s_\varepsilon s_\sigma}{\gamma_{in}(\gamma_{in} - \gamma_{kn})(\gamma_{in} - \gamma_{ln})}, \quad i \neq j \neq k; \tag{25}$$

$$\gamma_{1n} = \frac{s_{1n}}{3} + 2r_n \cos\frac{\theta_n}{3}, \quad \gamma_{2n,\,3n} = \frac{s_{1n}}{3} - 2r_n \cos\frac{\pi \mp \theta_n}{3}, \tag{26}$$

$$r_n = \pm \sqrt{|m_n|}, \quad \theta_n = \arccos(q_n r_n^{-3}), \tag{27}$$

here the sign of r_n is selected to be identical with the sign of q_n.

The terms describing vibrations are given by the expressions

$$u_{nz}^v = \frac{F s_\varepsilon}{3\mu} z + \frac{2F}{M} c_n(z) [A_{1n} \exp(-\alpha_{1n} t) + A_{2n} \exp(-\alpha_{2n} t) \sin(\beta_n t + \psi_n)], \tag{28}$$

where

$$A_{1n} = 2\alpha_{2n} \left(1 - \frac{1}{2} \frac{s_\varepsilon s_\sigma}{\alpha_{1n} \alpha_{2n}}\right) [(\alpha_{2n} - \alpha_{1n})^2 + \beta_n^2]^{-1}, \tag{29}$$

$$A_{2n} = A_{1n} \left(1 - \frac{1}{2} \frac{s_\varepsilon s_\sigma}{\alpha_{1n} \alpha_{2n}}\right)^{-1} \left[\frac{s_\varepsilon s_\sigma}{\alpha_{2n}^2 + \beta_n^2}\left(1 - \frac{1}{2}\frac{\alpha_{1n}}{\alpha_{2n}}\right) - 1\right] \operatorname{cosec} \psi_n, \tag{30}$$

$$\alpha_{1n} = s_{1n} - 2\alpha_{2n}, \quad \alpha_{2n} = \varkappa_n + \frac{s_{1n}}{3}, \tag{31}$$

$$\tan\psi_n = \frac{\beta_n [s_\varepsilon s_\sigma (2\alpha_{2n} - \alpha_{1n}) - 2\alpha_{2n}(\alpha_{2n}^2 + \beta_n^2)]}{(\alpha_{2n}^2 + \beta_n^2)(\alpha_{1n}^2 - \alpha_{2n}^2 + \beta_n^2) - s_\varepsilon s_\sigma(\alpha_{1n}\alpha_{2n} - \alpha_{2n}^2 + \beta_n^2)}. \tag{32}$$

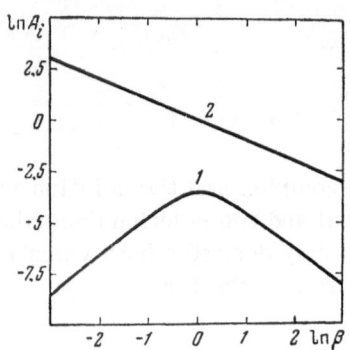

Fig. 1. Frequency dependence of the amplitudes of damped vibrations: 1) A_1; 2) A_2.

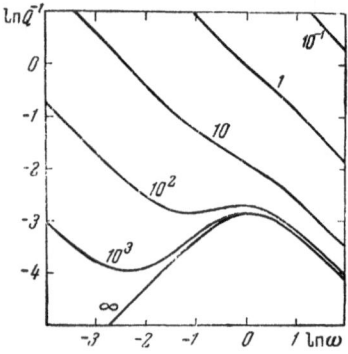

Fig. 2. Frequency dependence of the internal friction. Figures beside the curves indicate the values of the parameter τ_ε ($\tau_\sigma = 1$).

$\varkappa_n$ and β_n are found:

1) if $m_n > 0$

$$\varkappa_n = -r_n \sinh \theta_n, \qquad \beta_n = \sqrt{3}\, r_n \cosh \theta_n,$$

$$\theta_n = \sinh^{-1} (q_n r_n^{-3}); \qquad (33)$$

2) if $m_n < 0$, but $q_n^2 > m_n^3$,

$$\varkappa_n = -r_n \cosh \theta_n, \qquad \beta_n = \sqrt{3}\, r_n \sinh \theta_n,$$

$$\theta_n = \cosh^{-1} (q_n r_n^{-3}). \qquad (34)$$

The internal friction is found from Eq. (28) and the logarithmic decrement of vibrations, divided by π, is used as the measure of the internal friction:

$$Q^{-1} = \tan \delta \simeq \frac{\Delta_n}{\pi} = \frac{2\alpha_{2n}}{\beta_n}. \qquad (35)$$

If damping is not too strong, the temperature dependence of the internal friction can be calculated by means of Eq. (9) instead of Eq. (35). In the case of strong damping, it is necessary to allow for the variation of the natural frequency with temperature, and, therefore, it is necessary to use Eq. (35) in the calculations.

We shall now consider the frequency dependence of the internal friction of our system. Since the frequency can only have real values, the region of aperiodic motion will not be discussed. In this case the solution is given by Eq. (28). Using only the first harmonic, for which $c_1(z) = z/2l$, we obtain

$$u_z = Fz \left\{ \frac{1}{3\mu\tau_\varepsilon} + \frac{1}{Ml} \, [A_1 \exp (-\alpha_1 t) \right.$$

$$\left. + A_2 \exp (-\alpha_2 t) \sin (\beta t + \psi)] \right\}. \qquad (36)$$

Here, the first term gives the new equilibrium position for $t \to \infty$, the second term describes the shift of the equilibrium position with time, i.e., the elastic aftereffect, and the third term represents the damping of harmonics.

To estimate the contribution of the elastic aftereffect to the total motion, Fig. 1 gives the curves of A_1 and A_2 as a function of the natural frequency for $\tau_\sigma = 1$. The various values of β can be obtained not only by altering the geometrical dimensions of the rod, but also by altering the mass of the load attached to the sample.

It is evident from Fig. 1 that the condition $A_1/A_2 \ll 1$ applies over the whole range of frequencies, i.e., the elastic aftereffect is negligibly small and $u_z(z,t)$ describes damped vibrations about some equilibrium position.

Since, in the case of the frequency dependent internal friction, the relaxation and retardation times remain constant, it follows that the internal friction values calculated using Eqs. (9) and (35) will coincide and this makes it possible to use the simpler formula (9). The corresponding results are shown in Fig. 2, where τ_ε is used as the parameter and τ_σ is assumed to be equal to unity.

It is evident from Fig. 2 that a well-defined internal friction peak appears at $\tau_\varepsilon/\tau_\sigma \approx 10^2$-$10^4$. This condition corresponds to a higher rate of the volume relaxation processes than of the shear processes. In the opposite case the high internal friction background would have prevented the appearance of the second viscosity.

We shall consider some limiting cases of the internal friction.

1. No volume relaxation ($\tau_\sigma \rightarrow \infty$). Then, $Q^{-1} = 1/\omega\tau_\varepsilon$, i.e., there is an internal friction background, as in the usual Maxwell model.

2. Volume stresses relax at a very high rate ($\tau_\sigma \rightarrow 0$). In this case, $Q^{-1} = [\omega\tau_\varepsilon(1 + \chi)]^{-1}$, i.e., the background again is shifted somewhat compared with the preceding case, and, as before, $\chi = \mu/3K$.

The first case describes a rod which is absolutely rigid with respect to the volume deformations. In the second case, the volume deformations are elastic. The shear deformations in both cases are described by the Maxwell rheological model.

3. No relaxation of the shear stresses ($\tau_\varepsilon \rightarrow \infty$). In this case we obtain an internal friction peak

$$Q^{-1} = \Delta_E \omega\tau_\bullet (1 + \omega^2\tau_\bullet^2)^{-1},$$

where the new relaxation time and the new modulus defect are

$$\tau_\bullet = \tau_\sigma(1 + \chi)^{-\frac{1}{2}}, \quad \Delta_E = \chi(1 + \chi)^{-\frac{1}{2}}.$$

4. The case $\tau_\varepsilon \rightarrow 0$ corresponds to the full relaxation of the shear stresses, i.e., it describes the liquid state of matter.

Literature Cited

1. L.I. Mandel'shtam and M.A. Leontovich, Zh. Éksperim. i Teor. Fiz., 7(3):438 (1937).
2. E. Skudrzyk, Fundamentals of Acoustics [Russian translation], IL (1959), Vol. 2. [English edition: L-E. Kinster and A.R. Frey, eds.; 2nd ed., Wiley, New York (1962).]
3. L.D. Landau and E.M. Lifshits, Mechanics of Continuous Media, Moscow (1954).
4. T.D. Shermergor and S.I. Meshkov, Fiz. Metal. i Metalloved., 13(6):817 (1962).
5. V.S. Postnikov, Plasticheskie Massy, 11:60 (1960).
6. S.I. Meshkov and T.D. Shermergor, Zh. Prikl. Mekhan. i Tekhn. Fiz., No. 6:98 (1962); No. 3:20 (1963); No. 1:68 (1964).
7. S.I. Meshkov, V.S. Postnikov, and T.D. Shermergor, Izv. Akad. Nauk SSSR, Otd. Tekhn. Nauk, Mekhan. i Mashinostr., 3:90 (1964).
8. B. Gross, Mathematical Structure of the Theories of Viscoelasticity, Paris (1953).
9. Ya.I. Frenkel', Kinetic Theory of Liquids, Izd. Akad. Nauk SSSR (1945), Chap. IV.
10. A.I. Gubanov, Zh. Tekhn. Fiz., 19(1):34 (1949).
11. N.A. Slezkin, Prikl. Matem. i Mekhan., 9:233 (1945).
12. N.S. Fastov, Problems of Metallography and Physics of Metals, No. 5, Metallurgizdat (1958), p. 550.
13. B. Escaig, Acta Met., 10:829 (1962).
14. I.M. Lifshits, Zh. Éksperim. i Teor. Fiz., 44(4) (1963).

THEORY OF THERMOELASTIC INTERNAL FRICTION IN SOLIDS

Ya. S. Podstrigach and R. N. Shvets

The present paper deals with the dissipation of mechanical energy in thermoelastic processes in a finite solid, when heat is transferred through the surface of the solid into the surrounding medium, and with the establishment of a correlation between the thermoelasticity equations and the rheological relationships.

1. We shall calculate the value of the mechanical energy dissipated per unit volume of a finite solid, which is in mechanical and thermal contact with an external medium. From the first law of thermodynamics we find that the increase in the total energy of the body per unit time is [1, 2]:

$$\frac{d}{d\tau}\int_V \left(U + \frac{1}{2}\rho v^2\right)dV = \int_V (\overline{F}\overline{V})\,dV + \int_\Sigma (\overline{\sigma}_n v - I_n)\,d\Sigma, \tag{1}$$

where U is the internal energy per unit volume, ρ is the density, $\overline{v}$ and $\overline{u}$ are the velocity and displacement vectors, respectively; F is the mass force vector; I_n is the normal component of the heat flow vector $\overline{J}$; σ_n is the surface force vector

$$\overline{\sigma}_n = \sigma_i \cos(n,\widehat{x}_i), \quad \overline{\sigma}_i = \sigma_{ij}\overline{k}_j, \tag{2}$$

σ_{ij} are the components of the stress tensor; $\overline{k}_j$ are the coordinate unit vectors; $\overline{n}$ is the normal to the surface Σ.

Eliminating by means of the equations of motion

$$\frac{\partial\sigma_{ij}}{\partial x_i} = \rho\frac{dv_i}{d\tau} - F_i, \tag{3}$$

the kinetic energy from Eq. (1) and using Eq. (2), we find the change in the internal energy density in a thermoelastic body per unit time

$$\frac{dU}{d\tau} = -\operatorname{div}\overline{I} + \frac{1}{2}\sigma_{ij}\left(\frac{\partial v_i}{\partial x_j} + \frac{\partial v_j}{\partial x_i}\right). \tag{4}$$

Now, we obtain from Eqs. (1) and (4),

$$\frac{d}{d\tau} \int_V \left(\frac{1}{2} \rho \bar{v}^2\right) dV + \frac{1}{2} \int_V \sigma_{ij}\left(\frac{\partial v_i}{\partial x_j} + \frac{\partial v_j}{\partial x_i}\right) dV = \int_V (\overline{Fv})\, dV + \int_\Sigma (\bar{\sigma}_n \bar{v})\, d\Sigma. \tag{5}$$

Substituting into Eq. (5) the values of σ_{ij} taken from the equations of state

$$\sigma_{ij} = 2G\left[e_{ij} + \frac{\nu}{1-2\nu} \delta_{ij} e_{kk} - \frac{1+\nu}{1-2\nu} \delta_{ij} \alpha_t t\right] \tag{6}$$

and bearing in mind that, in the case of small deformations $\frac{de_{ij}}{d\tau} = \dot{e}_{ij} = \frac{1}{2}\left(\frac{\partial v_i}{\partial x_j} + \frac{\partial v_j}{\partial x_i}\right)$, we have

$$\frac{d}{d\tau} \int_V \left[G\left(e_{ij}e_{ij} + \frac{\nu}{1-2\nu} e_{kk}^2\right) + \frac{1}{2} \rho \bar{v}^2\right] dV = \frac{c_v \gamma_0}{T_0} \int_V t\dot{e}_{kk}\, dV$$
$$+ \int_V (\overline{Fv})\, dV + \int_\Sigma (\bar{\sigma}_n \bar{v})\, d\Sigma, \tag{7}$$

where $\gamma_0 = \frac{E\alpha_t T_0}{c_v(1-2\nu)}$, $2G = \frac{E}{1+\nu}$, E is the isothermal Young's modulus; T_0 is the absolute temperature; c_v is the specific heat; α_t and ν are the linear expansion coefficient and Poisson's ratio, respectively; $t = T - T_0$ is the change in the temperature of the body.

Eliminating from the right-hand side of Eq. (7) and from the heat-conduction equation [1, 3]

$$\lambda \Delta t - c_v \frac{\partial t}{\partial \tau} = c_v \gamma_0 \dot{e}_{kk} \tag{8}$$

the quantity $\dot{e}_{kk}$, and then using the Gauss formula and the law of heat transfer through the surface of a body,

$$\frac{\partial t}{\partial n} + \frac{\alpha_n}{\lambda}(t - t_c) = 0 \text{ on } \Sigma, \tag{9}$$

we find, after some transformations, the change in the mechanical energy W in a deformed thermoelastic body per unit time

$$\frac{dW}{d\tau} = -\int_V \left[\frac{\lambda}{T_0}(\text{grad}\, t)^2 - \overline{Fv}\right] dV - \int_\Sigma \left[\frac{\alpha_n}{T_0} t(t - t_c) - \bar{\sigma}_n \bar{v}\right] d\Sigma, \tag{10}$$

where λ, α_n are the thermal conductivity and heat transfer coefficient; $t_c = T_c - T_0$ is the change in the temperature of the medium; W, defined by the formula

$$W = \int_V \left[G\left(e_{ij}e_{ij} + \frac{\nu}{1-2\nu} e_{kk}^2\right) + \frac{1}{2} \rho \bar{v}^2 + \frac{c_v}{2T_0} t^2\right] dV, \tag{11}$$

represents the maximum work which can be done in a reversible transition from a given non-equilibrium state to a thermodynamic equilibrium state.

In the absence of external agencies ($\overline{F} = 0$, $\bar{\sigma}_n = 0$, $t_c = 0$), the change in the energy initially stored by a system

$$\left|\frac{dW}{d\tau}\right| = \frac{\lambda}{T_0}\int_V (\operatorname{grad} t)^2\, dV + \frac{\alpha_n}{T_0}\int_\Sigma t^2\, d\Sigma \tag{12}$$

represents the value of the energy dissipated per unit time.

We can see that the dissipation of mechanical energy in a system consisting of a body and a medium in the case of irreversible thermoelastic properties is due to the presence of a finite heat flow both within the body (λ finite) and through its surface (α_n finite). Equation (12) for the heat transfer coefficient $\alpha_n = 0$ is identical with the result given in [1], if the viscosity coefficient is $\eta = 0$.

It should be mentioned that the ratio $|\overline{\overline{W}}|/\overline{W}$ represents the internal (thermoelastic) friction of the system and the value of the absorption coefficient q [1, 5] for the propagation of a thermoelastic wave at a velocity v is given by

$$q = \frac{1}{2v}\frac{|\overline{\overline{W}}|}{\overline{W}}, \tag{13}$$

where $\overline{W}$ and $\overline{\overline{W}}$ are the time-average quantities given by Eqs. (11) and (12).

2. We shall consider the internal friction in a solid due to the thermoelastic dissipation caused by heat flow fluxes both within the body and through its surface and, at the same time, we shall establish a correlation between the equations of thermoelasticity and the rheological relationships.

For this purpose, we shall consider the longitudinal vibrations of a thin rod through the surface of which heat is transferred to the surrounding medium. The heat-conduction equation [4] has, in this case, the form

$$\lambda\frac{\partial^2 t}{\partial x^2} - r_0\frac{\partial t}{\partial \tau} - \mu_0 t - \gamma_0 c_v (1-2v)\dot{e}_\parallel = 0, \tag{14}$$

where $r_0 = c_v[1 + 2\alpha_t\gamma_0(1 + \nu)]$, $\mu_0 = \alpha_n/d$, $d = Q/P$ is the ratio of the cross-sectional area Q to the perimeter of the body P.

Eliminating the temperature t from Eqs. (6) and (14), we obtain the following relationship between the stress and deformation in a thermoelastic body:

$$r_0\dot{\sigma}_\parallel + \mu_0\sigma_\parallel - \lambda\frac{\partial^2\sigma_\parallel}{\partial x^2} = E\left[r_0(1+\gamma)\dot{e}_\parallel + \mu_0 e_\parallel - \lambda\frac{\partial^2 e_\parallel}{\partial x^2}\right], \tag{15}$$

where

$$\gamma = \frac{\alpha_t^2 E T_0}{r_0}.$$

The relationship (15) allows for the irreversibility associated with the flow of heat through the surface of the rod (finite α_n) as well as for the irreversibility due to the finite heat flow within the rod (finite λ).

For $\alpha_n = 0$ and finite λ, we obtain the relationship

$$r_0\dot{\sigma}_\parallel - \lambda\frac{\partial^2\sigma_\parallel}{\partial x^2} = E\left[r_0(1+\gamma)\dot{e}_\parallel - \lambda\frac{\partial^2 e_\parallel}{\partial x^2}\right], \tag{16}$$

which describes the inelastic behavior of the material solely due to the temperature relaxation within the rod.

In the case of a uniform thermomechanical state (or $\lambda = 0$), we obtain, from Eq. (15), the deformation law

$$\sigma_\parallel + r\dot\sigma_\parallel = E\,[e_\parallel + r(1+\gamma)\dot e_\parallel], \tag{17}$$

which represents a generalization of Hooke's law, allowing for the elastic aftereffect and relaxation [5, 6]. Here, $r = r_0/\mu_0$ and $r(1+\gamma)$ are, respectively, the stress and deformation relaxation times, whose values are completely determined by the known thermoelastic constants. We note that the assumption made in [6] about the coefficients in Eq. (17), in connection with the proof of the theorem of uniqueness of the boundary problem solution, can be justified directly from the physical point of view because $\gamma > 0$ always.

It follows that Eq. (15), which should govern the deformation of a thermoelastic body, is more general than Eq. (17), which describes a viscoelastic body. In a thin rod, whose thermomechanical state is described by Eqs. (3) and (15), two waves are propagated: a mechanical wave and a thermal wave [4, 7]. If $\lambda = 0$, i.e., the state of the rod is described by Eqs. (3) and (17), only the mechanical wave is propagated in the rod.

In the case of a longitudinal mechanical sinusoidal wave we have

$$\sigma_\parallel = \sigma_0 \exp i\,(sx - \omega\tau),$$
$$e_\parallel = e_0 \exp i\,(sx - \omega\tau), \tag{18}$$

where the complex quantity S is the root of the equation

$$\lambda v_0^2 s^4 - [\omega^2 \lambda - \mu_0 v_0^2 - iv_0^2 r_0 \omega\,(1+\gamma)]\,s^2 - \omega^2\,(\mu_0 + i\omega r_0) = 0,$$
$$v_0^2 = E/\rho, \tag{19}$$

which is obtained after substituting Eq. (18) into Eqs. (3) and (15). The velocity v and the absorption coefficient q of the wave are, respectively [4],

$$v = \frac{\omega}{\mathrm{Re}\,s}, \qquad q = \mathrm{Im}\,s. \tag{20}$$

Substituting Eq. (18) into Eq. (15), we find

$$\sigma_\parallel = \mu e_\parallel, \tag{21}$$

where the complex modulus is

$$\mu = E\,\frac{1 + i\omega\tau_\sigma}{1 + i\omega\tau_e}, \tag{22}$$

and the stress (τ_e) and deformation (τ_σ) relaxation times are, respectively,

$$\tau_e = \frac{r_0}{\lambda s^2 + \mu_0},$$
$$\tau_\sigma = (1+\gamma)\,\tau_e. \tag{23}$$

It follows from Eq. (23) that the relaxation time τ_e (τ_σ) is less than the corresponding relaxation time in the case of an isolated rod ($\mu_0 = 0$) or a rod in a uniform thermomechanical state ($\lambda = 0$). The quantities τ_e and τ_σ are complex and they depend on the frequency ω; this is due to the heat flow inside the rod ($\lambda \neq 0$).

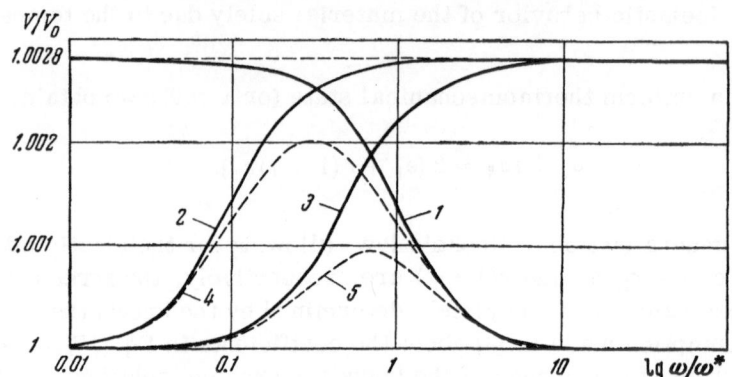

Fig. 1. Frequency dependence of the velocity of propagation of a wave. 1) $\mu = 0$, no heat flow; 2) $\mu = 0.1$, heat flow only through the rod surface; 3) $\mu = 0.5$, heat flow only through the rod surface; 4) $\mu = 0.1$, heat flow within the rod and through its surface; 5) $\mu = 0.5$, heat flow within the rod and through its surface.

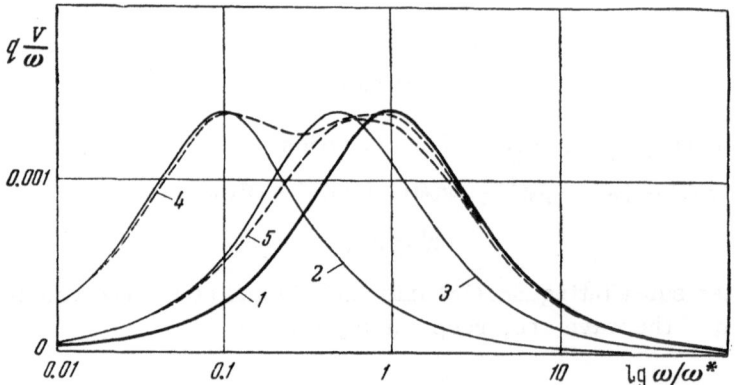

Fig. 2. Frequency dependence of the internal friction. The notation is the same as in Fig. 1.

The internal friction $\tan \delta$ is given by the expression

$$\tan \sigma = \frac{\operatorname{Im} \mu}{\operatorname{Re} \mu} = \frac{\gamma \omega \tau_e'}{(1 - \omega \tau_\sigma'')(1 - \omega \tau_e'') + \omega^2 \tau_e' \tau_\sigma'} , \tag{24}$$

where δ is the phase shift between $\sigma_{\parallel}$ and $e_{\parallel}$; $\tau_e = \tau_e' + i\tau_e''$, $\tau_\sigma = \tau_\sigma' + i\tau_\sigma''$. If $\omega \tau_e'' \ll 1$, $\omega \tau_\sigma'' \ll 1$, then we find from Eq. (24)

$$\tan \delta = \frac{\gamma \omega \tau_e'}{1 + (1 + \gamma) \omega^2 (\tau_e')^2} . \tag{25}$$

It should be mentioned that if the quantities $\omega \tau_e''$ and $\omega \tau_\sigma''$ in Eq. (24) are neglected compared with unity, then this corresponds to $\operatorname{Im} s = 0$ in Eq. (23). Then, using Eq. (20), we obtain real values of the relaxation times

$$\tau_e = \tau_e' = \frac{r_0 v^2}{\lambda \omega^2 + \mu_0 v^2} ,$$
$$\tau_\sigma = \tau_\sigma' = (1 + \gamma) \tau_e'. \tag{26}$$

Substituting Eq. (18) into the equation of motion (3) for a rod, we obtain

$$\sigma_{\|} = \rho \frac{\omega^2}{s^2} e_{\|}. \tag{27}$$

Comparing Eqs. (21) and (27), we find

$$\mu = \rho \frac{\omega^2}{s^2}. \tag{28}$$

From Eqs. (24) and (28) we find the expression for the internal friction in terms of the values s of the root of Eq. (19)

$$\tan \delta = 2 \frac{\operatorname{Im} s}{\operatorname{Re} s} \left[1 - \left(\frac{\operatorname{Im} s}{\operatorname{Re} s} \right)^2 \right]^{-1}. \tag{29}$$

Hence, on the assumption that $\operatorname{Im} s / \operatorname{Re} s \ll 1$, and using Eq. (20), we find the absorption coefficient in terms of the value of the internal friction

$$q = \frac{\omega}{2v} |\tan \delta|. \tag{30}$$

From the approximate formulas (25), (26), and (30) for $\omega \tau_e' \gg 1$, $\omega^2 > \mu_0 v_0$, and for $\omega \tau_e' \leq 1$, $v_0 \mu_0 < \omega$, we obtain the same values of the absorption coefficient as those calculated in [4] by a different method. If $r_0 = c_V$ and $\mu_0 = 0$, we obtain the results given in [8].

Figure 1 shows the velocity of propagation of a wave, and Fig. 2 gives the internal friction, both as a function of the quantity ω / ω^* ($\omega^* = v_0^2 r_0 / \lambda$) for $\gamma = 0.00561$ and different values of the dimensionless coefficient of heat transfer $\mu = (\mu_0 \lambda / v_0^2 r_0^2)$ from the surface of an aluminum rod.

Curve 1 represents the case when the surface of the rod is thermally insulated ($\mu = 0$). The internal friction reaches its maximum value in this case when $\omega = \omega^*$, where ω^* is the characteristic frequency [4, 7] of the rod, corresponding to a transition from adiabatic to isothermal vibrations.

The case when the energy dissipation during vibrations is solely due to heat flow through the surface of the rod is represented by curves 2 and 3 for $\mu = 0.1$ and 0.5, respectively. Here, the internal friction reaches its maximum value when $\omega = \mu \omega^*$.

Comparison of the curves 1, 2, and 3 in Fig. 1 shows that the dispersion tendencies are opposite in these two cases, i.e., the low-frequency waves are propagated faster than the high-frequency waves in a thermally insulated rod, while the opposite is true for a viscoelastic rod, whose material obeys the law of Eq. (17).

The dashed curves 4 and 5 represent the general case, allowing for the energy dissipation in a rod caused by heat flows both within the rod and through its surface ($\mu = 0.1$ and 0.5). The phase velocity in this case reaches its maximum value for $\mu = 0.1$ and 0.5 when $\omega = \sqrt{\mu} \, \omega^*$, but at low and high frequencies this velocity is equal to the isothermal velocity v_0.

Literature Cited

1.   L. D. Landau and E. M. Lifshits, Mechanics of Continuous Media, Gostekhizdat (1954).
2.   Ya. S. Podstrigach, Diffusion theory of deformation of an isotropic continuous medium, in collection: Problems of Mechanics of Real Solids, No. 2, Izd. "Naukova dumka" (1964).
3.   J. M. C. Duhamel, Second memoire sur les phenomenes thermo-mecanique, J. Ec. Polyt., 25:15 (1837).

4. Ya. S. Podstrigach and R. N. Shvets, Dynamic problem of thermoelasticity for a thin rod allowing for heat transfer from its surface, in collection: Problems of Mechanics of Real Solids, No. 2, Izd. "Naukova dumka" (1964).

5. C. M. Zener, Elasticity and Anelasticity of Metals [Russian translation], IL (1954). [English edition: University of Chicago Press, Chicago, Illinois (1948).]

6. A. Yu. Ishlinskii, Longitudinal vibrations of a rod in the case of a linear elastic after-effect and relaxation law, Prikl. Matem. i Mekhan., 4:1 (1940).

7. J. N. Sneddon, The propagation of thermal stresses in thin metallic rods, Proc. Roy. Soc. Edinburgh, A, 65(9) (1959).

8. T. D. Shermergor and V. S. Postnikov, Thermal relaxation in solids, in collection: Relaxation Phenomena in Metals and Alloys, Metallurgizdat (1963).

RELATIONSHIP BETWEEN FLUCTUATIONS, THE INTERNAL FRICTION AND THE ELASTIC MODULI OF A MEDIUM

L. A. Svergunenko

By considering the elastic and dielectric properties of a medium in external fields [1], it is possible to represent the dynamic characteristics of a medium in terms of thermodynamic fluctuations of the corresponding quantities. The method of fluctuations has been used to solve the problems of the diffuse scattering of electromagnetic waves in gases, liquids, and solids [2]. Bearing in mind that there is a direct relationship between the fluctuation and dissipative properties of a system, we shall consider below the relationship between the anelastic properties of a medium and the fluctuations of the corresponding physical quantities.

We shall consider a body, whose thermodynamic state is determined by the value of the deformation u, an internal parameter η, and temperature. As the internal parameter we can use such quantities as the long-range or short-range order, density, concentration of impurities, number of defects, etc. In a system in which elastic and anelastic phenomena are possible, changes in the value of the deformation are always accompanied by changes of at least one internal parameter. Neglecting changes in temperature of the whole system, we shall write the expression for the free energy per unit volume in the following form:

$$2\Delta F = \alpha u^2 + 2\gamma u\xi + \beta\xi^2,\tag{1}$$

where ξ represents the deviation of the parameter η from its equilibrium value η_0 in the absence of deformations; the coefficients α, β, and γ are the equilibrium values of the second derivatives of the free energy with respect to selected thermodynamic variables

$$\alpha = \left(\frac{\partial^2 F}{\partial u^2}\right)_0, \quad \beta = \left(\frac{\partial^2 F}{\partial \eta^2}\right)_0, \quad \gamma = \left(\frac{\partial^2 F}{\partial u\,\partial \eta}\right)_0.\tag{2}$$

The probability w of the appearance, in such an isotropic system, of fluctuations of the deformation u and of the internal parameter η is given, in accordance with the general theory of thermodynamic fluctuations [3], by the following expression:

$$w = \frac{V}{2\pi kT}(\alpha\beta - \gamma^2)^{1/2}\exp\left[-\frac{V}{2kT}(\alpha u^2 + 2\gamma u\xi + \beta\xi^2)\right],\tag{3}$$

if we can neglect variations of temperature T and volume V of the whole system. Using this expression, we find the mean-square fluctuations of the quantities of interest to us:

$$\overline{\langle u^2\rangle} = \frac{kT}{V}\frac{\beta}{(\alpha\beta - \gamma^2)}, \quad \overline{\langle \xi^2\rangle} = \frac{kT}{V}\frac{\alpha}{(\alpha\beta - \gamma^2)}, \quad \langle u\xi\rangle = \frac{kT}{V}\frac{\gamma}{(\alpha\beta - \gamma^2)}$$

$$\langle \xi F_u \rangle = \langle u F_\xi \rangle = 0, \quad \langle \xi F_\xi \rangle = \langle u F_u \rangle = \frac{kT}{V}, \quad \langle F_\xi F_u \rangle = \frac{kT\gamma}{V},$$

$$\langle F_\xi^2 \rangle = \frac{kT}{V}\beta, \quad \langle F_u^2 \rangle = \frac{kT}{V}\alpha,$$

$$\langle F_\xi \rangle \equiv \frac{\partial F}{\partial \xi}, \quad F_u \equiv \frac{\partial F}{\partial u} = \sigma. \tag{4}$$

Important characteristics of the fluctuations of the thermodynamic variables u and ξ and of the thermodynamic forces F_u and F_ξ are their correlation coefficients R_1 and R_2, defined by the relationships

$$R_1^2 = \frac{\langle u\xi \rangle^2}{\langle u^2 \rangle \langle \xi^2 \rangle},$$

$$R_2^2 = \frac{\langle F_\xi F_u \rangle^2}{\langle F_\xi^2 \rangle \langle F_u^2 \rangle}. \tag{5}$$

Since the thermodynamic forces F_ξ and F_u are linear combinations of u and ξ, we can show that the correlation coefficients R_1 and R_2 are equal, i.e., $R_1 = R_2 = R$. The correlation coefficient, defined in accordance with Eq. (5), satisfies [4] the inequality $|R| \leq 1$. If the thermodynamic variables u and ξ are statistically independent, then R = 0. The equality R = ±1 denotes the degeneracy of the probability relationship between u and ξ into a linear function of the u = ± ξ type. Using Eqs. (4) and (5), we can show that the correlation coefficient is $|R| < 1$ for those states of a system which are thermodynamically stable with respect to changes in the variables u and ξ, and tend to unity in the range of unstable thermodynamic equilibrium states, near phase transition points.

If the elastic deformations u in the system and the resultant changes in the internal parameter ξ are the consequence of the action of an external agency on the system, the change in the free energy ΔF of the system will again be given by Eq. (1), if the amplitude of the external agency is not too large. In this case the rate of change of the parameter ξ depends linearly on the thermodynamic force F_ξ, which leads to the equation

$$\dot{\xi} = -L(\beta\xi + \gamma u). \tag{6}$$

Now, the deformation u, which varies periodically at a frequency ω, is found to be related to the stress σ by the expression

$$\sigma = M(\omega) u,$$

where the complex elastic modulus $M(\omega)$ is expressed in terms of the coefficients α, β, and γ of the quadratic form (1) and in terms of the kinetic coefficient L of Eq. (6):

$$M(\omega) = \alpha - \frac{\gamma^2}{\beta}(1 + i\,\omega\tau_\xi)^{-1}, \quad \tau_\xi = (L\beta)^{-1}. \tag{7}$$

Using Eq. (7), we obtain the following expressions for the principal mechanical properties of the medium:

$$M_1(\omega) = \mathrm{Re}\, M(\omega) = \alpha - \gamma^2\beta^{-1}(1 + \omega_2\tau_\xi^2)^{-1}, \tag{8}$$

$$M_2(\omega) = \mathrm{Im}\, M(\omega) = \gamma^2\beta^{-1}\omega\tau_\xi(1 + \omega^2\tau_\xi^2)^{-1}, \tag{9}$$

$$M_v = M_1(\infty) = \alpha, \quad M_R = M_1(0) = \alpha - \gamma^2\beta^{-1}, \tag{10}$$

$$\Delta_1 = M_v - M_R = \gamma^2 \beta^{-1},$$

$$\Delta_2 = \frac{\Delta_1}{M_v} = \gamma^2 (\alpha\beta)^{-1}, \tag{11}$$

$$\tan \delta = \frac{M_2(\omega)}{M_1(\omega)} = \frac{\gamma^2}{[\alpha\beta(\alpha\beta - \gamma^2)]^{1/2}} \frac{\omega\tau}{1 + \omega^2\tau^2}, \tag{12}$$

$$\tau = \tau_\xi (\alpha\beta)^{1/2} (\alpha\beta - \gamma^2)^{-1/2}. \tag{13}$$

Here, $M_1(\omega)$ is the dynamic elastic modulus; M_V is the unrelaxed, and M_R is the relaxed (with respect to the parameter ξ) values of the elastic modulus; Δ_1 is the elastic modulus defect and Δ_2 is the relative value of this defect; $\tan \delta$ is the loss-angle tangent; τ_ξ is the relaxation time of the parameter ξ.

Use of the relationships given in Eq. (4) makes it possible to express the mechanical properties of a system of Eqs. (8)-(13), in terms of its fluctuation characteristics (the correlation parameter R, the mean-square values of the fluctuations of the deformation $<u^2>$, the stress $<\sigma^2>$, and the parameter $<\xi^2>$)

$$M_1(\omega) = \begin{cases} \dfrac{kT}{V} \dfrac{1}{\langle u^2 \rangle (1-R^2)} \left(1 - \dfrac{R^2}{1 + \omega^2\tau_\xi^2}\right) \\[2mm] \dfrac{V \langle \sigma^2 \rangle}{kT} \left(1 - \dfrac{R^2}{1 + \omega^2\tau_\xi^2}\right), \end{cases} \tag{14}$$

$$M_2(\omega) = \begin{cases} \dfrac{kT}{V \langle u^2 \rangle} \dfrac{R^2}{1 - R^2} \dfrac{\omega\tau_\xi}{1 + \omega^2\tau_\xi^2}, \\[2mm] \dfrac{V \langle \sigma^2 \rangle}{kT} R^2 \omega\tau_\xi (1 + \Delta^2\tau_\xi^2)^{-1} \end{cases} \tag{15}$$

$$M_v = \begin{cases} kT [V \langle u^2 \rangle (1-R^2)]^{-1}, \\ V \langle \sigma^2 \rangle / kT \end{cases}$$

$$M_R = \begin{cases} kT / V \langle u^2 \rangle \\ V \langle \sigma^2 \rangle (1-R^2) / kT \end{cases}, \tag{16}$$

$$\Delta_1 = \begin{cases} kTR^2 / V \langle u^2 \rangle (1-R^2) \\ V \langle \sigma^2 \rangle R^2 / kT \end{cases}, \quad \Delta_2 = R^2, \tag{17}$$

$$\tan \delta = R^2 (1-R^2)^{-1/2} \omega\tau (1 + \omega^2\tau^2)^{-1},$$

$$\tau = \tau_\xi (1-R^2)^{-1/2}, \tag{18}$$

$$\tau_\xi = \begin{cases} V \langle \xi^2 \rangle (1-R^2) / LkT \\ kT / LV \langle F_\xi^2 \rangle \end{cases}. \tag{19}$$

The relationships obtained show that the elastic properties and the value of the internal friction of a system in the case considered are determined completely by the mean-square values of the fluctuations of selected thermodynamic variables and by the coefficient of their correlation. It is evident from Eqs. (14)-(19) that the influence of the relaxation on the elastic properties and on the internal friction of the system depends considerably on the value of the coefficient of correlation R of the variables u or σ and the internal parameter η, characterizing a given relaxation mechanism, but is independent of the sign of R. The amplitude of an internal friction peak $(\tan \delta)_{max}$, reached when the condition $\omega \tau = 1$, is satisfied, is governed, according to Eq. (18), only by the value of the correlation coefficient R, increasing with increase in R and vanishing when R = 0. When R → 0, the amplitude of the peak $(\tan \delta)_{max}$ becomes infinite and the relaxed elastic modulus M_R decreases in accordance with Eq. (16) and vanishes when R = ±1. The situation when R = ±1 may arise, for example, in phase transitions and more correct results can be obtained in this case by an additional treatment allowing for factors such as the inhomogeneity of fluctuations, etc. Thus, over the whole range of variation of the correlation coefficient R $(0 \leq |R| \leq 1)$ with the exception of the values of R close to unity, the relationships (14)-(19) obtained above allow us to use the methods and results of the thermodynamic theory of fluctuations and of statistical physics in the treatment of the elastic properties and of the internal friction of the medium. On the other hand, the measurements of the internal friction and of the elasticity moduli of real systems, and the use of relationships of the (14)-(19) type make it possible to establish the nature of the thermodynamic fluctuations in such systems and the degree of correlation of the fluctuations.

Literature Cited

1.   L. A. Svergunenko, Izv. Vysshikh. Uchebn. Zavedenii, Fiz., No. 4 (1962).
2.   L. D. Landau and E. M. Lifshits, Electrodynamics of Continuous Media, Moscow (1957).
3.   L. D. Landau and E. M. Lifshits, Statistical Physics, Moscow (1951).
4.   G. Kramer, Mathematical Methods in Statistics [Russian translation], IL (1948).

THEORY OF THE INTERNAL FRICTION OF PLASTICALLY DEFORMED METALS

L. N. Aleksandrov, V. S. Mordyuk, and L. F. Savina

The plastic deformation of metals increases the internal friction level and displaces the deformation peak toward lower temperatures [1]. Postnikov's [2] microscopic treatment of the internal friction caused by the motion of defects in a stress field gave an expression for the internal friction component Q_p^{-1}, due to a preliminary deformation and explained the displacement of the deformation peak. From the approximate expression

$$\ln Q_p^{-1} = \ln n_p - \frac{U}{kT} - \frac{t}{\tau_0}\exp\left(-\frac{E_p}{kT}\right) + \ln A \tag{1}$$

it follows that increase in the internal friction with increasing degree of deformation of a metal can be explained by an increase in the defect concentration n_p. The observed rise in Q_p^{-1} is associated with a reduction of the time necessary to record the temperature dependence of Q^{-1}. Since the activation energy of the process of recovery, E_p, decreases with increasing degree of deformation, while the remaining quantities are practically constant, the acceleration of the processes of recovery impedes the determination of the nature of the dependence of the internal friction on the degree of deformation. Active defects have to be specified in order to compare the dependence $Q^{-1}(p)$ of Eq. (1) with the experimental data. The direct proportionality of the quantities Q^{-1} and n_p is not very likely (for example, according to [3], when the density of dislocations is increased by a factor of 9, Q^{-1} increases by a factor of 2) and the law of variation of E_p with p is unknown. According to Postnikov [2], it is difficult to allow for the influence of impurities on the value of Q_p^{-1}.

The Koehler–Granato–Lücke theory, modified by Swartz and Weertman [4], considers the oscillations of a dislocation line, pinned by impurity atoms. If the motion of a dislocation is restricte by a field of impurity atoms of concentration c, then the average displacement of the dislocation is governed by the Burgers vector b, divided by $c^{1/3}$. The internal friction is (for low stresses)

$$Q^{-1} = \frac{1.5 N b L_N}{\pi^2 \varepsilon c^{1/3}}, \tag{2}$$

where ε is the relative difference between the radii of the impurity and matrix (host) atoms; N is the dislocation density; L_N is the length of segments in a three-dimensional dislocation network (grid) ($L_N^2 \approx 3/N$). In the case of a concentration of interstitial atoms much lower than the concentration of substituent atoms, the distance between the nodal points of the network L_N in Eq. (2) should be replaced with b/c, but then the concentration of substitutional impurities is ignored in the calculation.

247

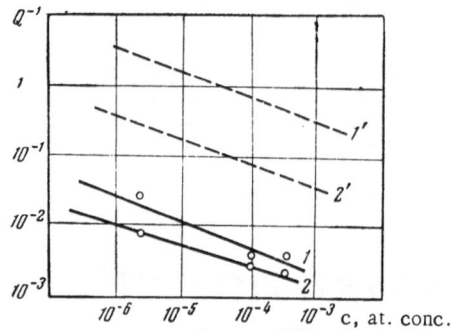

Fig. 1. Internal friction of molybdenum wire (0.4-mm diameter) containing various amounts of impurities, for dislocation densities of 10^{11} cm^{-2} (1, 1') and 10^9 cm^{-2} (2, 2'). 1, 2) Experimental data; 1', 2') calculated results.

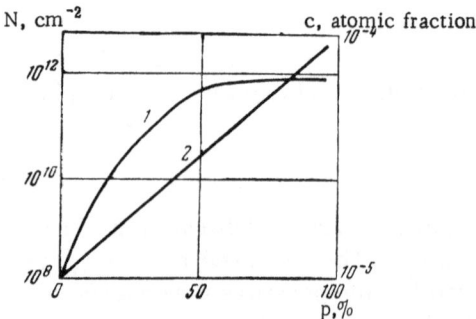

Fig. 2. Dependence of the dislocation density (1) and of the vacancy concentration (2) on the degree of deformation p.

The theory is applicable to metals with the bcc lattice and, in addition to the determination of the concentration dependence, it can be used to find the influence of the degree of deformation on Q^{-1} through the dislocation density. According to Eq. (2), $Q^{-1} \approx \sqrt{N}$ and since, in the annealing of a deformed metal N decreased by 3-4 orders of magnitude, the corresponding fall in Q^{-1} should be by 1-2 orders of magnitude. The value of the internal friction of a single crystal should be, according to Eq. (2), smaller than Q^{-1} of polycrystalline materials by 1-2 orders of magnitude. This difference is usually not more than one order of magnitude, for example in the case of tungsten [5]. Our investigation of the internal friction of deformed refractory metals made it possible to reduce to a minimum the influence of the recovery and to determine more clearly the influence of p on Q^{-1}. The concentration dependence of Q^{-1} was found to be well represented by Eq. (2) although the agreement between the absolute values was not very good. Figure 1 shows the results of experimental investigations and of calculations using Eq. (2) of the internal friction of molybdenum wire (0.4-mm diameter) of zone-purified (c = $2 \cdot 10^{-6}$ atomic concentration), pure (c $\approx$ $1 \cdot 10^{-4}$ atomic concentration), and industrial type Mch (c = $3 \cdot 10^{-4}$ atomic concentration) grades, the latter containing impurities in the form of Fe_2O_3, Ni, SiO_2 at 20°C; the two pairs of lines in Fig. 1 correspond to the dislocation densities N = 10^{11} cm^{-2} and 10^9 cm^{-2}. For substitutional impurities (Fe, Ni),

ε = 0.1, b = $2 \cdot 10^{-8}$ cm. The investigations were carried out using a directly heated relaxometer of the torsional pendulum type at a frequency of about 1 cps.

For single crystals with a low density of dislocations, the agreement between the theory and experiment was satisfactory, but for deformed polycrystalline samples the values of Q^{-1} calculated from Eq. (2) were found to be too high. Allowance for a higher volume concentration of vacancies did not alter greatly the results of the calculation. Since the atomic concentration c was not more than 10^{-3}, then evidently the damping influence of vacancies, acting as pinning points of dislocations, was due to their higher concentration at block and grain boundaries (according to [6], molybdenum contains $c_V \simeq 10^{-3}$ vacancies per host atom) and this influence was noticeable only at dislocation densities N < 10^{11} cm^{-2}, since at higher values of N the quantity L_n should be less than the length of the part of a loop b/c_V, which would be impossible. Analysis of Eq. (2) shows that at high degrees of deformation, due to interlocking of dislocations, the maximum loop length L_N should decrease (compared with its value found from the relationship 1.7/$\sqrt{N}$ by 1-2 orders of magnitude and the loop length should fall to the value of the interatomic spacing. Making this assumption, it was possible to explain the experimentally observed values of the internal friction in strongly deformed metals. Good agreement with the experiment could be obtained by allowing for the fact that in the theory of Swartz and Weertman

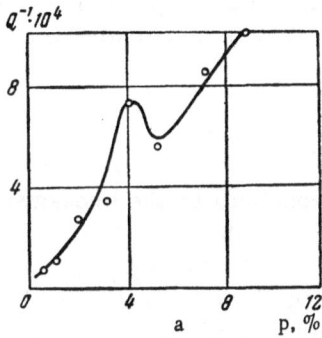

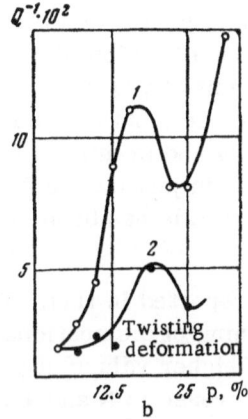

Fig. 3. Dependence of the internal friction of metals on the degree of deformation: a) niobium single crystal; b) polycrystalline tungsten (99.8% deformation, 250-μ diameter) of different plasticities (1, 2).

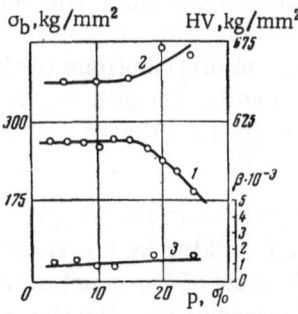

Fig. 4. Dependence of the tensile strength (1), microhardness (2), and the broadening of x-ray interference lines (3) on the degree of additional deformation of polycrystalline tungsten (99.9% compression, 100-μ diameter).

[4] only the contribution of the free dislocations (1-10% of the total number of dislocations) should be taken into account. The value of L_N was governed by the total value of N, but in Eq. (2) we should substitute the value $(1 - a)N$, where a was the dislocation pinning coefficient. The probability of a breakaway of a line of length L_N from its pinning points depended on the stress acting on the line. At low stresses and high dislocation densities, we find

$$Q^{-1} = \frac{NL_N^3 b\sigma_0}{9\pi^2 F_0},$$ (3)

where F_0 is the maximum stress necessary to break away a long loop; σ_0 is the amplitude of the applied stress. The length of the dislocation is $l < b/c$, where

$$l = \frac{2F}{\sigma_b b} = \frac{2F_0}{\sigma_0 b}.$$ (4)

If we take l to be the length of a loop b/c, then we obtain an expression for Q^{-1} which is applicable at high dislocation densities

$$Q^{-1} = NL_N^3 \frac{2c}{9\pi^2 b}.$$ (5)

In this case the concentration of impurities and vacancies should not be less than $10^{-8}\sqrt{N}$, since $b/c < L_N$. In the limiting case when the average distance between impurity atoms on a dislocation line is equal to the distance between the nodal points of a network, we obtain $Q^{-1} \approx 0.07$, bearing in mind that $NL_N^2 \approx 3$. When the concentration of point defects increases, Q^{-1} also increases. For a fixed density of dislocations, the increase of Q^{-1} can be explained in accordance with Eq. (5) by increasing vacancy concentration. When temperature is increased, further increase of c in accordance with the law $\exp(U/kT)$ begins to have an effect; U is the activation energy of vacancy formation.

To explain the low values of the internal friction ($\approx 10^{-3}$) at high degrees of deformation, it is necessary to allow for the breakup of the dislocation network when the degree of distortion of the lattice is increased, because of the interlocking [7] and the pinning of dislocations; the dislocation pinning coefficient for high values of p amounts to 0.9-0.99. The factor that retards the motion of dislocations is the presence of considerable static distortions in the lattice in the region of an oscillating dislocation segment.

The internal friction in strongly deformed metals, in the case of a complex-energy dissipation mechanism which includes resonance and hysteretic losses, can be calculated using the expression

$$Q^{-1} = \frac{0.25\,(1-a)\,b\;\sqrt{N}}{\varepsilon c^{1/2}}.$$

(6)

A possible increase in L_N (when the dislocation density is constant) by the breakaway at nodal points of a network can explain a further rise in the quantity Q^{-1} [8]. To eliminate the considerable contribution of the vacancies to the internal friction in the investigation of the dependence of Q^{-1} on p, it seemed of interest to consider the damping of vibrations at low temperatures but high degrees of deformation, giving rise to the saturation of the lattice with dislocations. It is known [9] that, when the degree of deformation is increased, the concentration of the vacancies increases linearly and the density of dislocations increases only up to p $\approx$ 50% and then it remains constant (Fig. 2). In the investigation of the internal friction of aluminum and other metals after cyclic deformation [10] it has been found that initially Q^{-1} rose with increase in the number of cycles and with increase in the number of defects, but this was followed by the stabilization or a fall of Q^{-1}, associated with the merging and precipitation of vacancies, as well as with the formation of microcracks; Q^{-1} rose again just before fracture. Similar rise in Q^{-1} with increasing p in the case of niobium single crystals has been reported in [11].

Figure 3a shows the dependence $Q^{-1}(p)$, plotted from the data reported in [11]. The density of dislocations was still very low and one could speak of the pinning of dislocations by impurities and vacancies, followed by the breakaway of dislocations at higher values of p. The increase of Q^{-1} on subsequent increase of p could be caused by an increase in N and L_N, but the dependence (6) described well the observed changes ($a = 0$).

Our comparison of the value of the internal friction calculated from Eq. (6), $Q^{-1}_{calc} = 5 \cdot 10^{-3}$, with the experimental value, $Q^{-1}_{exper} = 3 \cdot 10^{-3}$, for tungsten single crystals (N = 10^6 cm^{-2} and c < 10^{-4}%, according to mass-spectroscopic data) also confirmed the dependences obtained [12].

To investigate Q^{-1} when the lattice was saturated with dislocations, we used wire samples of different plasticities of the VA-3 grade, which were deformed by 99.8%. The samples contained the following impurities: 0.005% and 0.004% Fe_2O_3, 0.017 and 0.003% Mo. The subsequent shear deformation was carried out using a torsional machine K-2. The curves of the dependence $Q^{-1}(p)$ for two samples having different amounts and different distributions of the impurities are given in Fig. 3b. The stabilization of the value of the tensile strength σ_b, the microhardness HV, and of the broadening of the x-ray interference lines β (Fig. 4), all indicated saturation with dislocations, i.e., stabilization of the value of N.

Therefore, the increase in Q^{-1} could be associated with an increase in the length of dislocation loops. The stabilization or the fall of Q^{-1} due to dislocation locking by the vacancies corresponded to a subsequent small rise in HV and a fall of σ_b. In view of the constancy of N, the internal friction could be calculated in this case using Eq. (2), allowing only for the independent variation L_N or using Eq. (6), allowing for the change in the pinning coefficient a in the deformation limit. In both cases the faults of the theory of Swartz and Weertman [4] became apparent. The assumption of rigid nodal points in a dislocation network was found to be a rough approximation, and this was also true of the assumption of a uniform volume distribution of impurity atoms.

Literature Cited

1. M.A. Krishtal, Yu.V. Piguzov, and S.A. Golovin, Internal Friction in Metals and Alloys, Metallurgizdat (1964).
2. V.S. Postnikov, Collection: Relaxation Phenomena in Metals and Alloys, Moscow (1960), p.264. [English translation: B.N. Finkel'shtein, ed., Consultants Bureau, New York (1963), p. 199.]
3. L.N. Aleksandrov and V.N. Orlov, Collection: Relaxation Phenomena in Metals and Alloys, Moscow (1963), p. 294.
4. J.C. Swartz and J. Weertman, J.Appl.Phys., 32:1860 (1961) [Russian translation in: Ultrasonic Methods of Investigating Dislocations, IL, Moscow (1963), p. 58].
5. L.N. Aleksandrov, Collection: Crystallization and Phase Transitions, Izd. AN BSSR (1962), p. 294.
6. L.N. Aleksandrov, Collection: Mechanism and Kinetics of Crystallization, Izd. AN BSSR (1964), p. 371.
7. L.M. Utevskii, L.G. Orlov, and M.P. Usikov, Usp.Fiz.Nauk, 76:109 (1962).
8. A.S. Nowick, Collection: Creep and Recovery [Russian translation], Metallurgizdat (1961), p. 166. [English edition: R. Maddin, ed., American Society for Metals, Novelty, Ohio.]
9. S.D. Gertsriken and M.M. Novikov, Vestn.Kievsk.Gos.Univ., No. 1 (1958).
10. V.S. Postnikov and G.A. Gorshkov, Collection: Relaxation Phenomena in Metals and Alloys, Moscow (1963), p. 115.
11. R.H. Chamler and J. Schultz, Acta Met., 10(4):466 (1962).
12. L.N. Aleksandrov, V.S. Mordyuk, and L.F. Savina, Fiz.Tverd.Tela, 7(11):3153 (1965).

THERMAL RELAXATION OF MECHANICAL VIBRATIONS
IN POLYCRYSTALLINE MATERIALS

B. M. Darinskii and A. G. Fokin

The internal friction due to heat exchange between inhomogeneities in a medium has been considered by many workers [1-4]. However, usually the problem has been solved for a matrix medium and the first coefficient in a virial expansion calculated, allowing for heat exchange between an occlusion and the surrounding medium but neglecting the correlation between individual occlusions. The thermal relaxation in a disordered medium has been treated in [5], where it is shown that in polycrystalline materials of cubic symmetry thermal relaxation occurs in the case of shear deformations but is absent in the case of volume deformations. The present paper gives the solution of the problem of the thermal relaxation in polycrystalline materials of arbitrary crystal symmetry.

For a single-component polycrystalline sample, consisting of crystallites which differ from one another in their shape and the orientation of crystallographic axes, we have the following equations of thermoelasticity:

$$\nabla_k \lambda_{iklm} \nabla_m u_l - \nabla_k b_{ik} \theta = -f_i \quad \nabla_l \chi_{ik} \nabla_k \theta - \dot{\theta} - \gamma_{lm} \nabla_m \dot{u_l} = -\frac{q}{C_V} , \tag{1}$$

where u_l is the vector representing the displacement of the medium; λ_{iklm} is the tensor of the isothermal elastic moduli; $b_{ik} = \lambda_{iklm} a_{lm}$ is the thermal stress tensor; a_{lm} is the thermal expansion tensor; χ_{ik} is the thermal diffusivity tensor; θ is the deviation of the temperature of the polycrystalline sample from T_0, at which the stresses and deformations are zero; C_V is the specific heat per unit volume of matter; $\gamma_{lm} = (T_0/C_V) b_{lm}$, f_i, and q are, respectively, the density of the external mechanical forces and the density of the power flowing in the form of thermal energy from external sources of heat.

Applying to Eq. (1) the Fourier transformation $d/dt \rightarrow i\omega$ and introducing the notation $u_4 = \Theta$, $f_4 = q/C_V$,

$$L_{il} = \nabla_k \lambda_{iklm} \nabla_m, \quad L_{l4} = -\nabla_k b_{jk}, \quad L_{4l} = i\omega (T_0/C_V) L_{l4}, \quad L_{44} = \nabla_l \chi_{ik} \nabla_k - i\omega, \tag{2}$$

we obtain

$$L_{\alpha\beta} u_\beta = -f_\alpha.$$

The Greek subscripts assume values from unity to four, and the Latin subscripts assume values from unity to three.

The operator $L_{\alpha\beta}$ has a random dependence on r, governed by the orientation of the crystallographic axes of the crystallite points. We shall represent the operator $L_{\alpha\beta}$ in the form of a sum of the average value $\overline{L}_{\alpha\beta}$ and a fluctuating increment $\xi_{\alpha\beta}$

$$L_{\alpha\beta} = \overline{L}_{\alpha\beta} + \xi_{\alpha\beta}, \tag{3}$$

where $\overline{\xi}_{\alpha\beta} = 0$. The average value of the operator $\overline{L}_{\alpha\beta}$ is, by definition, an operator which is obtained by averaging the random coefficients of the operator $L_{\alpha\beta}$ over a region much greater than the dimensions of the crystallites. For a quasi-uniform and quasi-isotroptic polycrystalline body we shall have

$$\overline{L}_{il} = \overline{\lambda}_{iklm}\nabla_k\nabla_m, \quad \overline{L}_{i4} = -b\nabla_l,$$

$$\overline{L}_{4l} = -i\omega\frac{T_0}{C_V}b\nabla_l, \quad \overline{L}_{44} = \chi\Delta - i\omega, \tag{4}$$

where

$$\overline{\lambda}_{iklm} = \alpha\delta_{ik}\delta_{lm} + \beta(\delta_{il}\delta_{km} + \delta_{im}\delta_{kl}),$$

$$\overline{b}_{ik} = b\delta_{ik}, \quad \overline{\chi}_{ik} = \chi\delta_{ik}. \tag{5}$$

The random operator $\xi_{\alpha\beta}$ is defined by the following formulas:

$$\xi_{il} = \nabla_k\delta\lambda_{iklm}\nabla_m, \quad \xi_{l4} = -\nabla_k\delta b_{jk},$$

$$\xi_{4l} = i\omega\frac{T_0}{C_V}\xi_{l4}, \quad \xi_{44} = \nabla_l\delta\chi_{ik}\nabla_k, \tag{6}$$

where the symbol δ denotes the difference between the exact value of a random quantity at the point r and its average value, for example, $\delta b_{ik} = b_{ik} - b\delta_{ik}$.

The system of equations (1), together with Eq. (3), can be written in the form

$$\overline{L}_{\alpha\beta}\overline{u}_\beta = -f_\alpha - \overline{\xi_{\alpha\beta}u_\beta}. \tag{7}$$

We shall find the equation for the average value of the displacement $\overline{u}_i$ which is a macroscopic parameter, on the assumption that the fluctuations of the properties of the medium are small. Using the method given in [5, 6], we obtain, in the first nonvanishing approximation,

$$L^{(1)}_{\alpha\beta}\overline{u}_\beta = -f_\alpha, \tag{8}$$

where

$$L^{(1)}_{\alpha\beta} = \overline{L}_{\alpha\beta} + \overline{\xi_{\alpha\gamma}M_{\gamma\rho}\xi_{\rho\beta}}, \tag{9}$$

$M_{\gamma\rho}$ is an integral operator whose kernel $G_{\gamma\rho}(r)$ satisfies the equation

$$\overline{L}_{\alpha\gamma}G_{\gamma\rho}(r) = \delta(r)\delta_{\alpha\rho}, \tag{10}$$

which obeys the conditions $G_{\gamma\rho}(r) \to 0$ when $r \to \infty$. The solution of the system (10) in the Fourier space has the form

$$G_{il}(k) = \frac{\delta_{il}}{\beta}\frac{1}{k^2} - \frac{\widetilde{\alpha} + \beta}{\beta(\widetilde{\alpha} + 2\beta)}\frac{k_ik_l}{k^4},$$

$$G_{l4}(k) = -\frac{ib}{\widetilde{\alpha}+2\beta} \cdot \frac{k_l}{\chi k^2 + i\omega}, \quad G_{4l}(k) = i\omega \frac{T_0}{C_V} G_{l4}(k),$$

$$G_{44}(k) = \frac{1}{\chi k^2 + i\omega}\left[1 - i\omega \frac{T_0 b^2}{C_V (\widetilde{\alpha}+2\beta)(\chi K^2 + i\omega)}\right],$$

$$\widetilde{\alpha} = \alpha + i\omega \frac{T_0 b^2}{C_V (\chi k^2 + i\omega)}.$$

$$(11)$$

From the expressions given in Eq. (11), the Green's tensor $G_{\gamma\rho}(\mathbf{r})$ is found from the formula

$$G_{\gamma\rho}(r) = \frac{1}{8\pi^3}\int G_{\gamma\rho}(k) e^{-ikr} dk.$$

$$(12)$$

We shall consider first the fourth equation in the system (8). Using Eq. (9), we shall write it in the form

$$-i\omega \frac{T_0}{C_V} b\nabla_i \bar{u}_i + (\chi\Delta - i\omega)\,\bar{u}_4 + \int \overline{\xi_{4\gamma}(r) G_{\gamma\rho}(r-r')\,\xi_{\rho i}(r')}\,\bar{u}_i(r')\,dr' +$$

$$+ \int \overline{\xi_{4\gamma}(r) G_{\gamma\rho}(r-r')\,\xi_{\rho 4}(r')}\,\bar{u}_4(r')\,dr' = -f_4.$$

$$(13)$$

To calculate the integrals which occur in Eq. (13) and in similar expressions, we shall assume that the average values of the function $u_\alpha(r)$ differ little at distances of the order of the crystallite dimensions. Then, $\bar{u}_\alpha$ can be taken outside the integral sign and be treated as a constant. Using the integral Fourier transformation, we obtain an equation containing the effective values of the transport coefficients

$$-i\omega \frac{T_0}{C_v} b_{eff}\nabla_i \bar{u}_i + (\chi_{eff}\Delta - i\omega)\,\bar{u}_4 = -f_4,$$

$$(14)$$

where

$$b_{eff} = b - \frac{1}{9\beta} B_{lmii}^{lm}(0)\left(1 - \frac{2}{5}\frac{\alpha+\beta}{\alpha+2\beta}\right),$$

$$B_{lmii}^{lm}(0) = \overline{\delta b_{lm}\delta\lambda_{lmii}}.$$

$$(15)$$

In the derivation of Eq. (14) and formula (15) we have used the smallness of the physical parameter $T_0 b^2 / \beta C_V \ll 1$ of real solids.

The effective thermal diffusivity χ_{eff} in Eq. (14) is defined by the formula

$$\chi_{eff} = \chi - \frac{1}{9}\frac{1}{2\pi^2}\int_0^\infty \frac{\overline{\delta\chi_{lm}^2(k)}}{\chi k^2 + i\omega} k^2 dk,$$

$$(16)$$

where $\overline{\delta\chi_{lm}^2(k)}$ is the Fourier transform of the correlation function $\overline{\delta\chi_{lm}(0)\delta\chi_{lm}(r)}$.

The expression (16) shows that the effective thermal diffusivity of a polycrystalline material depends on the frequency. For low frequencies this dependence is unimportant, and in this case,

$$\chi_{eff} = \chi - \frac{1}{9}\frac{\delta\chi_{lm}^2}{\chi}.$$

$$(17)$$

A similar formula has been obtained in [7] for the effective coefficient of stationary diffusion in a polycrystalline medium.

We shall consider the first three equations of the system (8), writing them in the form

$$\overline{L}_{il}\overline{u}_l + \overline{L}_{i4}\overline{u}_4 + \int \overline{\xi_{i\gamma}(r)\, G_{\gamma\rho}(r-r')\, \xi_{\rho l}(r')}\, \overline{u}_l(r')\, dr' + \int \overline{\xi_{i\gamma}(r)\, G_{\gamma\rho}(r-r')\, \xi_{\rho 4}(r')}\, u_4(r')\, dr'. \tag{18}$$

Carrying out the calculations similar to those made earlier, and eliminating u_4 by means of Eq. (14), we obtain

$$\lambda_{iklm}^{\text{eff}} \nabla_m \nabla_k \overline{u}_l = -\widetilde{f}_i, \tag{19}$$

where $\lambda_{iklm}^{\text{eff}}$ has the form

$$\lambda_{iklm}^{\text{eff}} = \overline{\lambda}_{iklm} + \int B_{lmrs}^{ikpq}(r)\, \nabla_q \nabla_s G_{pr}(r)\, dr - \int B_{pqlm}^{ik}(r)\, \nabla_p G_{4q}(r)\, dr -$$

$$- \frac{i\omega T_0}{C_v} \int B_{ikpq}^{lm}(r)\, \nabla_q G_{p4}(r)\, dr + i\omega \frac{T_0}{C_v} \int B_{lm}^{ik}(r)\, G_{44}^{\circ}(r)\, dr + \frac{T_0 b_{\text{eff}}^2}{C_v}, \tag{20}$$

and b_{eff} is defined by Eq. (15). Here the symbol B denotes the various correlation functions, which are given by the expressions

$$B_{lmrs}^{ikpq}(r) = \overline{\delta\lambda_{ikpq}(0)\, \delta\lambda_{lmrs}(r)} = B_{lmrs}^{ikpq}(0)\, \varphi(r),$$

$$B_{lmpq}^{ik}(r) = \overline{\delta b_{ik}(0)\, \delta\lambda_{lmpq}(r)} = B_{lmpq}^{ik}(0)\, \varphi(r),$$

$$B_{lm}^{ik}(r) = \overline{\delta b_{ik}(0)\, \delta b_{lm}(r)} = B_{lm}^{ik}(0)\, \varphi(r). \tag{21}$$

The function $\varphi(r)$ depends on the geometrical structure of the polycrystalline sample. It has the following properties: $\varphi(0) = 1$, $\varphi(\infty) = 0$, $\varphi'(0) \neq 0$, which is the consequence of the sudden changes in the crystallite orientations.

Using the Fourier transformation, the expression (20) is reduced to the form

$$\lambda_{iklm}^{\text{eff}} = \overline{\lambda}_{iklm} - \frac{1}{3\beta} B_{lmpq}^{ikpq}(0) + \frac{1}{15\beta}\cdot\frac{\alpha+\beta}{\alpha+2\beta}\left(B_{lmqq}^{ikpp}(0) + 2B_{lmpq}^{ikpq}\right)$$

$$+ \frac{T_0 b_{\text{eff}}^2}{C_v} + \left[\frac{T_0 b^2}{15C_v(\alpha+2\beta)^2}\left(B_{lmpq}^{ikqp}(0) + 2B_{lmpq}^{ikpq}\right) - \frac{2T_0 b}{3C_v(\alpha+2\beta)} + \frac{T_0}{C_v} B_{lm}^{ik}(0)\right]\frac{1}{2\pi^2}\int_0^\infty \varphi(k)\,\frac{i\omega k^2}{\chi k^2 + i\omega}\, dk, \tag{22}$$

where $\varphi(k)$ is the Fourier transform of $\varphi(r)$.

The second and third terms in the expression (22) have been obtained first by Lifshits and Rozentsveig [8] by allowing for the correlation between the individual crystallites. The last term, which is a complex quantity, allows for the absorption of the mechanical energy in the process of heat transfer between the hot and cold parts of a polycrystalline sample.

Replacing the variables in (22),

$$k^2 = \frac{1}{\chi\tau}, \tag{23}$$

and transforming the expression for the average values of the products of the fluctuating coefficients, we shall find the expression for the effective elastic moduli tensor in the usual relaxation form

$$\lambda_{iklm}^{\text{eff}} = K\delta_{ik}\delta_{lm} + G\left(\delta_{il}\delta_{km} + \delta_{im}\delta_{kl} - \frac{2}{3}\delta_{ik}\delta_{lm}\right) + \Delta K\delta_{ik}\delta_{lm}\int f(\tau)\,\frac{i\omega\tau}{1+i\omega\tau}\, d\tau +$$

$$+ \Delta G \left(\delta_{il}\delta_{km} + \delta_{im}\delta_{kl} - \frac{2}{3}\delta_{ik}\delta_{lm} \right) \int f(\tau) \frac{i\omega\tau}{1 + i\omega\tau}. \tag{24}$$

The following notation is used in Eq. (24):

$$K = d + \frac{2}{3}\beta - \frac{3\alpha + 8\beta}{135\beta(\alpha + 2\beta)} B^{ikpp}_{ikqq},$$

$$G = \beta - \frac{3\alpha + 8\beta}{150\beta(\alpha + 2\beta)} B^{ikpq}_{ikpq} + \frac{6\alpha + 11\beta}{450\beta(\alpha + 2\beta)} B^{ikpp}_{ikqq},$$

$$\Delta K = \frac{2T_0 b^2}{135(\alpha + 2\beta)^2 C_v} B^{iipq}_{iipq}(0),$$

$$\Delta G = \frac{b^2 T_0}{75 C_v (\alpha + 2\beta)^2} \left(B^{lmpq}_{lmpq}(0) + \frac{1}{6} B^{mmpq}_{iipq}(0) \right) - \frac{b}{15(\alpha + 2\beta)} B^{lm}_{lmpp}(0) + \frac{1}{10}\frac{T_0}{C_v} B^{lm}_{lm}(0). \tag{25}$$

The distribution function of the relaxation times is found by means of the formula

$$f(\tau) = \frac{1}{4\pi^2}\varphi\left(\frac{1}{\chi^{1/2}\tau^{1/2}} \right) \frac{1}{\chi^{3/2}\tau^{3/2}}. \tag{26}$$

Obviously, the elastic moduli defects are, by definition, positive quantities.

The expression (24) shows that the thermal relaxation in the general case of an arbitrary crystallographic symmetry does take place in the case of shear and volume vibrations. In the special case of crystals of the cubic symmetry, we have $B^{iipq}_{iipq}(0) = 0$, which indicates the absence of a thermal relaxation of the volume vibrations. For the shear vibrations, we find from Eq. (25) a simpler expression for the modulus defect [6]

$$\Delta G = \frac{b^2 T_0}{75 C_v (\alpha + 2\beta)^2} B^{lmpq}_{lmpq}(0). \tag{27}$$

For a numerical estimate of the value of the absorption it is necessary to know the explicit form of the function $\varphi(r)$. We shall select it in the form

$$\varphi(r) = e^{-r\alpha}, \tag{28}$$

where a is some average characteristic of the crystallite dimensions. The Fourier transform of the function $\varphi(r)$ will be found by integration,

$$\varphi(k) = \frac{8\pi a^3}{(1 + a^2 k^2)^2}. \tag{29}$$

Using Eq. (29), we obtain

$$\int f(\tau) \frac{i\omega\tau}{1 + i\omega\tau} d\tau = \frac{i2x^2}{(\sqrt{2} + x + ix^2)^2}, \tag{30}$$

in which the dimensionless parameter $x = \sqrt{\omega a^2/\chi}$ is introduced. Separating the imaginary and real parts of Eq. (30), we obtain

$$\Delta G \int f(\tau) \frac{i\omega\tau}{1 + i\omega\tau} d\tau = \Delta G' + i\Delta G'', \tag{31}$$

$$\Delta G'' = \Delta G \frac{(1 + \sqrt{2}x) x^2}{(1 + \sqrt{2}x + x^2)^2},$$

$$\Delta k \int f(\tau) \frac{i\omega\tau}{1+i\omega\tau}\, d\tau = \Delta k' + i\Delta k''$$

$$\Delta k'' = \Delta k \frac{(1+\sqrt{2}x)\,x^2}{(1+\sqrt{2}x+x^2)^2}\,. \tag{32}$$

The imaginary parts of the elastic moduli $\Delta G''$ and $\Delta k''$ have each a maximum at

$$x_m^2 = 2 + \sqrt{3} \cong 3.73, \tag{33}$$

which is equal to $\tfrac{1}{4}$. This value will be assumed in the numerical estimate of the internal friction. We shall find expressions for the tangents of the angles between the stresses and the corresponding deformations. For the shear vibrations, we have

$$\tan\delta_1 = \frac{1}{4}\frac{\Delta G}{G} \cong \frac{1}{4\beta}\left[\frac{b^2 T_0}{75\,C_v\,(\alpha+2\beta)^2}\left(B_{lmpq}^{lmpq}(0) + \frac{1}{6}\,B_{llpq}^{mmpq}(0)\right) - \right.$$
$$\left. - \frac{1}{15}\frac{b}{\alpha+2\beta}\,B_{lmpp}^{lm}(0) + \frac{1}{10}\,B_{lm}^{lm}(0)\right]. \tag{34}$$

For the volume vibrations, we have

$$\tan\delta_2 = \frac{1}{4}\frac{\Delta K}{K} = \frac{1}{270}\frac{T_0 b^2}{(\alpha+2\beta)^2\left(\alpha+\frac{2}{3}\beta\right)C_v}\,B_{llpq}^{mmpq}(0). \tag{35}$$

The results of the numerical estimates of the value of the internal friction for some metals are given below.

Metal	Cd	Sn	Zn	Sb	Mg
$\tan\delta_1 \cdot 10^4$	1.3	2.8	1.0	0.16	$6.7\cdot10^{-2}$
$\tan\delta_2 \cdot 10^4$	1.1	$8.8\cdot10^{-3}$	0.28	$3.5\cdot10^{-2}$	$2.1\cdot10^{-3}$

In these calculations we used the values of the elastic moduli given by Leibfrid [9].

The corrections to the average values of the elastic moduli, allowing for the correlation between grains, amount to about 0.1 of the zeroth approximation, which justifies the application of the method of small vibrations to these metals.

The results given here show that in crystals of noncubic symmetry there is a relaxation of the volume vibrations, which in some cases may be detected experimentally by a sensitive relaxometer.

The relaxation of the shear vibrations is due to fluctuations of the elastic moduli tensor and of the stress tensor. These fluctuations increase the absorption of the vibration energy if the fluctuations of the tensors $\delta\lambda_{ikll}$ and δb_{ik} have opposite signs. Such a situation exists in polycrystalline tin, in which, as shown by calculations, the fluctuations of the thermal stress tensor make the principal contribution. In the other metals considered, the influence of fluctuations of the thermal stresses considerably reduces the internal friction.

The frequency of vibrations at which an internal friction peak is observed is found from the expression (33):

$$\omega_{\max} = 2.73 \frac{\chi}{a^2}\,. \tag{36}$$

A substitution into Eq. (36) of the numerical values of the thermal diffusivity coefficient and of the grain dimensions yields the frequency range 10^4-10^9 cps.

The authors are grateful to T.D. Shermergor for his comments on this investigation.

Literature Cited

1. C.M. Zener, Phys. Rev., 52:230 (1937); 53:90 (1938); 56:343 (1939).
2. M.A. Isakovich, Zh. Eksperim. i Teor. Fiz., 18:386, 907 (1948).
3. Yu. M. Sukharevskii, Dokl. Akad. Nauk SSSR, 51:229 (1947).
4. C.M. Zener, Collection: Elasticity and Anelasticity of Metals [Russian translation], IL (1954). [English edition: University of Chicago Press, Chicago, Illinois (1948).]
5. B.M. Darinskii and T.D. Shermergor, Fiz. Metal. i Metalloved., 5:645 (1964).
6. V.I. Tatarskii and M.E. Gertsenshtein, Zh. Éksperim. i Teor. Fiz., 44(2):676 (1963).
7. D.G. Dolgopolov, Fiz. Metal. i Metalloved., 13:209 (1962).
8. I.M. Lifshits and L.N. Rozentsveig, Zh. Éksperim. i Teor. Fiz., 16:967 (1946).
9. G. Leibfrid, Microscopic Theory of Mechanical and Thermal Properties of Crystals, Fizmatgiz (1963).

METHOD OF CALCULATING THE DAMPING OF MECHANICAL VIBRATIONS IN THE PRESENCE OF RELAXATION AND HYSTERESIS

S. I. Meshkov, V. V. Lukin, and V. S. Postnikov

It is known that the processes of relaxation and mechanical hysteresis are important sources of the internal friction. Usually the relaxation and hysteretic types of the internal friction are investigated independently of one another, because of the considerable differences between them. The former type exhibits a strong temperature and frequency dependence, while the latter increases considerably with increasing amplitude but is practically independent of the frequency [1].

A phenomenological description of the relaxation losses can be given within the framework of the linear rheology [2, 3], but in the case of the hysteretic losses it is necessary to use the nonlinear mechanics methods [1, 4, 5].

Shil'krut [5] has made an attempt to consider qualitatively the process of the damping of free vibrations due to the simultaneous action of relaxation and hysteresis. He used the "energy method" of Panovko [4], according to which the energy ΔW, dissipated in one period during any dissipative process, is equal to

$$\Delta W = -cTa\frac{da}{dt},\tag{1}$$

where c is the reduced rigidity modulus, which depends on the elastic properties of the material, the type of sample construction, and the vibration mode; T is the period of vibrations; a (t) is the equation for the envelope of the damped amplitudes.

In solving the problem of the damping, Panovko assumed (in the first approximation) that the energy ΔW was equal to the area of the hysteresis loop, while Shil'krut took ΔW to be equal to some function of time F(t). This result was associated with the fact that Shil'krut calculated the damping using a nonstationary regime of forced vibrations, which is a very dubious method of allowing for the energy dissipation in a medium.

In view of this, we shall consider the problem from a somewhat different point of view. Let us assume that in the one-dimensional case the relationship between the stress σ and the deformation ε has the form

$$\sigma = M_\bullet\varepsilon \pm f(\varepsilon, a),\tag{2}$$

where the modulus M_* is introduced with the aid of an integral Boltzmann—Volterra operator having a relaxation kernel $\varphi(t)$

$$M_* \varepsilon(t) = \int_{-\infty}^{t} \varphi(t - t') \dot{\varepsilon}(t') \, dt', \tag{3}$$

and the function $f(\varepsilon, a)$, where a is the deformation amplitude, describes the shape of the mechanical hysteresis loop in the coordinates $\sigma - \varepsilon$.

However, according to Panovko [4], in the case of the hysteretic internal friction in the first approximation, not the shape of the loop, but its integral characteristic — the area representing the energy dissipated in one period — is the important parameter. Thus, a loop of the elliptical shape, which is the most convenient shape in calculations, makes it possible to isolate the fundamental frequency using any other nonlinear hypothesis, provided the energies are equivalent, i.e., the loop areas are equal.

Therefore, following this form of "linearization," we shall assume

$$f(\varepsilon, a) = \beta a^n \sqrt{1 - \left(\frac{\varepsilon}{a}\right)^2}. \tag{4}$$

To derive the equations for the envelope of the vibration amplitudes in the case of a moderate rate of damping, we shall use Eq. (1), in which, according to Eq. (2), the quantity c should be replaced by the dynamic elastic modulus $M_*(\omega)$.

Although the expression (1) has been obtained for the damped vibrations, which do not have a closed loop in the $\sigma - \varepsilon$ coordinates, the value of the energy ΔW can, in the first approximation (because the damping is weak), be equated to the area of a closed loop. This area can be easily found from Eq. (2), assuming in it a periodic deformation $\varepsilon = a \sin \omega t$.*

As a result, we obtain

$$\sigma = [\omega L_s \varphi(\omega)] \varepsilon \pm [\omega L_c \varphi(\omega) + \beta a^{n-1}] a \sqrt{1 - \left(\frac{\varepsilon}{a}\right)^2}, \tag{5}$$

where L_S and L_C denote, respectively, the integral operators of the sine and cosine Fourier transformations, i.e.,

$$L_s \varphi(\omega) = \int_0^{\infty} \varphi(\xi) \sin \omega \xi \, d\xi,$$

$$L_c \varphi(\omega) = \int_0^{\infty} \varphi(\xi) \cos \omega \xi \, d\xi; \quad \xi = t - t'. \tag{6}$$

From Eq. (5), we find the area of a closed loop ΔS and the dynamic modulus $M_*(\omega)$

$$\Delta S = \pi a^2 [\omega L_c \varphi(\omega) + \beta a^{n-1}], \tag{7}$$

$$M_*(\omega) = \omega L_s \varphi(\omega). \tag{8}$$

It is evident from Eq. (7) that the area of a closed loop in the $\sigma - \varepsilon$ coordinates represents, according to Eq. (2), a superposition of two terms. The first of these terms, associated with

_____
*The lower limit in Eq. (3), $t = -\infty$, automatically eliminates the transient nonstationary regime of forced vibrations.

the hereditary properties of the medium, depends on the frequency and is proportional to the square of the amplitude. The second term, which is due to the hysteresis, depends only on the amplitude and, in this case, is proportional to the $(n-1)$-th power of the amplitude.

Assuming that $\Delta W \approx \Delta S$, we obtain the following equation for the determination of the envelope $a(t)$:

$$- [\omega L_s \varphi(\omega)] T \frac{da}{dt} = \pi a [\omega L_c \varphi(\omega) + \beta a^{n-1}]. \tag{9}$$

Separating the variables and integrating Eq. (9), we find

$$t = -\omega L_s \varphi(\omega) \frac{T}{\pi} \int_{a_0}^{a(t)} a^{-1} [\beta a^{n-1} + \omega L_c \varphi(\omega)]^{-1} da, \tag{10}$$

where a_0 and $a(t)$ are the initial $(t=0)$ and later (t) deformation amplitudes, respectively.

In the calculation of the integral in the first part of Eq. (10), it is necessary to distinguish two cases: $n = 1$ and $n \neq 1$.

When $n = 1$ (E.S. Sorokin's hypothesis), the equation for the envelope has the form

$$a(t) = a_0 \exp \left\{ - \frac{t}{T} \frac{\pi [\omega L_c \varphi(\omega) + \beta]}{\omega L_s \varphi(\omega)} \right\}. \tag{11}$$

When $n \neq 1$, the integrand represents a binomial differential, and, therefore, the integral is calculated by means of the substitution $a = \sqrt{x^{n-1}}$. Carrying out the necessary calculations, we obtain the following expression for the envelope:

$$a(t) = \frac{a_0 \exp\left(-\frac{t}{T}\gamma\right)}{\sqrt[n-1]{1 + \frac{a_0^{n-1}\beta}{\omega L_c \varphi(\omega)} \left\{1 - \exp\left[-\frac{t(n-1)}{T}\gamma\right]\right\}}}, \tag{12}$$

where

$$\gamma = \pi L_c \varphi(\omega) / L_s \varphi(\omega). \tag{13}$$

We can easily show that γ represents the logarithmic decrement of vibrations in a hysteresis-free medium. In fact, for $\beta = 0$, we have

$$a(t) = a_0 \exp\left(-\frac{t}{T}\gamma\right). \tag{14}$$

It is evident from Eq. (12) that the damping of mechanical vibrations in a medium whose dissipative properties are described by Eq. (1), depends both on the frequency and on the amplitude of vibrations. Quantitative agreement with the experimentally observed frequency and amplitude dependencies can be obtained by a suitable selection of the rheological and hysteretic parameters (n, β).

The thermodynamic theory of relaxation phenomena in solids gives, for a continuous spectrum, the following formula for the relaxation kernel:

$$\varphi(t) = \int_0^\infty \psi(s) \exp(-st) ds \equiv L\psi(t), \tag{15}$$

where $\psi(s)$ is the distribution function of the relaxation frequencies and $s \equiv 1/\tau$; τ is the relaxation time; L is the operator of the integral Laplace transformation. For a discrete spectrum, $\psi(s)$ is expressed in terms of the δ-functions

$$\psi(s) = \sum_i \Delta M_i \delta(s - s_i), \tag{16}$$

where ΔM_i is a quantity which has the dimensions of the elastic moduli and characterizes the contribution of the i-th relaxation process.

As a result, we obtain

$$L_c \varphi(\omega) = \sum_i \Delta M_i s_i \, (s_i^2 + \omega^2)^{-1}, \tag{17}$$

$$L_s \varphi(\omega) = \sum_i \Delta M_i \omega \, (s_i^2 + \omega^2)^{-1}. \tag{18}$$

The formula for the logarithmic decrement for a standard linear body is obtained by assuming that $i = 2$, $s_i = s$, $s_2 = 0$, $\Delta M_i = M_\infty - M_0$, $\Delta M_2 = M_0$:

$$\gamma = \pi \frac{M_\infty - M_0}{M_0} \frac{\omega s}{s^2 + \dfrac{M_\infty}{M_0} \omega^2}. \tag{19}$$

It is evident from Eq. (19) that γ is π times larger than the loss-angle tangent [6], which confirms the well-known correlation between these two measures of the internal friction in the case of weak damping.

Obviously, if there are no relaxation processes, i.e., all $s_i = 0$, then the damping of vibrations is solely due to the mechanical hysteresis. In fact, according to Eqs. (17) and (18), when $s_i \to 0$, $\gamma \to 0$. Then, expanding the formula (12) as a series in terms of γ, and taking only the first terms in the expansion, we obtain

$$a(t) = \frac{a_0}{\sqrt[n-1]{\dfrac{1 + a_0^{n-1}\beta t\,(n-1)\,\pi}{MT}}}, \tag{20}$$

where $M = \sum_i \Delta M_i$ is the unrelaxed elastic modulus. This special result has been obtained by Panovko [4].

Literature Cited

1. G.S. Pisarenko, Energy Dissipation in Mechanical Vibrations, Izd. AN UkrSSR, Kiev (1962).
2. C.M. Zener, Collection: Elasticity and Anelasticity of Metals [Russian translation], IL (1954). [English edition: University of Chicago Press, Chicago, Illinois (1948).]
3. T.D. Shermergor, Collection: Relaxation Phenomena in Metals and Alloys, Metallurgizdat (1963).
4. Ya.G. Panovko, Internal Friction in Vibrations of Elastic Systems, Fizmatgiz (1960).
5. D.I. Shil'krut, Collection: Energy Dissipation in Vibrations of Elastic Systems, Izd. AN UkrSSR (1963), pp. 93, 97.
6. H. Kol'sky, Stress Waves in Solids [Russian translation], IL (1955). [English edition: 2nd ed. (paper), Dover, New York.]

DIFFUSION INTERNAL FRICTION IN
POLYCRYSTALLINE MATERIALS

T. D. Shermergor and B. M. Darinskii

The deformation of a polycrystalline sample gives rise to a random elastic field, which is due to the random distribution of crystallographic axes of anisotropic crystallites. The bending of the crystallites, described by dilatation gradients, should be accompanied by an enhanced diffusion, which should lead to the diffusion internal friction in the case of periodic deformations [1]. Moreover, polycrystals of low crystallographic symmetry should exhibit a relaxation effect, caused by the anisotropy of the concentration-stress tensor $b_{ik} = \partial\sigma_{ik}/\partial c$, where σ_{ik} is the stress tensor, $c = n/N_0$ is the concentration of impurity atoms, n and N_0 are, respectively, the numbers of impurity atoms and all other atoms per unit volume. We shall consider, for example, two neighboring crystallites of hexagonal symmetry, to which a tensile stress is applied; the axis of one of these crystallites will be assumed to be parallel and that of the second perpendicular to the tensile stress. Then, if $b_{11} \ll b_{33}$, the diffusion of impurity atoms from the second crystallite to the first will give rise to the relaxation of the stresses and to a reduction in the elastic energy of the system. In the periodic deformation case this mechanism gives rise to the internal friction which we shall calculate below.

The initial system of equations, which describes a polycrystal, has the form [2-4]

$$\nabla_k \lambda_{iklm} \nabla_m u_l - \nabla_k b_{ik} c = -f_i, \tag{1}$$

$$\nabla_i D_{ij} \nabla_j c - \frac{\partial c}{\partial t} - \frac{1}{kTN_0} \nabla_i c D_{ij}^0 \nabla_j b_{lm} \nabla_m u_l = -q, \tag{2}$$

the first of which is the equilibrium equation, which allows for the relaxation due to diffusion, and the second is the equation for the diffusion arising from these conditions. The parameters have the following meanings: λ_{iklm} is the tensor of the elastic constants; D_{ij} is the diffusion coefficient; D_{ij}^0 is the value of the diffusion coefficient when $c \rightarrow 0$; u_l is the displacement vector; f_i and q are, respectively, the density of the forces acting and the intensity of the source of impurity atoms. The coefficients in these equations are random functions of the coordinate point, but the variables u_l and the c have regular and random components.

The parameter D_{ik} is a transport coefficient. It governs the relaxation time but does not affect the modulus defect and the amplitude of the internal friction peak. Therefore, D_{ik} can be replaced by its effective value $D_{ik} = D\delta_{ik}$. Moreover, we shall assume that u_l, $c \propto \exp(i\omega t)$, we shall make the substitution $\partial/\partial t \rightarrow i\omega$, and we shall linearize Eq. (2) by the substitution $c \rightarrow c_0$ in the third term.

Then, the equation for the resultant diffusion will have the form

$$Lc - gD\Delta b_{lm}\nabla_m u_l = -q, \tag{3}$$

where

$$L \equiv D\Delta - i\omega, \quad g \equiv c_0 D_0 / kT N_0 D. \tag{4}$$

Introducing the reciprocal operator $M = L^{-1}$, we shall solve Eq. (3) with respect to c and substitute the result in Eq. (1). This gives

$$L_{il} u_l = -f_i'. \tag{5}$$

Here, the operator L_{il} is given by the equation

$$L_{il} = \nabla_k \lambda_{iklm}\nabla_m - g\nabla_k b_{ik} b_{lm}\nabla_m + i\omega g\nabla_k b_{ik} M b_{lm}\nabla_m, \tag{6}$$

and f_i' is a renormalized source of the displacement field.

The calculation of the internal friction in such a polycrystalline sample reduces to the averaging of the system of equations (5) and to the calculation of the tensor of the effective elastic moduli. We shall carry out the averaging over the volume including a sufficient number of crystallites, but still small compared with the dimensions over which $\nabla_m \bar{u}_i$ and $\bar{c}$ vary greatly. After such averaging the parameters in Eq. (5) will be constant and the functions will be regular.

We shall consider only the direct averaging of the operator L_{il} and the replacement of u_l by its regular value $\bar{u}_l$. Then, the averaging of the first term in Eq. (6) gives the average elastic modulus. Thus, in this approximation the correlation effects, due to the spatial fluctuations of the elastic-constants tensor λ_{iklm}, are not included.

To average the second term in Eq. (6), we shall bear in mind that the symmetrical tensor b_{ik} for a medium of arbitrary crystallographic symmetry can be reduced to the principal axes and represented in the form

$$b_{ik} = b_1\delta_{i1}\delta_{k1} + b_2\delta_{i2}\delta_{k2} + b_3\delta_{i3}\delta_{k3}. \tag{7}$$

Its average value is

$$\langle b_{ik}\rangle = \frac{1}{3}(b_1 + b_2 + b_3)\delta_{ik} \equiv b\delta_{ik}. \tag{8}$$

Using Eqs. (7) and (8), we find that the correlation tensor

$$A_{iklm} \equiv \langle \delta b_{ik}\delta b_{lm}\rangle = \frac{1}{10} AD_{iklm}, \tag{9}$$

where the following notation is used:

$$\delta b_{ik} = b_{ik} - \langle b_{ik}\rangle, \tag{10}$$

$$A \equiv A_{ikik} = \frac{2}{3}(b_1^2 + b_2^2 + b_3^2 - b_1 b_2 - b_1 b_3 - b_2 b_3), \tag{11}$$

$$D_{iklm} = \delta_{il}\delta_{km} + \delta_{im}\delta_{kl} - \frac{2}{3}\delta_{ik}\delta_{lm}. \tag{12}$$

Thus,

$$\langle b_{ik} b_{lm} \rangle = b^2 \delta_{ik} \delta_{lm} + A_{iklm}. \tag{13}$$

We shall now consider the averaging of the last term in Eq. (6). According to the definition of the operator M, its action on an arbitrary function is a convolution of the Green's function G of the direct operator L and this function. Therefore,

$$\langle b_{ik} M b_{lm} \nabla_m \rangle \bar{u}_l = b^2 \delta_{ik} \int G(\mathbf{r} - \mathbf{r}') \, \bar{u}_{ll}(\mathbf{r}') \, dV' + $$
$$+ \int B_{ik_m l}(\mathbf{r} - \mathbf{r}') \, G(\mathbf{r} - \mathbf{r}') \, \bar{u}_{lm}(\mathbf{r}') \, dV'. \tag{14}$$

Here, the binary tensor correlation function is given by the expression

$$B_{iklm}(\mathbf{r} - \mathbf{r}') = \langle \delta b_{ik}(\mathbf{r}) \, \delta b_{lm}(\mathbf{r}') \rangle, \tag{15}$$

and $\bar{u}_{lm} \equiv \Delta_m \bar{u}_l$. The first term in Eq. (14) describes the diffusion which would take place in a uniform isotropic medium in the presence of dilatation gradients $\nabla_i \bar{u}_{ll} \neq 0$. Since we shall consider below only the diffusion flux between the grains of a polycrystal, it is necessary to assume that $\bar{u}_{ik} = $ const. This leads to the following expression for the effective elastic-constants tensor:

$$\overset{*}{\lambda}_{iklm} = \bar{\lambda}_{iklm} - g A_{iklm} + i\omega g I_{iklm}, \tag{16}$$

where

$$I_{iklm} = \int B_{iklm}(\mathbf{r}) \, G(\mathbf{r}) \, dV. \tag{17}$$

To calculate the integral (17), we shall assume the following dependence of the tensor B_{iklm} on the coordinates

$$B_{iklm}(\mathbf{r}) = A_{iklm} \exp(-r/a), \tag{18}$$

where a is of the order of the average dimensions of the crystallites. Then, using the Parseval theorem for the integral Fourier transformation, we obtain

$$I_{iklm} = A_{iklm} \cdot \frac{1}{\pi^2} \int \frac{a^3 dV_k}{(1 + a^2 k^2)^2 (Dk^2 + i\omega)}. \tag{19}$$

We shall carry out the integration over the angles and use the value of the integral [5]

$$\frac{4}{\pi} \int_0^\infty \frac{a^3 k^2 dk}{(1 + a^2 k^2)^2 (k^2 + i\rho^{-2})} = -i\rho^2 \gamma(x), \tag{20}$$

$$\gamma' \equiv \mathrm{Re}\,\gamma = \frac{x(x + \sqrt{2x})}{(1 + x + \sqrt{2x})^2}, \quad \gamma'' \equiv \mathrm{Im}\,\gamma = \frac{x(1 + \sqrt{2x})}{(1 + x + \sqrt{2x})^2}, \tag{21}$$

where $x = a^2/\rho^2$, and $\rho = \sqrt{D/\omega}$ is the characteristic distance traveled by a flux of diffusing atoms in one period. Then, for the effective tensor λ^*_{iklm} we obtain

$$\overset{*}{\lambda}_{iklm} = \bar{\lambda}_{iklm} + \frac{1}{10} A g (\gamma - 1) D_{iklm}. \tag{22}$$

Hence, it follows that the bulk modulus $\overline{K}$ retains its previous value, but the effective shear modulus is

$$\mu^{\bullet} = \overline{\mu} + \frac{1}{10} Ag (\gamma - 1).$$

(23)

Using Eq. (23), we find the internal friction in the case of the shear Q_μ^{-1} and of the longitudinal Q_E^{-1} vibrations

$$Q_\mu^{-1} = \frac{Ag}{10\mu} \gamma'', \quad Q_E^{-1} = \frac{EAg}{30\mu^2} \gamma''.$$

(24)

Here, E is Young's modulus. It follows from Eq. (24) that at low frequencies $Q^{-1} \propto \omega$, but at high frequencies $Q^{-1} \propto 1/\sqrt{\omega}$. Estimating the dependence of Q^{-1} on the crystallographic symmetry, we note that for crystals of the cubic structure $b_1 = b_2 = b_3$ and $A \equiv 0$, while for the hexagonal crystals $b_1 = b_2 \neq b_3$ and $A = (2/3) (b_1 - b_3)^2$. For a lattice of lower symmetry, A is given by the full formula (11).

The components of the tensor b_{ik} can be calculated from the tensor or the concentration deformations $a_{ik} = \partial \varepsilon_{ik}/\partial c$, where ε_{ik} is the deformation tensor, using the relationship $b_{ik} = \lambda_{iklm} a_{lm}$. The quantities a_{lm} are determined by x-ray diffraction from the change in the lattice parameter when the concentration of the impurity is increased. By way of example, we shall find the amplitude of the internal friction peak for the hexagonal system titanium–nitrogen. Representing a_{ik} in the form analogous to Eq. (7), we find that $a_1 = 0.0013$, $a_3 = 0.0085$ [6]. Since the numerical values of the elastic constants c_{ik} are unknown, we shall use the isotropic approximation for λ_{iklm}. We can show that in this case $b_3 - b_1 = 2\mu (a_3 - a_1)$. Bearing this in mind, we find from Eq. (24) that $(Q_\mu^{-1})_{max} = 0.83 \cdot 10^{-4}$ at T = 1100°K and c = 1%. Since the solubility of nitrogen in titanium at this temperature is very high (c $\approx$ 15%), then, even when the nonlinearity of the tensor a_{ik} is allowed for, the internal friction peak may be fairly high.

Literature Cited

1.   C. M. Zener, Collection: Elasticity and Anelasticity of Metals [Russian translation], IL (1954). [English edition: University of Chicago Press, Chicago, Illinois (1948).]
2.   B. Ya. Lyubov and N. S. Fastov, Dokl. Akad. Nauk SSSR, 84(5) : 939 (1952).
3.   G. L. Krasko and B. Ya. Lyubov, Dokl. Akad. Nauk SSSR, 142(2) : 326 (1962).
4.   M. A. Krivoglaz, Fiz. Metal. i Metalloved., 17(2) : 161 (1964).
5.   B. M. Darinskii and T. D. Shermergor, Fiz. Metal. i Metalloved., 18 : 5 (1964).
6.   A. E. Vol, Structure and Properties of Binary Metal Systems, Fizmatgiz (1962). [English edition: Longmans, Green, and Co.]
7.   T. W. Kaye and T. H. Laby, Tables of Physical and Chemical Constants, 12th edition [Russian translation], Fizmatgiz (1962).
8.   R. Wasilenski and G. Kehl, J. Inst. Metals, 83(3) : 94 (1954).